ROLAND GLASER

Biophysics

Roland Glaser

Biophysics

With 162 Figures

 Springer

Prof. Dr. ROLAND GLASER
Institut für Biologie
Experimentelle Biophysik
Humboldt-Universität zu Berlin
Invalidenstraße 42
10115 Berlin
Germany
e-mail: roland.glaser@rz.hu-berlin.de

Original title: BIOPHYSIK by Roland Glaser
Published by: Gustav Fischer Verlag, Jena 1996 (fully revised 4th edition)
Copyright © Spektrum Akademischer Verlag GmbH, Heidelberg Berlin 1999

ISBN 978-3-642-08639-7

Library of Congress Cataloging-in-Publication Data
Glaser, Roland. [Biophysik. English] Biophysics/Roland Glaser. – Rev. 5th ed. p. cm. Rev. ed. of:
Biophysik. 4th ed. 1996. Includes bibliographical references (p.).

1. Biophysics. I. Glaser, Roland Biophysik II. Title. QH505.G5413 2000 571.4-dc21

Springer is a part of Springer Science and Business Media
springeronline.com

© Springer-Verlag Berlin Heidelberg 2010
Printed in Germany

Cover design: design & production, 69121 Heidelberg, Germany

31/3150 5 4 3 2 1 – Printed on acid free paper

Preface

When I started to teach biophysics to biology students at the Friedrich-Schiller University in Jena in 1965 the question arose: What actually is biophysics? What should I teach? Only one thing seemed to be clear to me: biophysics is neither "physics for biologists" nor "physical methods applied to biology" but a modern field of science leading to new approaches for our understanding of biological functions.

Rashevsky's book on *Mathematical Biophysics* (1960), the classical approaches of Ludwig von Bertalanffy (1968), as well as the excellent book by Katchalsky and Curran on *Nonequilibrium Thermodynamics in Biophysics* (1965), showed me new ways of looking at biological processes. Thus, I came to the conclusion that it would be worthwhile trying to integrate all these various physical and physicochemical approaches to biological problems into a new discipline called "biophysics". The first German edition of this textbook, published in 1971, was developed from these considerations.

Meanwhile, I had moved from Jena to the Humboldt-University in Berlin where I organized courses for biologists specializing in

* Translated by H. T. Lowe-Porter, Penguin Books, 1985, p. 275-276

biophysics. The idea was, why should only physicists find their way to biophysics? Why not help biologists to overcome the "activation energy" barrier of mathematics and physics to discover this fascinating discipline?

In Berlin, a special group was established (1970) in the Department of Biology with the aim of teaching biophysics. This led to a full university degree course of biophysics which has developed successfully and attracts an increasing number of students today.

Consequently, my co-workers and I had the responsibility of organizing not only introductory courses to biophysics for biology students, but also advanced courses in molecular biophysics, biomechanics, membrane biophysics, bioelectrochemistry, environmental biophysics and various aspects of theoretical biophysics.

The evolution of this textbook in the following years was the result of these courses. Innumerable discussions with students, colleagues and friends led to continuous refinement and modification of the contents of this book, resulting in a second, third, and in 1996, a fourth German edition. New topics were added, others updated or even deleted. The only sentences that remained unchanged are those of Thomas Mann at the beginning of the Preface.

The philosophy of this book is that biophysics is not a simple collection of physical approaches to biology, but a defined discipline with its own network of ideas and approaches, spanning all hierarchical levels of biological organization. The paradigm of a holistic view of biological functions, where the biological system is not simply the sum of its molecular components but is rather their functional integration, seems to be the main concept of biophysics.

While it is easier to realize such an integrated view in a 'one-man book', this has, of course, the disadvantage that the knowledge and experience of many specialists cannot be incorporated. However, to a certain degree this problem has been compensated for by discussions with colleagues and friends and by their continuous support over a period of more than three decades. Further problems are the selection of the topics to be included in the book and the emphasis placed on the different aspects, avoiding underestimation of others. Although the author has tried to balance the selection and emphasis of topics by looking at the development of biophysics over the last three decades, he is not sure that he has succeeded. Even if this is the case, this book will at least help to answer the question: What is biophysics? It provides a solid introduction to biophysics. For further reading, books and reviews are recommended at the end of each chapter. The extensive index at the end of the book ensures an easy orientation and will enable this book to be used as a reference work. As mentioned above, this book is written primarily for biologists and biophysicists with a background in biology. Therefore, some basic knowledge of biology is required, but less knowledge of physics and mathematics is needed. It should encourage biologists to enter the field of biophysics and stimulate further research. The German editions have shown that physicists also will profit from reading this book.

This first English edition is not just a translation of the fourth German edition, but is rather a fully revised fifth edition. For an author, it is impossible to translate his book without substantial rewriting and refining. All chapters have been more or less revised, and results which have been published since the last edition have been integrated. Many figures have been redrawn, some are new; some totally new chapters have also been included.

Last, but not least, I wish to express again my sincere gratitude to all of my colleagues and friends, throughout the world, who helped me before with all previous editions and especially for helping me with this English edition. Thanks are also extended to the staff of Springer-Verlag for encouraging me to write this English version and for correcting my imperfect English.

Berlin, July 2000   ROLAND GLASER

Contents

List of Fundamental Constants and Symbols

The numbers in parentheses indicate equations in the text, where the symbols are explained or defined.

$\equiv$	identical
$=$	equal
$\overset{!}{=}$	equal by definition
$\approx$	approximately equal
$\sim$	proportional

A	Arrhenius-coefficient [Eq. (2.3.17)]
A	affinity [Eq. (3.1.75)]
A	area
a	chemical activity [Eq. (3.1.34)]
B	magnetic flux density [Eq. (4.4.1)]
b	electrophoretic mobility [Eq. (2.5.13)]
C	electric capacity [Eq. (3.5.6)]
C^*	complex electric capacitance [Eq. (3.5.11)]
C_l	clearance-constant [Eq. (5.2.28)]
C_V	isochoric heat capacity [Eq. (2.4.25)]
c	molar concentration
c_0	speed of light in vacuum $= 2.998 \cdot 10^8$ m s^{-1}
D	diffusion coefficient [Eq. (3.3.6)]
E	energy (general expression)
E^0	standard redox potential [Eq. (2.2.1)]
$\mathbf{E}$	electric field strength [Eq. (2.4.4)]
e	basis of natural logarithm $= 2.71828$
e	absolute amount of charge on electron $= 1.60218 \cdot 10^{-19}$ C
$\mathbf{F}$	mechanical force
F	Faraday $= 9.6485 \cdot 10^4$ C val^{-1}
F	Helmholtz free energy [Eq. (3.1.22)]
f	symbol for an arbitrary function
f	generalized coefficient of friction [Eq. (3.1.52)]
f	activity coefficient [Eq. (3.1.34)]
G	Gibbs free energy [Eq. (3.1.23)]
G	electrical conductivity [Eq. (3.5.2)]
g	specific conductivity [Eq. (3.5.2)]
g	osmotic coefficient [Eq. (3.2.26)]

H	enthalpy [Eq. (3.1.21)]
H	magnetic field strength [Eq. (4.4.1)]
h	Planck's constant $= 6.626 \cdot 10^{-34}$ J s
I	sound intensity [Eq. (4.3.3)]
I	information [Eq. (2.3.7)]
I	ionic strength [Eq. (2.4.16)]
I_A	second moment of area [Eq. (3.6.13)]
I_P	polar second moment of area [Eq. (3.6.16)]
i	unit vector in x-direction
j	imaginary unit $= \sqrt{-1}$
j	unit vector in y-direction
j	electric current density [Eq. (4.4.7)]
J	flux [Eq. (3.3.1)]
J	unidirectional flux in kinetic equations
K_p	equilibrium constant of isobaric chemical reactions [Eq. (3.1.71)]
K	bending ($K = 1/R$), (Sect. 3.6.4)
k	Boltzmann's constant $= 1.380658 \cdot 10^{-23}$ J K^{-1} $= 8.6174 \cdot 10^{-5}$ eV K^{-1}
k	rate constant
L_i	phenomenological coefficient relating a flow to a force [Eq. (3.1.49)]
L	decibel intensity of sound [Eq. (4.3.3)]
l	second (or: azimuthal) quantum number (Sect. 2.1.1)
l	distance, length
M	moment of force [Eq. (3.6.12)]
M	molar mass
m	mass
m	magnetic quantum number (Sect. 2.1.1)
N	Avogadro's number $= 6.0221367 \cdot 10^{23}$ mol^{-1}
n	primary (or: principal) quantum number (Sect. 2.1.1)
n	number of particles, individuals etc.
P	mathematical probability (Sect. 2.3.1)
P	permeability coefficient [Eq. (3.3.9)]
P	electrical power density [Eq. (4.4.7)]
p	pressure
Pa	Pascal
Q	heat
q	electric charge
R	molar gas constant $= 8.314510$ J K^{-1} mol^{-1}
R	radius of curvature ($R = 1/K$), [Eq. (3.6.8)]
R	resistance coefficient relating a flow to a force [Eq. (3.1.50)]
R	Ohm's resistance (reactance), (Sect. 3.5.3)
Re	Reynolds number [Eq. (3.7.1)]
r	radius, radial distance

r	Donnan ratio [Eq. (3.2.38)]
S	entropy [Eqs. (2.3.4) and (3.1.10)]
s	spin quantum number (Sect. 2.1.1)
T	temperature
t	time
U	internal energy [Eq. (3.1.9)]
V	volume
$\bar{V}$	partial molar volume [Eq. (3.1.8)]
$\mathbf{v}$	velocity
W	thermodynamic probability [Eq. (2.3.4)]
W	work [Eq. (3.1.9)]
$\mathbf{X}$	generalized force [Eq. (3.1.42)]
x	coordinate in an orthogonal system
x	mole fraction [Eq. (3.1.35)]
Y	Young's modulus [Eq. (3.6.7)]
Y^*	electric admittance [Eq. (3.5.1)]
y	coordinate in an orthogonal system
z	coordinate in an orthogonal system
z_i	number of charges
α	electrical polarizability [Eq. (2.4.8)]
β_T	isothermic compressibility [Eq. (2.4.26)]
γ	velocity gradient or shear rate [Eq. (3.6.1)]
γ	surface tension (Section 2.5.1)
δ	difference of length
Δ	sign, indicating a difference between two values
ε	mechanical strain [Eq. (3.6.6)]
ε	dielectric constant or permeability number [Eq. (2.4.1)]
ε_0	dielectric permittivity of vacuum $= 8.854187817 \cdot 10^{-12}$ C V^{-1} m^{-1}
ζ	electrokinetic potential [Eq. (2.5.14)]
η	viscosity [(Eq. (3.6.2)]
κ	Debye-Hückel constant [Eq. (2.4.15)]
λ	thermal conductivity [Eq. (4.1.1)]
λ	wavelength
μ	magnetic permeability [Eq. (4.4.1)]
μ_0	magnetic permeability of vacuum $= 1.2566370 \cdot 10^{-6}$ V s A^{-1} m^{-1} [Eq. (4.4.1)]
μ	electric dipole moment [Eq. (2.4.7)]
μ_i	chemical potential of the component i [Eq. (3.1.33)]
$\tilde{\mu}_i$	electrochemical potential of the salt i [Eq. (3.1.41)]
ν	stoichiometric number [Eq. (3.1.65)]
ν	kinematic viscosity ($\nu = \eta/\rho$)
ν	frequency in Hz ($\nu = \omega/2\pi$)
ξ	degree of advancement of a chemical reaction [Eq. (3.1.73)]

π osmotic pressure [Eq. (3.2.14)]

ϱ density

ϱ charge density in space [Eq. (2.4.12)]

σ Stefan-Boltzmann constant [Eq. (4.1.2)]

σ mechanical stress [Eq. (3.6.5)]

σ entropy production [Eq. (3.1.63)]

σ_0 surface charge density [Eq. (2.5.15)]

σ Staverman's reflection coefficient [Eq. (3.2.28)]

τ time constant

τ sheer stress [Eq. (3.6.3)]

Φ Rayleigh's dissipation function [Eq. (3.1.64)]

ϕ fluidity ($\phi = 1/\eta$) (Chapter 3.6.1)

χ magnetic susceptibility [Eq. (4.4.2)]

ψ electrical potential [Eq. (2.4.3)]

ω angular frequency ($\omega = 2\pi v$)

ω coefficient of mobility [Eq. (3.1.52)]

CHAPTER 1

Nature and Subject of Biophysics

The subjects of biophysics are the physical principles underlying all processes of living systems. This also includes the explanation of interactions of various physical influences on physiological functions, which is a special sub-area, called environmental biophysics.

Biophysics is an interdisciplinary science somewhere between biology and physics, as may be concluded from its name, and is furthermore connected with other disciplines, such as mathematics, physical chemistry, and biochemistry. The term "biophysics" was first used in 1892 by Karl Pearson in his book *The Grammar of Science.*

Does biophysics belong to biology, or is it a part of physics? Biology, by definition, claims to be a comprehensive science relating to all functions of living systems. Hence, biophysics, like genetics, biochemistry, physiology etc., should be considered as a specialized sub-area of biology. This view has not remained undisputed by physicists, since physics is not confined to subjects of inanimate matter. Biophysics can be considered, with equal justification, as a specialized part of physics. It would be futile to try to balance those aspects against each other. Both of them are justified. Biophysics cannot flourish unless cooperation is ensured between professionals from either side.

Delimitation of biophysics from clearly unrelated areas has appeared to be much easier than its definition. Biophysics, for example, is by no means some sort of a melting pot for various physical methods and their applications to biological problems. The use of a magnifying glass, the most primitive optico-physical instrument, for example, has just as little to do with biophysics as the use of most up-to-date optical or electronic measuring instruments. Biophysical research, of course, requires modern methods, just as other fields of science do. The nature of biophysics, however, is actually defined by the scientific problems and approaches rather than by the applied methods.

Biophysical chemistry and bioelectrochemistry can be considered as specialized sub-areas of biophysics. Medical physics, on the other hand, is an interdisciplinary area which has its roots in biophysics but has ramifications of far-reaching dimensions, even with medical engineering.

In terms of science history, biophysical thought, according to the above definition, can be traced back to early phases of philosophical speculations on nature, that is back to antiquity. This applies to the earliest mechanistic theories of processes of life and to insights into their dynamics, for example of Heraclitus in the 5th century B.C. The promotion of scientific research in the Renaissance also includes biophysical considerations. Leonardo da Vinci (1452–1519), for example, investigated mechanical principles of bird flight in order to use the information for engineering design; research which would be termed bionics today. A remarkably comprehensive biomechanical description of functions, such as mobility of limbs, bird's flight, swimming movement, etc., was given in a book by Alfonso Borelli (1608–1679) *De motu animalium* published in Rome, as early as 1680. The same Borelli founded a school in Pisa of *iatro-mathematics* and *iatro-physics* in which the human body was perceived as a mechanical machine, and where attempts were made to draw medical conclusions from that perception (Iatric – Greek term for medical art). Iatro-physics has often been considered as a mechanistic forerunner of medical biophysics.

Parallels to processes of life were established not only in the area of tempestuous progress of mechanics but at all levels throughout the development of physics. Reference can be made, in this context, to the frog experiments undertaken by Luigi Galvani (1737–1798). The physics of electricity was thus studied in direct relationship with phenomena of electrophysiology. Worth mentioning is the strong controversy between Luigi Galvani and Alessandro Volta (1745–1827) about the so-called elettricità animale (animal electricity), which had serious personal consequences for both.

It is well-known that medical observations played a role in the discovery of the first law of thermodynamics by J. R. Mayer (1814–1878). Calorimetric studies of heat generation of mammals were conducted in Paris by A. L. Lavoisier (1743–1794) and P. S. de Laplace (1749–1827) as early as about 1780. Reference should also be made, in this context, to investigations of Thomas Young (1773–1829), and later Hermann v. Helmholtz (1821–1894) on the optical aspects of the human eye and on the theory of hearing. These activities added momentum to the development of physiology which thus became the first biological platform for biophysics.

The development of physical chemistry around the turn of this century was accompanied by applications of these discoveries and insights in understanding various functions of living cells. There have also been many instances in which biologically induced problems had stimulating effects upon progress in physics and physical chemistry. Brown's motion, discovered in pollen grains and subsequently calculated by A. Einstein, is an example. Research on osmotic processes, as well, were largely stimulated by the botanist W. Pfeffer. The temperature dependence of rate constants of chemical reactions was initially formulated in terms of phenomenology by S. Arrhenius (1859–1927), and has, ever since, been applied to a great number of functions of life, including phenomena as sophisticated as processes of growth. Studies of physiochemical foundations of cellular processes have continued to be important in biophysical research, especially after the introduction of the principles of nonequilibrium thermodynamics. In particular, biological membranes, as highly organized anisotropic structures, are always attractive subjects for biophysical investigations.

A decisive impetus has been given to biophysical research through the discovery of X-rays and their application to medicine. It was attributable to close cooperation between physicists, biologists, and medical scientists which paved the way for the emergence of radiation biophysics which not only opened up possible new approaches to medical diagnosis and therapy but also made substantive contributions to the growth of modern molecular biology.

The year 1948 saw the publication of Norbert Wiener's book *Cybernetics* dealing with control and communications in men and machines. While regulation and control of biological systems had been subjects of research before, biocybernetics has given further important inspiration to biophysics. In the 1970s, biological system theory moved very close to thermodynamics. It should be borne in mind, in this context, that the expansion of classical thermodynamics to cover nonequilibrium systems with non-linear equations of motion was strongly stimulated by biological challenges. Supported by the works of A. Katchalsky, I. Progogine, H. Haken and many others, elements of thermodynamics are found to be closely interconnected to those of kinetics within the theory of non-linear systems.

The word "bionics" was coined by a synthesis of "biology" and "technics" at a conference in Dayton, USA, in 1960. More specific shape was thus given to the millennial quest of man to look at nature's complete technological design. Biophysics, and especially biomechanics, play a substantial role in these attempts.

This brief view of the history and the development of biophysics allows us now to draw the following conclusions about its nature and relevance: biophysics seems to be quite a new branch of interdisciplinary science, but, in fact, biophysical questions have always been asked in the history of science. Biophysics relates to all levels of biological organization, from molecular processes to ecological phenomena. Hence, all the other biological sub-areas are penetrated by biophysics, including biochemistry, physiology, cytology, morphology, genetics, systematics, and ecology.

Biological processes are among the most intricate phenomena with which scientists find themselves confronted. It is, therefore, not surprising that biologists and other scientists have repeatedly warned against schematism and simplifications. Such warning is justified and is a permanent reminder to the biophysicist of the need for caution. Yet, on the other hand, there is no reason to conclude that biological phenomena are too sophisticated for physical calculation. Despite the fact that at present we are not able to explain all biological reactions, no evidence has ever been produced that physical laws were no longer valid, when it comes to biological systems.

Further reading: Rowbottom and Susskind (1984)

CHAPTER 2

Molecular Structure of Biological Systems 2

This section starts with quantum-mechanical approaches, which allow us to explain molecular bonds and processes of energy transfer. Later we will explain the physical basis of thermal noise, which, finally, will lead us to the problem of self organization, or self assembly of supramolecular structures. It is the intention of this section to make the reader familiar with some specific physical properties of biological systems at the molecular level. The leading idea of this section is the controversy between thermal fluctuation against the forces of molecular orientation and organization.

Two kinds of physical behavior meet at the molecular level of biological structures: on the one hand, there are the characteristic properties of *microphysical* processes, based on the individual behavior of single small particles like atoms, molecules or supramolecular structures. These processes are mostly stochastic. On the other hand, there are reactions which resemble "*macrophysical*" properties, the kind of behavior of "large" bodies. The "macrophysics" is ruled by the laws of classical physics, for example classical mechanics. Our daily experiences with macrophysical systems teach us that their behavior is generally deterministic.

To explain this difference, let us consider a simple mechanical wheelwork. The knowledge of its design and construction allows a precise prediction of the behavior of the system. This prediction is based on the laws of classical mechanics. In contrast to this, a chemical reaction with a small number of molecules in a homogeneous phase depends on stochastic collisions of the individual molecules with each other. Since this process is stochastic, it is only predictable in a statistical way.

This stochastic behavior of molecular systems can be transformed into a deterministic one, if the number of participating stochastic events is large, or if the degrees of freedom of the single reactions are extremely limited. The increase of stochastic events can be realized either by an increasing number of participating molecules, by enlarging the volume for example, where the reaction takes place, or by an increase of the time interval of observation. This consideration indicates an interesting interplay between volume, time constants, and reliability of a biochemical reaction.

The limitation of the degree of freedom of a biochemical reaction is realized by a property of the system which is called *anisotropy*. In contrast to isotropic systems, like simple solutions, in anisotropic systems the mobility of molecules in various directions is not identical, but is restricted in some directions, and promoted in others. This, for example, is the case for enzymatic reactions, where the participating enzymes are orientated in membranes, or if the reactions of charged or polar reactants occur in strong electric fields of electrical double layers.

In many fields the biological organism works as an amplifier of the microphysical stochastics. A molecular mutation, for examples, leads to a reaction chain, which finally ends with a phenomenological alteration of the organism. Or, as another example: a few molecular events in the pigments of optical receptors can lead to perception and to reaction in behavior.

During the first step in considering molecular mechanisms of biological systems, a further aspect is taken into consideration. Unfortunately, biologists often ignore the fact that a qualitative jump has to be made in the transition from the "visible" macrophysical structures, to the microphysical systems such as atoms or molecules. This includes not only the above-mentioned transition from the deterministic behavior of macroscopic systems to the stochastic behavior of single molecules, but many more aspects as well. The biologists, for example, must acknowledge that the term "structure" receives a new meaning. The visible "biological structure", as known in the fields of anatomy, morphology and histology, now appears as concentration profiles or as systems of electric charges or electromagnetic fields. Instead of visible and measurable lengths, diameters or distances, as common in the visible world, in the microphysical world so called *effective parameters* are used. These sorts of parameters are exactly defined and they can be measured with arbitrary exactness, but they do not correspond to some visible boundaries. A single ion, for example, has no diameter in the sense of the diameter of a cell, or a cell nucleus, which can be measured by a microscopic scale. In the following sections we will define effective parameters like crystal radius, hydration radius and Debye-Hückel radius, which really are important parameters for functional explanations.

It is not the intention of this book to describe the topics of molecular biology. However, the theoretical foundations and principles will be explained to make possible a link between structure and function at the molecular level and current biological thinking in these dimensions.

2.1
Intramolecular Bonds

Any representation of the dynamics of molecular and supramolecular structures has to begin with the atom, its organization and energy states and with interactions between atoms in a molecule. The molecule, as described in the next chapter, is initially assumed to be thermally unaffected. The thermal energy of movement will be introduced as an additional parameter in Section 2.7.

2.1.1
Some Properties of Atomic Orbitals

The Schrödinger equation is the theoretical basis for calculation of the wave functions of electrons and the probability of their presence at a particular point in space. This is a link between wave mechanics and the atomic model postulated by Niels Bohr. The latter concept, however, is substantially extended by the incorporation of the wave properties of elementary particles.

The quantification of electric energy by Planck's theory which has already been postulated ad hoc in the atomic model of Niels Bohr, results directly from the solution of Schrödinger's equation in the wave mechanical model. For some considerations, it will be more convenient as a kind of approximation, and quite legitimate to postulate the particular nature of the electron for the purpose of model construction. The limitation of this model is defined by Heisenberg's Uncertainty Principle.

The energy state of electrons and their distribution in space are expressed by so-called *quantum numbers*. Every single electron is characterized by four quantum numbers which are interconnected with each other in a well-defined way.

The *primary* (or *principal*) *quantum number* (n), according to Bohr's model, expresses which electron shell the electron belongs to and can assume the following values: $n = 1, 2, 3, 4, \ldots$

The *second* (or *azimuthal*) *quantum number* (l) determines the distribution of charge density in space. It is dependent on n, as it can assume only the following values: $l = 0, 1, 2, 3, \ldots (n - 1)$. This implies that the possible values of l are limited by n. Only value 0 and 1, therefore, can be assumed by l, if $n = 2$. Electrons with azimuthal quantum numbers $l = 0, 1, 2$ are defined as: *s-electrons*, *p-electrons*, and *d-electrons* respectively.

The *magnetic quantum number* (m) results from the fact that a moving electron, similar to an electric current in a coil, generates a magnetic field and, consequently, can also be influenced by an external magnetic field. The following values can be assumed by the magnetic quantum number: $m = -l, \ldots 0 \ldots +l$. With $n = 2$, and $l = 0$ or 1; m, therefore, can only be $-2, -1, 0, +1, +2$.

The *spin quantum number* (s) of the electron describes its direction of rotation around its own axis. There are only two conditions possible: clockwise or anti-clockwise rotation. These situations are denoted by $s = +1/2$, and $s = -1/2$. The quantum number has no effect on electron energy, unless magnetic fields are involved.

It can be easily calculated by the combination of quantum numbers, especially those with high n-values, that a large number of electron states can be achieved. In 1926 W. Pauli made a postulation which, so far, has not been invalidated. According to this, the so-called *Pauli exclusion principle*, it is impossible for two electrons with identical quantum numbers to occur in the same atom. This principle limits the number of possible electron orbits and is of great importance for various applications in quantum mechanics.

The points discussed so far can best be demonstrated using the hydrogen atom. Only the values: $n = 1$, $l = 0$, and $m = 0$ are possible. The energy of the electron can be calculated for this condition by the following equation:

$$E = -\frac{e^4 m}{8\varepsilon_0^2 n^2 h^2} = -2.18 \cdot 10^{-18} \, J = -13.6 \, eV, \qquad (2.1.1)$$

where m is the mass of an electron ($m = 9.109 \cdot 10^{-31}$ kg), e is the charge of the electron ($e = 1.60218 \cdot 10^{-19}$ C), h is Planck's constant (h $= 6.626 \cdot 10^{-34}$ J s $= 4.136 \cdot 10^{-15}$ eV s). The electric constant (also referred to as the permittivity constant), $\varepsilon_0 = 8.854 \cdot 10^{-12}$ C V^{-1} m^{-1}, is just a factor which converts the charge as an electrical variable into the mechanical variables: length and force. It will be explained in detail in Section 2.4.1. Here, eV, as a unit of energy is introduced. It will be used frequently in the following text. It stands for "electron volt" and is identical with the kinetic energy of an electron, which has been accelerated by a potential difference of 1 V. Its relation to other energy units is as follows:

$$1 \, eV = 1.602177 \cdot 10^{-19} \, J = 4.45049 \cdot 10^{-23} \, Wh$$
$$= 1.634 \cdot 10^{-20} \, kpm = 1.602 \cdot 10^{-12} \, erg \qquad (2.1.2)$$

The energy of an electron E, as calculated for example by Eq. (2.1.1) for the valence electron of hydrogen, is defined as the amount of work which is required to lift it from its orbit, against the attractive force of the positively charged nucleus, and to move it to an infinite distance from the nucleus. It is also possible to speak of an energy potential, called *ionization potential* because the formation of an ion results when an electron is lost from an atom. In Sections 2.2.2 and 4.5 reference to this idea will be made, describing ionization through interaction with electromagnetic radiation.

The wave function describing the stationary state of an electron is called the *orbital*. Sometimes the term *orbital cloud* is used, which gives a better illustration of the statistical nature of the electron distribution. Spherical orbitals, as in the case of the hydrogen atom, occur when $l = 0$; $m = 0$. Such charge clouds are called *s-orbitals*.

For: $n = 2$, not only $l = 0$, but additionally, the azimuthal quantum number $l = 1$ is possible. In this case, the corresponding orbital resembles a double sphere. The position of the common axis of this double sphere will be determined by the magnetic quantum numbers $m = -1$; $m = 0$ and $m = +1$. These shapes are called *p-orbitals*, with the variants p_x, p_y, and p_z (see Fig. 2.1).

When the azimuthal quantum number is $l = 2$ then there are five possibilities for m, namely the numbers -2, -1, 0, $+1$, and $+2$. Accordingly, in this case five different orbitals are possible. These so-called *d-orbitals* play an important role in the ligand field theory of coordinative bonds which will be explained in Section 2.1.4.

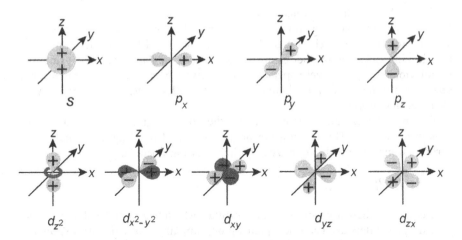

Fig. 2.1. Schematic representation of the *s*-, *p*-, and *d*-orbitals. For clear demonstration of the geometric conditions, the clouds are cut at certain amounts of probability. Therefore, the sharp borders of the orbitals do not reflect their realistic structure

It must be underlined that the spin quantum number does not influence either the shape, or the size of the orbitals. This parameter is important in connection with the pairs of valence electrons when considering the Pauli principle.

2.1.2
Covalent Bonds, Molecular Orbitals

The calculation of molecular orbitals provides the basis for the theoretical interpretation of atomic interactions. A mutual approach of atoms is accompanied by an overlapping of their electromagnetic fields and, consequently, by a change in the wave function of their electrons. The energy levels of the wave functions, modified in such a way, can sink lower than the sum of the levels of energy of the undisturbed atoms. In this case the connection of the two atoms becomes a stable chemical bond.

The intramolecular bonding energy of atoms consists of several components. The kinetic energy of the electrons, the electrostatic interaction among the electrons, and the interactions between electrons and nuclei, for example, are to be included in this calculation. These components have different signs, and their interaction forces are determined by various functions of distance. If at a certain distance the sum of all of these energy functions becomes a minimum then this distance will determine the bonding distance between two atoms in the molecule. This function has been accurately calculated just for H_2, the simplest molecule, which consists of two proton nuclei and one common electron. Yet serious problems already occur when calculating other, more complicated, diatomic molecules.

The most important chemical bond is the *covalent bond*. It can only occur if two mutually approaching atoms have unpaired electrons in their valence shells. This means that both relevant orbits are each occupied by one electron only, thus leaving space for another electron. In this case the Pauli exclusion principle is not violated. Figuratively speaking, an electron pair with anti-parallel spin quantum numbers is formed, being common to both atoms, and forming a type of molecular orbital.

This may be explained using carbon, the atom of greatest importance for biological systems. The distribution of the valence of bonding electrons of the carbon atoms can be characterized by the following formula:

$$2s^2\, 2p_x\, 2p_y.$$

Consequently, there are three occupied orbitals which belong to shell $n = 2$. The p_x- and p_y-orbitals are each occupied by one unpaired electron (no exponent, i.e., exponent = 1), but the s-orbital is occupied by an electron pair (exponent = 2). Hence, only two of the four bond, or valence electrons are able to form bonds. It is, however, a peculiarity of the carbon atom that it jumps to another energy state immediately prior to a reaction by a relatively low energy input (some 250 kJ/mol). For this, one of the two electrons of the s-orbital moves to a completely unoccupied p_z-orbit:

$$2s^2\, 2p_x\, 2p_y \rightarrow 2s\, 2p_x\, 2p_y\, 2p_z.$$

This is a specific property of the carbon atom, and it is just one of the many other physiochemical peculiarities which enabled the emergence of life at all. The nitrogen atom, in contrast to this, is not capable of such modification. Its valency electrons may be formulated as follows:

$$2s^2\, 2p_x\, 2p_y\, 2p_z.$$

Consequently, the nitrogen atom remains trivalent in its reaction.

As a first approach, molecular geometry can be represented by a simple overlapping of the orbitals of the involved atoms. The configuration of the H_2O molecule, using this principle, is depicted in Fig 2.2A. Oxygen has two unpaired electrons with $2p_x$- and $2p_y$-orbitals. Hence, the $1s$-orbitals of the two hydrogen atoms can only form bonds with the two electrons if they come from two defined directions. This leads to an electron cloud built up by one electron from each atom (with different spin numbers!) which is common to both atoms. Accordingly, this is defined as an *sp-bond*.

The above example illustrates the angular orientation of the covalent bond. Measurements have shown, however, that significant deviations can occur from the models which had been obtained by simple geometrical considerations. For example, in the water molecule, the bonding angle between the two hydrogen atoms is 104.5° (Fig. 2.2A) and not 90°, as might have been expected from the orientation of the $2p_x$-orbital of oxygen toward its $2p_y$-orbital (Fig. 2.2A). Such

Fig. 2.2. A H_2O-molecule (simplified construction using the overlapping of the atomic orbitals); B sp^3-hybrid-orbital of CH_4; C π-orbital of a double bond; D π-orbitals of benzene (after Christen 1968)

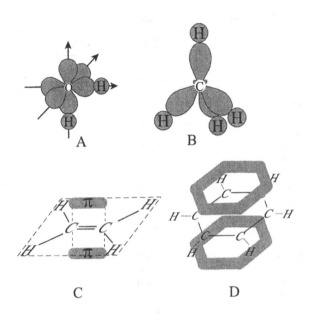

shifts of bonding angles are the result of mutual electrostatic repulsion of the valence electrons. These divergences can be even stronger in other molecules. The four bonding orbitals of the carbon atom in the promoted state have already been discussed. Correspondingly, in methane (CH_4), the three hydrogen atoms should form sp-bonds at right angles to one another, while the ss-bond should be fully undirected. In reality, however, the four hydrogen atoms are found to be in a precise tetrahedral arrangement (Fig. 2.2B).

These examples show that simple geometrical constructions are not sufficient to indicate a realistic picture of the molecule orbitals. For accurate analysis, the Schrödinger equation must be resolved for the entire molecule, or at least for some parts of it. Approximative calculations of this kind result in so-called *hybrid orbitals*, which reflect all of the interactions within the given molecule. The tetrahedrally arranged bond angles of methane, for example, correspond to so-called sp^3-orbitals. The number 3 in the exponent, in this context, indicates that a hybridization of one s-orbital occurred with three p-orbitals.

The sp-orbital, and the sp^3-hybrid orbital indicate rotational symmetry. Bonds of this kind are called σ-bonds. They may be subjected to thermal rotations, as described in detail in Section 2.3.6. The electrons involved in such bonds are called σ-electrons. In case of double bonds, so-called π-orbitals occur with corresponding π-electrons. Such orbitals are not symmetrical with regard to their bonding orientation, as may be seen from Fig. 2.2C.

From Table 2.1 it can be seen that, in a double bond, the interatomic distance is shortened when compared with that of a single bond. The total energy of a double bond is smaller than the sum of energies of two single bonds. In contrast to the classical model of Kekulé, the benzene ring must not be perceived as a

Table 2.1. Properties of biologically important bonds. (After Pullman and Pullman 1963)

Kind of bond	Bond distance (nm)	Bond energy (kJ mol^{-1})
C—C	0.154	348
C=C	0.133	615
C≡C	0.121	812
N—N	0.145	161
N=N	0.124	419
N≡N	0.110	946
C—N	0.147	292
C=N	0.126	615
C≡N	0.115	892
C—O	0.143	352
C=O	0.114	716 (for aldehydes)
		729 (for ketons)
C—S	0.182	260
C=S		477
C—Cl	0.176	329
O—H	0.096	463
N—H	0.101	391
S—H	0.135	340
C—H	0.109	414

static arrangement of alternate single and double bonds, but rather as a continuos π-orbital in which the π-electrons can move around the ring.

Molecule orbitals may sometimes result in a considerable charge displacement within the molecule. Simply, this can be illustrated by considering that the electron pairs common to two atoms in a covalent bond are distributed asymmetrically. In other words, the molecular orbital of the bond is shifted toward one of the two covalently bonded atoms. The atom with the greater probability of the presence of the electron pair is more strongly "electronegative" than the other one.

In this respect, a series of atoms with an increasing degree of *electronegativity* can be constructed. In the periodic system of elements this series is directed toward increasing atomic numbers within the periods as well as within the groups. Hence, the following relation applies:

H < C < N < O < F, and

J < Br < Cl < F.

In this way, the displacement of bonding angles within a water molecule can be explained. Because of the strong electronegativity of the oxygen atom in relation to the hydrogen atom, a dipole O—H results and, consequently, the two hydrogen atoms will repel each other (for more details on the structure of water and the importance of this polarization effect, see Sect. 2.4.2). This polarization effect contributes directly to the bonding energy. About 22% of the bonding energy in a C—O bond, and 39% in a H—O bond are attributable to this effect.

2.1.3
Ionic Bonds

If the polarization effect of covalent bonds, as described in the previous section, is pushed to the extreme, it is no longer possible to refer to a molecular orbit, nor to bonding electrons. There occurs a total transmission of an electron from the valency orbital of one atom to that of the other. This causes a full separation of charges. Ions are generated, which attract each other electrostatically. With the loss of the molecular orbitals there also occurs a loss of the molecular identity. Strictly speaking, it makes no sense to use the term NaCl-molecule. In solutions, the molecular character of this salt is just expressed by the stoichiometric relation of the number of anions and cations. A crystal of NaCl, on the other hand, can be considered as a super molecule, because in this case the ions are arranged in an electrostatic lattice.

The ionic bond, in contrast to the covalent bond, can be considered simply from an electrostatic point of view. The basic formula of the electrostatics is *Coulomb's law*. It defines the force (F) with which two points, carrying electric charges (q_1) and (q_2), repel each other at a distance (x) in a vacuum.

$$\mathbf{F} = \frac{q_1 q_2}{4\pi\varepsilon_0 x^2}. \tag{2.1.3}$$

(The meaning of the electric field constant $\varepsilon_0 = 8.854 \cdot 10^{-12}$ C V^{-1} m^{-1} will be explained in Chap. 2.4.1). This equation also makes it possible to calculate the bonding force, as the negative value of the force (−F) which is required to separate two ions. This force, however, is not measurable directly. It is better therefore to transform this equation in such a way that it can provide information on the bonding energy or, which is the same, on the energy of ionization. The *bonding energy* is equivalent to the work which is necessary to move a given ion from its bonding place up to an infinite distance from the counter ion. A work differential (dW), which has a very small, but finite value, can be calculated from the product of force (F) and a corresponding distance differential (dx):

$$dW = -\frac{q_1 q_2}{4\pi\varepsilon_0 x^2}\, dx. \tag{2.1.4}$$

The sign of dW depends on the point of view or consideration, and thus is a matter of definition. Work applied to a system (e.g., by combination of the ions) means a gain of work by the system, and at the same time a loss of work by the environment. As in the majority of textbooks, we will consider here positive dW as a gain of work by the system. Furthermore, positive dx means an enlargement, negative dx, on the other hand, a diminution of the distance between the charges. In the case under review [Eq. (2.1.4)], dW becomes positive if the signs of q_1 and q_2 are equal, and both charges are moved toward each other (this means: $dx < 0$).

The overall work (W) which would be necessary to move the particles from the distance $x = \infty$ closer to the bond distance $x = x_1$ can be obtained by integration of Eq. (2.1.4) within these limits.

$$W = \int_{\infty}^{x_1} dW = \int_{\infty}^{x_1} -\frac{q_1 q_2}{4\pi\varepsilon_0 x^2}\, dx = \frac{q_1 q_2}{4\pi\varepsilon_0 x_1}. \qquad (2.1.5)$$

As Eq. (2.1.5) applies to any value of x, no index is required for these variables. Bonding energy (E) we defined above as the energy needed to interrupt the bond, this means, to move the particle from $x = x_1$ to a distance $x = \infty$. Hence, $E = -W$.

Let us illustrate this situation by the following consideration. What is the value of the bonding energy between a Na^+, and a Cl^- ion? The charge of either ion is based on the excess or the lack of one electron respectively. The electrostatic charge of a single electron is $1.602 \cdot 10^{-19}$ C. This value, furnished with the appropriate sign, must be inserted in Eq. (2.1.5). Now, information on the bonding distance (x) is still required. On account of their electrostatic attractive force, both ions would fully fall into each other, however, at the minimum distance of approximation a strong electrostatic repulsion is generated. This repulsive force is related to the structure of their inner electronic orbits. The minimum distance of approximation of two ions is, in fact, the sum of their Van der Waals radii. The *van der Waals radius* (sometimes also called *Bohr's radius*) can be determined by means of X-ray diffraction diagrams from the position of the atoms in the crystal. It is also referred to as the *crystal radius*. For the sodium ion this value is 0.098 nm (see Table 2.3, page 60), for chloride: 0.181 nm. Consequently, the distance between both ions in the NaCl molecule (i.e., in the crystal) amounts to: $x = 2.79 \cdot 10^{-10}$ m.

Substituting these values in Eq. (2.1.5), we obtain:

$$E = -W = -\frac{(-1.602 \cdot 10^{-19}) \cdot (1.602 \cdot 10^{-19})}{4 \cdot \pi \cdot (8.85 \cdot 10^{-12}) \cdot (2.79 \cdot 10^{-10})} = 8.27 \cdot 10^{-19}\, \text{J}, \qquad (2.1.6)$$

or, according to the conversion, given in Chap. 2.1.1

$$E = (8.27 \cdot 10^{-19}) \cdot (6.242 \cdot 10^{18}) = 5.31\, \text{eV}.$$

Usually, the unit electron volt (eV) is used to characterize the energy of single bonds in a molecule. For macroscopic considerations the unit Joule (J) is used. For calculations of the molar bonding energy, the *Avogadro number* is required, which gives the number of molecules per mol:

$$E = (8.27 \cdot 10^{-19}) \cdot (6.02 \cdot 10^{23}) = 4.98 \cdot 10^5\, \text{J mol}^{-1},$$

$E = 498 \, \text{kJ mol}^{-1}$.

The electrostatic nature of ionic bonds as described in this section, automatically indicates that in contrast to the covalent bonds, no valence angle is predicted.

2.1.4
Coordinative Bonds, Metallo-Organic Complexes

The stability of a series of biologically important molecules is mediated by polyvalent metals and transition elements. This stability cannot be explained either by single covalent bonds, or by ionic bonds. For these types of bonds, which occur mainly in inorganic chemistry, the terms *coordinative bonds* or *complex bonds* have been introduced.

These complexes, which may also be charge-carrying ions, consist of a *central atom* which is surrounded by ligands, i.e. molecular complexes, in geometrically defined arrangement. There are also polynuclear complexes with several central atoms. Ligands are directly bound to the central atom but form no bonds among each other. If a ligand has two or more bonds to a central atom it is called *chelate complex*.

Most of the central atoms are elements with incompletely occupied d-orbitals, i.e. elements with bonding electrons having principal quantum numbers $n > 2$. Some trace elements of biological systems are of special interest in this context, for example Fe (in the porphyrin complex of hemoglobin), Mg (in chlorophyll), Co (in B_{12}-vitamin). The same phenomenon is of importance for the Ca-Mg-antagonism in cell physiological processes.

An explanation for these bonding mechanisms is given by the *ligand field theory*. The approach of the ligands to the central atom leads to a re-orientation of the electron orbitals. This interaction of the central atom with the ligands is caused by the so-called *ligand field*. As may be seen from Fig. 2.1, according to the differences in the magnetic quantum number (m), five different d-orbitals are possible. Considering that each orbital can be occupied by two electrons with different spin quantum numbers, the total number of possible electron states will be higher than the number of electrons actually present.

An isolated Fe^{3+}-ion, for example, occupies in free solution each of these orbitals simply with one unpaired electron (Fig. 2.3). Hence, all five electrons are in the same energy state. When a ligand now approaches this atom, some of the electrons will assume a higher energy level, and some of them a lower one, depending on the direction of the approach, relative to the orbital coordinates. For reasons of quantum mechanics, the mean energy of all these levels has to remain constant. This means that the amount by which one level is elevated must be the same as that by which the other level has been lowered.

If the field is still weak, the differences in the energy levels are too low to induce changes of the spin quantum numbers which would be necessary to avoid conflicts with Pauli's exclusion principle (see Sect. 2.1.1). Such a situation is called a *high spin state*, since the number of unpaired electrons remains at a

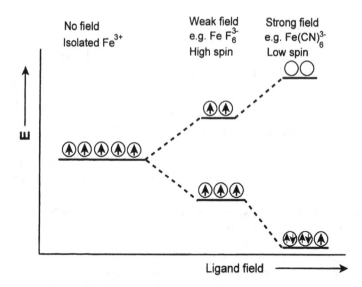

Fig. 2.3. The occupation of the orbitals of Fe^{3+} influenced by weak or strong ligand fields. The *arrows* in the *circles* indicate the orientation of the electron spins. *Circles* without *arrows* indicate empty orbital positions

maximum. Strengthening the field will cause all of the electrons to collect together at the lowest level of energy, with some of them undergoing spin flip. This results in a *low spin state*. A favorable energy state has thus been reached for the central atom, as may be seen in Fig. 2.3. The stability of the complex is guaranteed by the existence of the difference between the new and the old energy levels.

The hem-complex, which is a prosthetic group of a number of important proteins (hemoglobin, myoglobin, cytochrome, etc.), as depicted in Fig. 2.4 is an example of a chelate complex. Here, the central iron atom is under the influence of the ligand field of porphyrin. This molecule consists of four pyrrole nuclei, located in one plane and kept together by methine bridges. Four nitrogen atoms are coordinated to the iron atom. The remaining bond orbitals of the iron are perpendicular to this plane and can be occupied by other ligands. Chlorophyll is a porphyrin complex with magnesium as the central atom.

Free bonding sides of the central atom can also be occupied by water molecules. A major role in the ligand binding process is played by the competition between ligands with different affinities. This competition explains many inhibitory effects in metabolism. The action of cyanide, a respiratory poison, may be mentioned as an example. In this case, water is displaced by CN^- from its coordination bonds with the iron atom of hemoglobin.

The highly specific nature of reactions involving complex formation in biological systems cannot be explained, however, simply by differences in the bond affinity. In some cases, stearic aspects of macromolecular structures are

Fig. 2.4. The structure of the hem complex. The Fe^{2+} is bound four times with the pyrrole ring (chelate bond). Both free valences of the central atom (Fe^{2+}) are perpendicular in relation to the plane of the pyrrol-ring system

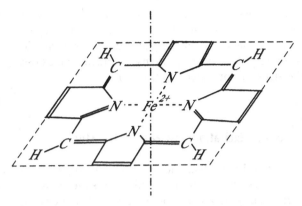

responsible for this. Because of the dynamic properties of biological structures, in other cases the rate constants of the processes of complexation are essential. For example, the equilibrium constant of the complexation of ATP (adenosine triphosphate) with magnesium is nearly the same as with calcium. Nevertheless, the Ca-ATP complex occurs in larger concentration, because the rate constant of its formation is about 100 times faster than for the Mg-ATP complex.

2.1.5
Hydrogen Bond

A brief reference will be made at this point to a special kind of dipole-dipole interaction which is of great importance in molecular biology. The contact of a hydrogen atom with an electro-negative partner leads to the formation of a polar molecule in which the hydrogen atom is the positive pole (see Sect. 2.1.2). This dipole can attract another polar molecule which will then turn its negative pole towards the bonded hydrogen atom. The two dipoles may approach each other very closely. These distances between 0.26 and 0.31 nm are even below Van der Waals radii. This suggests that in the formation of such bonds there is a covalent contribution in addition to the electrostatic interaction. Once the two molecules have moved sufficiently close to each other, the hydrogen atom can no longer be unambiguously assigned to one of them. It belongs, quasi, to both molecules simultaneously and, consequently, constitutes a so-called *hydrogen bridge*.

To calculate the bonding energy of a hydrogen bridge, aspects of wave mechanics have to be taken into consideration. This energy is between 13 and 25 kJ mol^{-1} and is a function of bond distance. The hydrogen bond , therefore, falls into the category of "weak" bonds, which can easily be split by thermic collision in the range of biological temperatures (cf. Sect. 2.3.5). The two molecules connected by a hydrogen bridge face each other with their negative poles and therefore repel each other; thus the hydrogen bond between them becomes stretched. Thus, hydrogen bonds which are formed on large molecules are orientated at defined angles.

Hydrogen bridges may be formed within large molecules as well as between them. More details on the importance of this type of bond will be given in the following sections, in the context of the formation and stabilization of the secondary structure of proteins and nucleic acids (Sect. 2.4.5) and with regard to the structure of water (Sect. 2.4.2).

2.2
Molecular Excitation and Energy Transfer

From the thermodynamic point of view, the living organism can be considered as an open system in nonequilibrium state. This nonequilibrium is maintained permanently by a continuous uptake of free energy. The biosphere assimilates this energy from photons of sunlight through the process of photosynthesis. This assimilated energy is then stored and transferred in molecules with energy rich bonds at first within plants, and then through the food chain into animals.

This long process, starting with energy assimilation in the thylakoid membranes of chloroplasts and ending, for example, in energy transformation and dissipation in humans and other animals, is associated with numerous elementary processes which can be described in terms of quantum mechanics. In the following chapter we will briefly discuss this variety of processes.

2.2.1
Mechanisms of Photon-Induced Molecular Excitation

Whenever an assessment of the effect of any radiation energy is made, it should always be borne in mind that the effective amount of energy is only that which is actually absorbed by the system, and not that part which penetrates it. This is the so-called *Grotthus-Draper principle*. The Grotthus-Draper principle is applicable for all kinds of radiation. It plays a particular role in photobiology because of the specific spectral absorption properties of light.

The wavelength of visible light lies between 400 and 800 nm. This corresponds to quantum energies between 3.1 eV and 1.6 eV respectively. (cf. Sect. 4.4, Fig. 4.13). The quantum energy of the photons of visible light, therefore, is much lower than the ionization energy of water (12.56 eV), a value which is arrived only in short-wave ultraviolet light (UVC). The energy of thermic noise (kT), which we will consider in detail in Section 2.3.4, in relation to this, is much smaller. At temperatures of biological importance it amounts only to about 0.025 eV. For photosynthesis the organism uses exactly this gap between the quantum energy of thermic noise on the one hand, and the quantum energies causing ionization of water and organic molecules on the other hand, using the energy of "visible" light. The photons of light are strong enough to be used for an effective process of energy conversion, but they do not endanger the stability of biomolecules. How narrow this biologically usable part of the electromagnetic spectrum in fact is, can be seen from Fig. 4.13.

Lets consider first the process of molecular excitation in general. If a molecule absorbs a quantum of energy of light, a sequence of processes will be induced. This is illustrated in Fig. 2.5. In a first step, the absorption of a photon leads to the raising of an electron to an orbit of higher quantum number. This can occur in the framework of the following two series of energetic states, which are qualitatively different: $S_0, S_1, S_2, \ldots$, and $T_1, T_2, T_3, \ldots$ Between these steps with additionally increasing quantum numbers, small energetic steps of thermic excitations are positioned. In the case of *singlet states* (S), the electrons of a pair have anti-parallel orientated spins; the spin quantum numbers, therefore, have different signs. In the case of *triplet states* (T), the spins of the electrons of the pair have antiparallel orientation, their spin quantum numbers are identical. As already stated, the occurrence of electrons in which all quantum numbers are equal is ruled out by Pauli's exclusion principle (cf. Sect. 2.1.1). Thus, if the triplet state represents an electron pair with identical spin quantum numbers, these two electrons must differ with regard to other energy parameters although their orbitals can be energetically very similar. A triplet is a so-called *degenerated state*.

These two states of excitation differ substantially in their life span. Figure 2.5 shows that a setback, $S_1 \rightarrow S_0$ occurs under emission of *fluorescent light* within 10^{-9} to 10^{-5} s, whereas a setback, $T_1 \rightarrow S_0$, recordable as *phosphorescence*, will occur at a much slower rate. Consequently, triplet states are characterized by an increased stability when compared with excited singlet states.

2.2.2
Mechanisms of Molecular Energy Transfer

All processes of life need a driving energy, which originally comes from the quantum energy of visible light, emitted by the sun and absorbed by pigments of photosynthetic units. After this process of molecular excitation, the absorbed

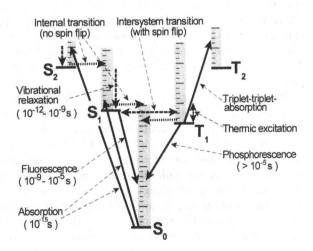

Fig. 2.5. Energy diagram of ground and excited states of organic molecules

energy must be accumulated and transmitted to other parts of the cell, to other parts of the plant, and finally to other organisms which are not able to obtain energy by photosynthesis directly.

For this, the energy of molecular excitation will be transformed to the chemical energy of so-called *high-energy* compounds. The most common accumulator of chemical energy in the cell is adenosine triphosphate (ATP), formed in the process of photosynthesis and used in nearly all processes of energy conversion in other cells. The hydrolysis of ATP, producing adenosine diphosphate (ADP) and catalyzed by special enzymes, the ATPases, allows the use of this stored energy for ionic pumps, for processes of molecular synthesis, for production of mechanical energy and many others. The amount of energy, stored by the ADP $\rightleftarrows$ ATP reaction in the cell, however, is limited just because of osmotic stability. Therefore, other molecules, like sugars and fats, are used for long-term energy storage. The free energy of ATP is used to synthesize these molecules. Subsequently, in the respiratory chain, these molecules will be decomposed, whereas ATP is produced again.

These metabolic steps are the subject of biochemistry. Here, the processes of energy transfer will be discussed in their most general form. The following nomenclature of electrochemistry will be used: the energy-supplying molecule will be referred to as *donor*, the energy receiving molecule as *acceptor*, no matter what the actual mechanism of energy transfer is.

In general, the following mechanisms of intermolecular energy transfer must be considered:

– energy transfer by radiation
– energy transfer by inductive resonance
– energy transfer by charged carriers

Energy transfer by radiation can be envisioned in the following way: the excited molecule emits fluorescent radiation which matches precisely the absorption spectrum of the neighboring molecule and, consequently, excites it. Such mechanisms are capable of transferring energy over distances which are large compared with the other processes described in this context. However, the efficiency of this process is quite low, it declines sharply with increasing distance. In fact, such mechanisms do not play a significant role in biological processes.

Of particular importance, primarily in the process of photosynthesis, is the transfer of energy by inductive processes, namely the so-called *resonance transfer*. This form of molecular energy transfer is a *non-radiant energy transfer*, since fluorescent light does not occur in this process. This mechanism can be envisioned as some sort of coupling between oscillating dipoles. The excited electron of the donor molecule undergoes oscillations and returns to its basic state thus inducing excitation of an electron in the acceptor molecule. This process requires an overlapping of the fluorescent bands of the donor with the absorption band of the acceptor, i.e., on the resonance of both oscillators. The smaller the difference between the characteristic frequencies of donor and acceptor, the faster the transfer will be. Singlet, as well as triplet states can be involved in such processes. So-called strong dipole-dipole couplings are possible to distances of up to 5 nm. This is a process in which a $S_1 \rightarrow S_0$ transition in the donor molecule induces an $S_0 \rightarrow S_1$ excitation in the acceptor. Inductions through triplet states are only effective over shorter distances.

Energy transfer by charge carriers is the most common reaction in metabolic processes. It is a process which can take very different courses. The redox process is a classical example of this. It consists basically in the transfer of one or two electrons from the donor to the acceptor molecule. In this way, the donor becomes oxidized and the acceptor reduced. This apparently simple scheme however conceals a number of complicated sub-routines which have not yet been completely resolved.

For this process of electron transfer, donor and acceptor molecules must be in exactly defined position to each other and at a minimum distance, so that overlapping of respective electron orbitals can occur. In the first place, donor and acceptor will form a complex of highly specific stearic configuration, a so-called *charge transfer complex*. This process of complex formation which occasionally requires stearic transformations of both molecules, causes the actual transfer. It takes place at lower rates than energy transfer by induction as discussed earlier. Hence, the charge-transfer complex is an activated transition state which enables redox processes to take place between highly specific reaction partners in the enzyme systems of the cellular metabolism. Because of the oscillating nature of electron transfer, this coupling of two molecules is strengthened by additional electrostatic forces sometimes called *charge-transfer forces*.

In the process of energy transfer, differences between energetic potentials of donor and acceptors play an important role. An uphill transfer of electrons is only possible through an input of external radiation energy. Actually, these differences are slight, when compared with the absolute values of the ionization energy. They are only in the region of about 1.5 eV.

What is the scale of these energy gradients? Szent-Györgyi in his remarkable "Study in Cellular Regulations, Defense, and Cancer" in 1968 proposed to introduce a scale of so-called *biopotentials*. This scale is based on the ionization energy of water (12.56 eV), the basic molecule of life, and has the opposite direction to the scale of ionization energy (see Fig. 2.6). This scale has never really come to the fore, probably because the name is somewhat confusing as the same term had already been used for a completely different parameter, namely for the electrical transmembrane potential. On the other hand, it is quite illustrative.

In electrochemistry, a more common scale is used, namely the scale of redox potentials. This is based on measurements with hydrogen electrodes, i.e. platinum electrodes, surrounded by hydrogen gas. A potential of 0.82 V exists between a hydrogen electrode and an oxygen electrode, generating water. This therefore corresponds to the reference point of Szent-Györgyi's biopotenials.

The energies, depicted in Fig. 2.6, refer to bonds which are involved in the corresponding reactions. If chains of biochemical reactions are studied, the term

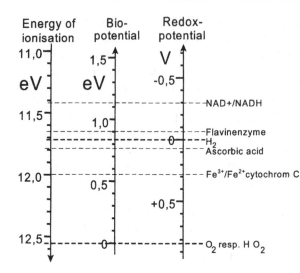

Energy of ionisation   Bio-potential   Redox-potential

Fig. 2.6. Correlations of the scales of ionization energy, redox-potential and Szent-Györgyi's biopotential

transfer potential is used. The redox potential would thus be an electron transfer potential. Similarly, the transition of ATP → ADP is referred to as a phosphate transfer potential. In general terminology, these are all group transfer potentials. As we will see later, in the terminology of phenomenological thermodynamics it is a standard free Gibbs energy of formation (ΔG^0) with the following relationship to the standard redox potential (E^0):

$$\Delta G^0 = -n\mathrm{F}E^0, \tag{2.2.1}$$

where n is the number of transmitted electrons and F is the Faraday constant.

Redox processes are of great importance in biological metabolism, since oxygen has through the ages progressively invaded what was initially an anaerobic world. The electron transfer from O_2 to its end product of reduction – H_2O is a complicated process and is attributed to the formation of free radicals. A radical is an electrically neutral species with an unpaired electron. These radicals can occur either on a photodynamic way, or by catalysis of transition metals or enzymes. A special role in cell metabolism is played by the so-called *reactive oxygen species* (ROS). They trigger chain reactions capable of damaging the different constituents of the cell. The organism uses different anti-oxidant defense systems against this toxicity of oxygen. We will come back to the role of ROS when discussing primary processes of radiation chemistry of water (Sect. 4.5.2).

In connection with the processes of charge transfer, the *semi-conductor effects* shall be mentioned. In solid state physics, a semi-conductor is defined as a material the electric resistance of which is higher than that of a metallic conductor indicating a very specific temperature dependence. Semi-conductor phenomena can be based on mechanisms which differ significantly from each other. This is particularly true for organic semi-conductors.

Evidence for the existence of semi-conductor phenomena in biological structures is very difficult to obtain. Objects which contain water usually exhibit an ion conductivity which masks the semi-conductor effect. Water-free macromolecules, on the other hand, may be considered to be denatured. In spite of these difficulties, semi-conductor properties have been verified from proteins, nucleic acids and lipids. Its role in biological functions, however, is still unclear. In some cases also the effect of supra-conductivity was also proposed to occur in biological molecules. Supra-conductivity is to be considered as an electron transition with negligible loss of energy.

In relation to the above mentioned mechanisms, reference must also be made to *piezoelectric effects* in biological matter. These comprise translocations of charges in crystals or crystalloid macromolecular structures resulting from mechanical loading, such as compression, elongation or bending. Piezoelectric properties have been found in bones, wood and various tissues (including tendons), as well as in isolated biomacromolecules. It should be noted that the piezoelectric properties of bones are not the result of mechanical tension in the inorganic crystals, but rather of the protein components of the bone. The biological role of this piezoelectricity is unclear. In connection with the streaming potentials occurring in living bones, there may be a connection with processes of bone remodeling (cf. Sects. 3.5.2 and 3.5.6).

Further Reading: For free radicals and biochemical redox-systems: Buettner 1993; Bucala 1996; for supra-conductivity of biological materials: Tributsch and Pohlmann 1993; for piezoelectricity: Fukada 1983; Guzelsu and Walsh 1993

2.2.3
Photosynthesis as Process of Energy Transfer and Energy Transformation

Molecular processes of energy transformation take place in many metabolic systems. Of special interest, however, are the primary reactions in which the quantum energy of sunlight is transformed into the energy of chemical bonds. This is the case in photosynthesis, as well as in photoreception. Photosynthesis is the first step of energy gain in the biosphere. Photoreception on the other hand, is a process of signal transduction.

About 0.05% of the total 10^{22} kJ energy which reaches the earth every year from the sun is assimilated by photosynthesis. This is the general energetic pool for all living processes of the earth. The efficiency of photosynthesis is very high, compared with our recent technical equipment. A large amount of the energy of sunlight, absorbed by the photosynthetic reaction centers of the green plants is transformed by the primary process of photosynthesis. During subsequent metabolic processes of energy transfer, however, an additional loss of energy occurs. The total efficiency of the process of photosynthesis, in fact, is assumed to be about 5%.

In eucaryotic plants the process of photosynthesis occurs in the chloroplasts, especially in the thylakoids located there. *Thylakoids* are flat vesicles with a diameter of about 500 nm, stacking together to form so-called *grana*. About 10^3 thylakoids are located in one chloroplast. Every thylakoid contains about 10^6 pigment molecules. A precondition for the efficiency of energy transfer in the process of photosynthesis is the high degree of supramolecular organization.

In general, photosynthesis can be considered as a reaction during which water is split, driven by the energy of photons, producing O_2, and transferring hydrogen to the redox system NADPH/NADP$^+$, the nicotinamid-adenin-dinucleotid phosphate. Simultaneously, a proton gradient across the thylakoid membrane is generated. Within a separate process, this leads to the synthesis of ATP. Subsequently, an ATP-consuming synthesis of carbohydrates occurs. This process is usually called the *dark reaction* of photosynthesis occurring in the stroma of chloroplasts. In general, photosynthesis can be considered as the reversal of respiration and can be characterized roughly by the following scheme:

$$6\,CO_2 + 6\,H_2O \xrightarrow{h\nu} C_6H_{12}O_6 + 6\,O_2.$$

This process is characterized by a standard free Gibbs energy of reaction of $\Delta G^0 = +2868$ kJ mol^{-1} [the basic thermodynamic parameters will be explained in detail in Section 3.1.2; see Eq. (3.1.31)].

As indicated in Fig. 2.7, the absorption of the photons $h\nu_1$ and $h\nu_2$ occur at two points: at photosystem I, and at photosystem II. Each of these photosystems has its own reaction center, the place where the photochemical processes actually occur, and an antenna system, that occupies the largest part of the complex. These antennae contain polypeptides with light absorbing pigments. They are responsible for the absorption of light of various wavelengths and, correspondingly, quantum energies, and for their transport in the form of so-called *exitons* to the reaction center. This energy transmission is realized by processes of non-radiant energy transfer. One reaction center corresponds to nearly 300 antenna molecules. This antenna system, in fact, enlarges the diameter of optical effectiveness of the photoactive pigments nearly by two orders of magnitude. Depending on the species of plant, chlorophyll and various pigments (e.g., carotenoids like xantophylls) may be part of these antennae. The size, composition and structure of the antennae are quite different for bacteria, algae and various higher plants. This is understandable, considering the large differences of intensities and of spectral characteristics of the light leading to photosynthesis in organisms ranging from submarine algae, to tropical plants. Probably, the antenna molecules of photosystem I and photosystem II are interconnected to each other. If photosystem II, for example, is overloaded, absorbed energy will be transferred to photosystem I by a so-called *spillover process*. There are further mechanisms to protect the photosystem from over exposition.

In Fig. 2.7. the primary process of photosynthesis is depicted schematically. The complexes: photosystem I (PSI), photosystem II (PSII), as well as the cytochrome b/f complex (Cyt b/f) are seen as components of the thylakoid

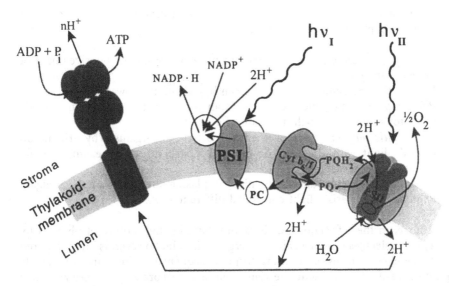

Fig. 2.7. Structural and functional organization of primary processes of photosynthesis in the thylakoid membrane. Explanation in text. (From Renger 1994, redrawn)

membrane. The light-induced splitting of water occurs in the so-called water oxidizing multienzyme complex, structurally connected to photosystem II, where the molecular oxygen becomes free, but the hydrogen, in contrast, is bound by plastoquinone (PQ), forming plastohydoquinone (PQH$_2$). The cytochrome b/f complex mediates the electron transport between PSII and PSI, reducing two molecules of plastocyanin (PC), using the hydrogen of PQH$_2$. This light-independent process at the same time extrudes two protons into the internal volume of the thylakoid. In a second photochemical process, the photosystem I (PSI), using the redox potential of plastocyanin, transfers one proton to NADP$^+$, producing NADPH, one of the energy rich products of photosynthesis.

As a general result of this process, starting with molecular excitations by light, including a system of extremely complicated redox reactions, protons are pumped from the stroma into the internal volume of the thylakoids. This electrochemical gradient of protons is the energy source of an ATP-synthetase, which finally releases ATP as a mobile accumulator of energy into the stroma.

These mechanisms of absorption and transport of energy in the system of photosynthesis, were recently investigated using modern methods of photometry with laser flashes, the durations of which were as short as 10^{-14} s. To study the mechanisms of primary photobiological reactions, many investigations were performed on bacteria rhodopsin, a pigment-protein complex which is also located in a special system of membranes.

In addition to photosynthesis as the primary process of energy conversion, a number of other photobiological processes exist, which are responsible for light reception and therefore use photons as a source of information. This concerns the following physiological functions:

- *Visual perception* of photons to obtain a more or less comprehensive image of the environment
- *Phototaxis* – movement of an organism by orientation towards light
- *Phototropism* – movement of parts of an organism by orientation towards light (for example: growth direction of a plant)
- *Photonastic movement* – movement of parts of an organism by orientation towards light with the help of performed structures (for example: daily orientation of blossoms)
- *Photodinesis* – light-induced alterations of plasma flow (for example: in algae)
- *Photomorphogenesis* – light controlled differentiation.

Of course, in these functions, as in photosynthesis, the capture of photons by receptor molecules means a gain in energy. This gain, however, is low compared with the energy turnover resulting from subsequent reactions. Just one single photon is sufficient to excite the visual cell of a vertebrate eye. These reactions are possible only by processes of amplification, which need a considerable input of metabolic energy. This high sensitivity of the receptors is associated with a high tolerance of the system to overexposure. The visual cell which is even responsive to the absorption of one single photon tolerates a maximal irradiation of 10^5 quanta. We will mention this problem during the discussion of formal theories of radiation effects in Section 4.5.5.

Further reading: Barber 1992; Deisenhofer and Norris 1991; Govindjee 1975; Hall and Rao 1994

2.3
Thermal Molecular Movement, Order and Probability

In this section, the biophysics of molecular organization of biological systems will be discussed in the context of processes of thermal movements. Having presented atomic processes, which are best described by the equations of quantum and wave mechanics, we now come to topics where a statistical mechanics approach can be applied. Stochastic phenomena are of great importance in molecular biophysics. Here, we are immediately confronted with the dialectics of arbitrary distribution on the one hand, and organized order on the other. This touches on problems of biological self-organization and stability of the resulting structures.

2.3.1
Thermodynamic Probability and Entropy

In 1854, Rudolf J. E. Clausius introduced the *entropy* (S) as a parameter of phenomenological thermodynamics, and defined it as the heat, added to a system in a reversible way in relation to the temperature [see Eq. (3.1.10)]. Later, in 1894, Ludwig Boltzmann used this parameter in the framework of statistical thermodynamics. In this context, entropy and the second principle of thermodynamics become more imaginable. Entropy appears as a kind of measure of disorder, or as a degree of random distribution, i.e. of missing order. The correlation between order and probability and, as will be explained later – information – is of great importance for the understanding of the principles of biological organization.

Let us start with the assumption that entropy is a measure of randomization of a given distribution. We will consider a system of maximal entropy as a system in maximal disorder. Furthermore, let us demand that the entropy be an extensive parameter. Therefore, like volume, or mass, but in contrast for example to temperature or density, the entropies S_1 and S_2 of two systems can be added, if these systems come together:

$$S_1 + S_2 = S. \tag{2.3.1}$$

How can we now define a parameter which indicates the degree of randomization or, on the contrary, the degree of order? What does order of organization mean? Of course, our daily experience shows that an ordered system spontaneously transforms into a disordered one, but not vice versa. This, actually, is the consequence of the second principle of thermodynamics.

Let us consider a very simple structure, just the distribution of four distinguishable spheres in two compartments of a box (Fig. 2.8). Let each of these spheres, independently of the three others, fall just by chance into one or the other compartment of the box. All of the 11 possibilities of the distribution, as indicated in Fig. 2.8, therefore, have the same degree of probability, because the probability of each sphere individually falling into compartment 1 or into compartment 2 is equal. Summarizing the patterns of distribution, there is only one way to realize the distributions 0:4 and 4:0. In contrast, there are four ways to realize the distribution 3:1 and 1:3, and, finally, six ways for equal distribution: 2:2.

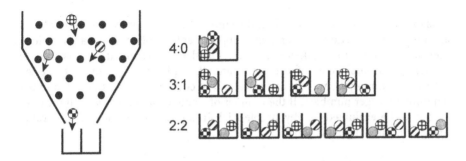

Fig. 2.8. All possibilities of the statistical distribution of four spheres in two compartments of a box

Let us now ignore the fact that the spheres are distinguishable. Let us simply ask: how large is the probability that just by stochastic distributions one of the relations 4:0, 3:1, 2:2, 1:3, or 0:4 occurs? Apparently, the probability of any kind of distribution will be larger if it can be realized in a larger number of ways. The distribution mode 2:2, for example, is six times more likely than the distribution 4:0, or 0:4. The number of ways which lead to the realization of a definite situation, in fact, seems to be a measure of the probability of the occurrence of it. We will designate this number of ways by the parameter W which we will call *thermodynamic probability*. The amount of W can be at least 1 and at maximum ∞, in contrast to the mathematical probability (P), which we will use in Chapter 2.3.2, and which ranges between 0 and 1.

Now we come to the following sequence of conclusions: if W really is a measure of the probability of getting a definite distribution, and if an increase of the degree of order is the most uncertain result of a stochastic distribution and finally, if the entropy (S) is a parameter, indicating the degree of disorder, then S should be a function of W. If two situations with relative probabilities W_1 and W_2 are connected together, then the probability of this combined situation results from the product $(W_1 \cdot W_2)$. Using Eq. (2.3.1), this means:

$$S = f(W) = S_1 + S_2 = f(W_1) + f(W_2) = f(W_1 \cdot W_2). \qquad (2.3.2)$$

This demand is met by the logarithmic function:

$$\ln A + \ln B = \ln(A \cdot B). \qquad (2.3.3)$$

Hence entropy is proportional to logarithm of W:

$$S = k \ln W. \qquad (2.3.4)$$

This is the Boltzmann equation of entropy. Boltzmann's constant k was defined as a universal constant later by Max Planck. It must have the same unit of measurement as entropy, and is as follows:

$$k = 1.380658 \cdot 10^{-23} \, \text{J K}^{-1} = 8.6174 \cdot 10^{-5} \, \text{eV K}^{-1}.$$

This explanation has been based on the simplest experiment where four spheres are distributed randomly over two compartments. One step towards serious thermodynamics can be taken if, for example, the compartments of this box are thought of as molecules of a system, and the spheres as quanta of energy, distributed among them. This complication, of course, means a transition to the handling of larger numbers. If the number of elements and classes are increased, W cannot be evaluated just by simple trial. It is possible to calculate this value using the following equation:

$$W = \frac{n!}{n_1! \cdot n_2! \cdot n_3! \cdots \cdots n_m!}, \qquad (2.3.5)$$

where n is the total number of all elements in the system (in case of Fig. 2.8, the total number of spheres: $n = 4$); n_i (for $i = 1, \ldots, m$) is the number of elements in each class of state (this means the number n_1 in compartment 1 and n_2 in compartment 2); and m is the number of classes of state (namely: number of compartments in the box).

2.3.2
Information and Entropy

C. E. Shannon introduced a parameter into information theory which was to some extent, by its very substance, related to entropy; this and was named *information*. The information of a message depends on the effort required to guess it by a highly set system of questions. Hence, information is some sort of degree of the actuality of a message.

It is not difficult to guess the result of the toss of a coin, since there are only two possibilities of equal probability. To guess a certain card in a full deck of playing cards is much more difficult. In this case, a much greater uncertainty factor has to be taken into account. Using a more systematic approach, a large number of yes-no questions have to be answered. Hence, the information content of a playing card is higher than that of a tossed coin. Should a deck consist of cards which are all the same, and this is known by the challenged person, guessing will not make sense at all. The information content of each of these cards is zero. The probability by which possibilities are turned into reality, consequently, seems to become a measure of information.

In contrast to thermodynamics, in information theory the mathematical term probability (P) is used, which is defined as follows:

$$P = \frac{\text{number of favorable cases}}{\text{greatest possible number of cases}}.$$

On average, coins tossed a hundred times will land with heads up in 50 instances. Hence, the probability of heads facing up may be expressed as follows:

$$P = \frac{50}{100} = \frac{1}{2}.$$

On the other hand, the probability of throwing a "six" with some dice is only $P = 1/6$, whereas the probability of throwing one of the three even numbers would be higher: $P = 3/6 = 1/2$.

Whereas the thermodynamic probability (W) is always larger than 1, (cf. Sect. 2.3.1), the value of the mathematical probability lies between 0 and 1 ($0 \leq P \geq 1$). In this context, $P = 0$ means an impossibility, while $P = 1$ expresses absolute certainty.

The logical conclusions which led to the derivation of the Boltzmann equation (cf. Sect. 2.3.1) are the same as those on which the Shannon relation is based. Information (I) is a function of mathematical probability (P):

$$I = f(P). \tag{2.3.6}$$

The conditions for the function f in Eq. (2.3.6) are again satisfied by a logarithmic function, since here too, the multiplication rule for the calculation of probabilities must be valid [cf. Eq. (2.3.3)]. Therefore:

$$I = K \ln P. \tag{2.3.7}$$

This is the *Shannon equation* of information theory. The unit of I is determined by the unit of the factor K. The bit ("binary digit") is most commonly used. It expresses the number of binary yes-no decisions which are needed to determine a given message. For example, the one side of the coin can be guessed by one single decision, its information value, consequently, is 1 bit. Five binary decisions will be sufficient to guess a card from a deck. Hence, the information value of one card is 5 bits. The factor $K = -1/\ln 2 = -1.443$ must be used to calculate I in bits. In information theory the logarithm to the base 2 ($\log_2$) is occasionally used:

$$I = -1.443 \ln P = -\log_2 P \quad (I \text{ in bit}). \tag{2.3.8}$$

If Boltzmann's constant (k) is used ($K = -k$), then information is obtained formally in entropy units:

$$I = -k \ln P \quad (I \text{ in J K}^{-1}) \tag{2.3.9}$$

The similarity of Eqs. (2.3.9) and (2.3.4) is evident. The formal interconnection of Shannon's definition of information with entropy triggered off an intense discussion about the relation between these parameters. The starting point for this discussion was the characterization of entropy as the measure of disorder according to the second principle of thermodynamics. Isolated systems spontaneously try to reach a state of maximum disorder. This can be seen in the process of information transmission where a message loses some of its information value.

The idea of a link between information and entropy was first suggested by Boltzmann and was then developed by other authors. E. Schrödinger (1944) made the frequently quoted statement: "The living system feeds on negative entropy". This is the reason why sometimes the term "negentropy" is used.

What is the real importance of Shannon's information equation in biophysics? In principle, it is possible to calculate the information content of a protein by the above approach. The requirements are, firstly, a statistical record of the frequency of the occurrence of the individual amino acids in proteins. This will provide the probability (P) for the presence of a given amino acid at a certain locus in the protein. Subsequently, using Eqs. (2.3.8) or (2.3.9) the information content of each monomer can be calculated. Subsequently, the information content of a whole protein can be obtained by addition of the values of its monomers.

The information content of a nucleic acid can be obtained in the same way. One mammalian DNA molecule consists on average of 15 000 pairs of nucleotides. Assuming that the four possible types of nucleoside bases have an equal probability of occurrence, then the information content of each single nucleotide will, consequently, have a value of 2 bits. The information capacity of one DNA molecule, in this case, amounts to 30 000 bits.

The problems of this kind of calculation may be illustrated in the following example: the information content of an English text can be calculated from the frequency of the use of individual letters. In this way one can derive the information content of a word, a sentence, and, consequently, even of this textbook. It is, however, obvious that this parameter does not reveal anything about the "information value" of the book as generally understood. The same information (I) would be given by any other book with the same number of meaningless strings of English words.

This situation, however, does not invalidate Shannon's information concept. Everybody knows how important the calculation of information is today in the field of computer science. So, for example, for the author of this book it is quite important to know which sort of discs he needs to store the text and the figures of this book. But, obviously, this is just a question of its volume, of the codes used, but in no case of its "information value" in the common sense.

Does this mean that it is impossible to quantify biologically important information? Does it mean that the information concept is not applicable to biological systems at all? In fact, there is no reason for scepticism. Consequently, a distinction has to be made between a syntactic measure of information, and a semantic measure.

The *syntactic* information content of a DNA molecule, as calculated above, provides some information on the maximum storage capacity of the genome. The amount of information actually stored is even lower, if the redundance in the storage of genetic information which is required for the protection of the information is taken into account. Estimates of information capacity in the genome vary between $3 \cdot 10^2$ bit and 10^{12} bit.

The *semantic* information, in contrast to the syntactic information, really should contain some kind of validation of the content. Despite many attempts, quantification of semantic information has not yet been achieved. In spite of the problems of giving an exact and meaningful quantification of information, it is doubtless a quality of matter and has a place in information theory and thermodynamics.

The interconnection of information with entropy in terms of thermodynamics may be illustrated best by a thought experiment conceived by James Clerk Maxwell in 1881 and still discussed today (Fig. 2.9). Maxwell proposed a room, which is connected with another by an opening. This opening can be closed by means of a slide. Both rooms are filled with a gas which is, in the beginning, in equilibrium, e.g. of equal pressure and temperature. An intelligent creature, later called "Maxwell's demon", is able to handle the slide between the two rooms with ease. This "demon" can observe accurately the direction and the velocity of the molecules in his room. Velocity and direction of these particles are

Fig. 2.9. Maxwell's demon

statistically distributed. If a particle flies accidentally toward the opening, the demon opens the slide to let the particle pass. As a result of such sorting, the pressure in the lower room would rise. The demon could also take another approach. For example, he could separate fast from slow particles. In this case, a difference in the temperature between the two rooms would occur. In both cases, the entropy of the whole system would be reduced and energy might be generated by an appropriate turbine. The result would be a "perpetuum mobile", a perpetual motion machine of the second order, as it would contradict the second principle of thermodynamics.

This apparent contradiction was subsequently the subject of a large number of scientific publications. The problem, however, may be resolved by the following consideration: the "demon" requires information to carry out the sorting. He collects this information by "watching" the molecules. In order to "see" he needs light. For this, the demon in Fig. 2.9 symbolically carries a lit candle. Yet, a body will only be able to emit light in a state of nonequilibrium relative to its environment. This, however, contradicts the equilibrium condition at the beginning of the experiment. The same would apply to any other approach to acquisition of information. This resolves the apparent contradiction with the second law of thermodynamics.

Why do we discuss this thought experiment here, if it is clear that it will not work as a perpetual moving machine? What has this to do with biophysics? In fact, independently of this discussion about the virtual violation of the second law of thermodynamics, Maxwell's demon demands particular interest in biophysics because of its analogy to various functions of living systems. The living cell, too, reduces its entropy at the expense of its environment, using information processing. Yet, in this case it is not the energy of statistical fluctuations which is used. The biological system selects molecules from its environment which are rich in free Gibbs energy of formation, and correspondingly have a low content of entropy. It uses this energy and extrudes molecules with lower free energy and larger entropy. The basic information for this process of selection, in other words, the "software" for this process, is stored in the structure information of the proteins, which are responsible for the recognition of these molecules, and eventually for their metabolism. These proteins get their information during the process of synthesis via the RNA, from the DNA of the genome.

This example raises the question: what is the threshold value of information which is required to control the processes of living systems? Or translated into the language of modern computer science: how large must the simplest software for this sort of a biological Maxwell demon be? What is the threshold information that carries out not only the metabolic function of the primordial organism but additionally its replication? How did the first accumulation of information in a molecule occur? A purely accidental combination of monomers to build their first functional macromolecule must be ruled out. The probability for this occurrence is too low by far. Today, so-called *prebiotic evolution* is assumed, i.e. chemical selection of developing polymers even before the stage of the biopolymer (see also Sect. 5.3.3).

Further reading: Shannon and Weaver 1962; bioinformatics in molecular biology: Kauffman 1993; Strait and Dewey 1996; Volkenstein 1994; Yockey (1992); Maxwell's demon: Leff and Rex (1990)

2.3.3
Biological Structures: General Aspects

In the previous section we introduced expressions like *order*, *structure* and *organization* and discussed them in relation to entropy and information, as well as with the statements of the second law of thermodynamics. This touches on a set of questions which are of central interest in biophysics and which will be mentioned in many sections of this textbook. Therefore it is necessary at this point to explain some basic definitions and principal ideas.

What, really, is a structure? To the biologist, the term "structure", is usually related to the macroscopically or microscopically visible organization of an organism. This means, for example, the structure of an animal skeleton, structure of a cell, of a mitochondrion etc. The term "molecular structure" already lies outside the limits of this view. It refers to a certain arrangement of atoms, without defined contours, which can be described just by means of wave mechanics. The same applies for the concentration profile of an electrical double layer (Fig. 2.42), and also for so-called time structures, namely special time courses, like oscillations of a biological system (Figs. 5.4 and 5.19), like the shape of an electrocardiogram (Fig. 3.33), or the sonogram of a bat's cry (Fig. 4.12). This means that the definition of the term "structure", which is used in biophysics has to be broader than that of the morphologists and cytologists. It must include these structures as well as those of metabolic networks, ecosystems, or others.

The best, and most generalized definition of this term is given by the set theory of mathematics. Therefore: *a system is an aggregate of elements with certain interrelations between them.* The totality of these interrelations is called the *structure* of the system. This definition does not prescribe at all what kind of elements, and what kind of interrelations these are. It is applicable to all kinds of systems including biological systems and structures. In biophysics, we are especially interested in *dynamic systems*, i.e. where the interrelations between their elements are *interactions*. In contrast to this, in *static systems* the elements have no interaction at all, but are just interrelated by formal *relations*. Examples for static systems, are the system of natural numbers in mathematics, or the system of animal species or of plants in biology.

The elements of a metabolic network are the metabolites, and the interrelations between them, i.e. their interactions, are the steps of biochemical reactions. Correspondingly, an ecosystem is to be considered as an interaction of individuals and populations depending on abiotic conditions.

Biological macromolecules can be considered as systems too. In this case, the following levels or organization occur, depending on what will be assumed as their elements.

- *Primary structure* – linear sequence of monomers (=elements) in the linear molecular chain (=system). For example, a polypeptide chain: ...-serine-alanine-lysine-arginine-...
- *Secondary structure* – positioning in space of monomers (=elements) in a part of the molecule (=system) relative to each other. For example, the α-helix, or the β-sheet structure of an amino acid sequence in a protein
- *Tertiary structure* – position in space of molecular regions of homogeneous secondary structures (=elements) in a molecule (=system). For example, intramolecular coordination of the position of several helical regions relative to each other or to a β-sheet
- *Quaternary structure* – position in space of macromolecules (=elements) in a supramolecular complex (=system). For example, single proteins in a multi-enzyme complex

When a salt is crystallized, a *periodic structure* occurs, which is characterized by a periodic arrangement of the elements. E. Schrödinger (1944) called the biological structure an *aperiodic crystal*. This means a highly organized structure, the elements of which are not simply repeated periodically. Sometimes one tries to evaluate this structural organization as *structural information*. As we pointed out in the previous section, however, it is hard to quantify this parameter. Probably, the structural information should be measured as the effort which is necessary to describe such a structure perfectly.

Consequently, the processes of structure formation of biological systems, be it the development of life, its reproduction, or simply the biosynthesis of a macromolecule, are all accompanied by reduction of entropy. This appears to be contrary to the second law of thermodynamics and has given rise to heated philosophical discussions in the past. The second law of thermodynamics actually postulates that in spontaneous processes occurring in isolated systems, the entropy strives towards a maximum. Yet, neither an organism nor its environment, i.e. the earth as a whole can be considered as an isolated system. The earth is constantly absorbing energy from the sun and is emitting this energy again. That continuous flow of energy maintains a permanent nonequilibrium state which manifests itself not only in a direct way in photosynthesis with subsequent metabolism of heterotrophic organisms, but also in the environment of life, for example, in flowing water, in alternation of light and darkness, and in changes in temperature, humidity etc.

In fact, structures are also formed in inanimate nature under loss of entropy. Basically, a distinction must be made between two kinds of structure formation which at the same time explain the two ways of self-organization in living systems.

- *Equilibrium structures*, for example a salt crystal, formed spontaneously during evaporation of solvent
- *Nonequilibrium structures* (or: *dissipative structures*), for example an arrangement of cirrus clouds as the result of processes of air convection under special meteorological conditions

The genesis of life which is based on the prebiotic formation of a first bio-macromolecule, and the subsequent evolution of all organisms, can be understood as the result of complicated processes which occur far from equilibrium. In this context, a major role is played by dissipative structures of the inanimate environment, such as periodic temperature variations, tides etc. Substantial importance also has to be attributed to such nonequilibrium structures in the course of life itself. They are frequently represented by time structures, such as for example the heart rate, or other kinds of oscillations, which in some cases are associated with the so-called biological clock. (for further explanations see Sects. 3.1.4, 5.3.2 and 5.3.3).

Equilibrium structures, such as inorganic salt crystals with a very primitive geometry, become very complex and variable in shape, when based on the sophisticated pattern of the primary structure of bio-macromolecules rather than on the relatively simple field of interaction of spherico-symmetric ions. The spontaneous folding of proteins and their arrangement to supramolecular structures, such as complex enzyme systems of even ribosomes, must be viewed in this way. Such processes are also referred to as *self assembly*. More specific aspects relating to the formation of equilibrium and nonequilibrium structures will be given in subsequent chapters of this textbook.

Further reading: Eigen (1971, 1992); Meinhardt (1982); Kauffman (1993); Strait and Dewey (1996)

2.3.4
Distribution of Molecular Energy and Velocity at Equilibrium

The Boltzmann equation of entropy [Eq. (2.3.4)] as derived in Section 2.3.1, helps to illustrate the second law of thermodynamics, according to which isolated systems spontaneously approach a state of maximum entropy. We demonstrated there that one can also express this in the following way: at equilibrium isolated systems reach a state of highest realization probability (maximum of W). Now we will ask the question: what are the mean properties of the molecules at this equilibrium state? This is of fundamental importance for further considerations.

It is possible to answer this question with the help of statistical thermodynamics, which helps to make predictions on the probability of distribution of energy among the elements of a system in thermodynamic equilibrium. It should, however, be emphasized that a deterministic description of the properties of single particles lies outside the limits of this discipline. Nevertheless, it will be seen that even a statistical statement allows the drawing of valuable conclusions regarding reaction rates, stability of molecules and more.

Let us imagine a space which contains gas molecules of uniform mass (m), which at the beginning all have the same velocity (v), and consequently, the kinetic energy (E) which can be calculated by the following equation:

$$E = \frac{m}{2} v^2. \tag{2.3.10}$$

This equality of the kinetic energy of all molecules is, in fact, a highly improbable state. In this case, using Fig. 2.8, all molecules would belong to a single box, or a single class of properties n_i. Thus, according to Eq. (2.3.5): $W = n!/n_i! = 1$. This situation will change instantaneously. The molecules would exchange their energy by elastic collisions with each other, and soon a great number of energy states would be occupied. A simple mathematical example shows that W, and according to Eq. (2.3.4) also the entropy (S), will increase with a growing number of (m) of state classes, provided the following relation applies:

$$\sum_{i=1}^{m} n_i = n.$$

Because of the law of conservation of energy (first principle of thermodynamics), the following condition must be satisfied at the same time:

$$\sum_{i=1}^{m} n_i E_i = \text{const.}$$

The energy of this system, therefore, can be distributed randomly among all of its molecules. The total energy of the system, however, must always remain constant.

Now, let us look for a state with maximum probability; this means with maximum entropy, and corresponding to the second principle of thermodynamics this is actually the thermodynamic equilibrium. Translated for our example: how will the energy and, correspondingly, the velocity be distributed between the particles, if the system is isolated, and if it is left to itself for a sufficiently long period of time? Even under equilibrium conditions, of course, the energy of individual molecules will change permanently, but nevertheless the statistical mode of energy distribution, or of the distribution of various degrees of molecular velocity at equilibrium becomes stationary, i.e. time independent (for a detailed description of various kinds of stationary states see Sect. 3.1.4).

Considerations of this type lead to *Maxwell's equation of velocity distribution*:

$$\frac{dn(v)}{n_0 dv} = \frac{4}{\sqrt{\pi}} \left(\frac{m}{2kT}\right)^{3/2} v^2 e^{-\frac{mv^2}{2kT}}. \tag{2.3.11}$$

The left part of this equation contains a relation which expresses the relative number (dn/n_0) of those molecules which are related to a particular velocity interval (dv). Its unit is s m^{-1}. This is a function expressing the probability of the distribution of the velocity, where m is the mass of a molecule [not to be confused with the numeral m in Eq. (2.3.5)], and k is the Boltzmann constant. Multiplying in Eq. (2.3.11) the denominator and numerator of the expression within the brackets as well as that in the exponent with the Avogadro number (N = 6.023 · 10^{23} mol^{-1}) introduces molar, instead of molecular parameters:

$$M = N \cdot m, \tag{2.3.12}$$

and

$$R = N \cdot k, \tag{2.3.13}$$

where M is the molar mass, and R = 8.314 J K^{-1} mol^{-1} gas constant.

In Fig. 2.10, as an example, the velocity distribution of molecules of oxygen is depicted $(M = 0.032$ kg mol$^{-1})$. This curve is not symmetrical. The mean velocity (weighed arithmetic mean value) in general will be higher than the maximum value. The following relation applies:

$$v_{\max} = \sqrt{\frac{2kT}{m}} = \sqrt{\frac{2RT}{M}}; \quad \bar{v} = \sqrt{\frac{8kT}{\pi m}} = \sqrt{\frac{8RT}{\pi M}}. \tag{2.3.14}$$

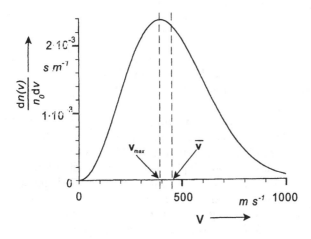

Fig. 2.10. Velocity distribution of O_2 molecules at 37 °C based on Eq. (2.3.11)

These values increase with increasing temperature (T) and decrease with increasing molar mass (M).

Maxwell's velocity distribution is based on the Boltzmann function of energy distribution. It determines the number (n_i) of those particles which under equilibrium conditions of the system have an energy level of E_i:

$$n_i = Ce^{-\frac{E_i}{kT}}, \tag{2.3.15}$$

(C is a factor of standardization). The exponent of this equation, as in Eq. (2.3.11), expresses the relation between energy E_i of the particle and the mean energy of thermal noise (kT). The Boltzmann constant (k) should be replaced by the gas constant (R) in cases where E_i expresses not the energy (in: J) of a single molecule but is a molar parameter (in: J mol^{-1}), as is customary in chemical thermodynamics.

The Boltzmann equation predicts the distribution of any form of energy. It expresses the situation wherein lower states of energy tend to occur more often than states of higher energetic levels. Related to the total number of particles, it may also be written in the following form:

$$\frac{n_i}{n_{\text{total}}} = \frac{e^{-\frac{E_i}{kT}}}{\sum_{i=0}^{\infty} e^{-\frac{E_i}{kT}}}. \tag{2.3.16}$$

This equation provides the basis for many calculations in statistical thermodynamics. For example, it can be used to determine the percentage of molecules in a given system which are able to overcome a certain energy barrier (E_i). This aspect will be discussed in greater detail in the following chapter.

The use of this equation is allowed only in cases of thermodynamic equilibrium. At first glance it seems to limit its applicability to biological systems which are often far from equilibrium. The process of the adjustment of the Boltzmann distribution is, however, so fast that an imbalance of this kind has to be considered only in rare cases. On many occasions in this book we will indicate that in fact subsystems could be in equilibrium even in cases where the whole system is far from it (see especially Sect. 3.1.4).

2.3.5
Energy of Activation, Theory of Absolute Reaction Rate

The temperature dependence of the rate of a chemical reaction was described by S. Arrhenius by the following equation:

$$k_R = A e^{-\frac{E_A}{RT}}. \tag{2.3.17}$$

Here, k_R is the rate constant. The energy E_A was defined as *activation energy*; A is an empirical factor.

This equation describes not only the temperature dependence of chemical reactions. It can also be used to describe reaction rates of other physiochemical processes, such as diffusion, kinetics of phase transitions etc. At certain temperature intervals, complicated biological processes, like growth rate, or heart rate of poikilothermic animals etc., can also be fitted in this way.

To evaluate experiments in which time dependent parameters like rate constants, reaction rates, frequencies of oscillations etc. were measured as functions of temperature, a so-called *Arrhenius plot* may be applied (see. Fig. 2.11). In this case the logarithmic form of Eq. (2.3.17) is used:

$$\ln k_R = \ln A - \frac{E_A}{RT} \tag{2.3.18}$$

Applying this kind of evaluation, $y = -\ln k_R$ is plotted against $x = 1/T$. If in the investigated process the activation energy itself is temperature independent, then the measured points are located on a straight line:

$$y = ax + b \tag{2.3.19}$$

The slope of this straight line is, according to Eq. (2.3.18): $a = -E_A/R$. The relation: $b = \ln A$ and, consequently, the factor of A can be read from the extrapolated value of $x = 1/T = 0$. Direct derivation of the logarithmic function to dT is more complicated. We need it later:

$$\frac{d \ln k_R}{dT} = \frac{E_A}{RT^2}. \tag{2.3.20}$$

Real biological processes usually consist of a large number of single reactions, each with different activation energies. Hence, a temperature dependence must

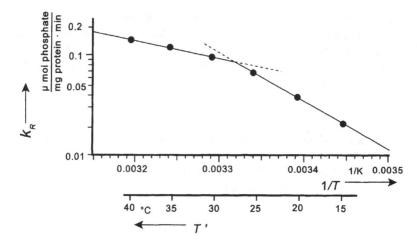

Fig. 2.11. An example for an Arrhenius-plot: the logarithm of the activity of isolated ATPase (A) is plotted versus the reciprocal value of the temperature in a region between 15 and 40 °C. (Parameters from Blank and Soo 1992)

be expected which is much more complicated than that expressed by the Arrhenius equation [Eq. (2.3.17)]. Surprisingly, however, even in these cases functions will still be obtained which can be described by straight lines in the Arrhenius plot. This is caused by the fact that in complicated processes the rate of the overall process is determined by a simple rate limiting reaction. The Arrhenius plot just reflects the activation energy of this rate limiting process.

In some physiological papers, a so called *Van't Hoff rule* of temperature dependence is used. According to this the rate of a reaction increases by two to three times, if the temperature rises by 10 °C. This factor is referred to as Q_{10} value. This rule, however, is not supported by thermodynamic considerations. Even in cases where the energy of activation is constant, according to the Arrhenius equation, the Q_{10} value will increase with increasing temperatures.

In some reactions the energy of activation (E_A) itself is a function of temperature. In these cases the straight line in the Arrhenius plot gets kinks (Fig. 2.11). The reasons for this can vary. In complex processes it is possible that at different temperatures, different reactions with different energies of activation will become rate limiting. In other cases such transitions may occur as the result of conformational changes of one of the components of the reaction, probably a conformational change in the enzyme itself. In reactions of membrane-bound enzymes, sometimes phase transitions of the enzyme near lipids may cause such effects.

The statistical mechanics allow a direct derivation of the Arrhenius equation. This is based on the *theory of absolute reaction rate* (or *theory of transmission states*) as developed by H. Eyring. Here we can explain just some basic ideas of this approach.

Chemical reactions presuppose splitting of chemical bounds. To analyze such processes it is necessary to know the interaction energy of a given atom in the molecule as a function of the distance to each of its neighbors. This is an intricate function which can be expressed mathematically as an n-dimensional space corresponding to n possible ways of geometrical approach. In this space a line may be found which shows the way from state A to state B over the smallest peaks of the activation energies. This line is called the *reaction coordinate*. It is possible to represent the energy level along this line in a two-dimensional graph.

In Fig. 2.12 such a function with two minima is plotted, representing the stationary states A and B. In these positions, the molecule will be in a stable state, which is however continuously deflected in a direction of the reaction coordinate by thermic collisions (the term *stability* will be explained in detail in Sects. 3.1.4 and 3.5). The effectiveness of those collisions depends on their energy i.e. on the relation of their energy to the activation energy E_A. The energy of collisions, in fact, is thermic energy, which can be expressed by the factor kT (or in molar units, correspondingly in RT). The Boltzmann equation [Eq. (2.3.16)] allows the calculation of the relative number of those molecules (n_i/n_{total}) which are able to overcome the energy barrier (E_i), which in our case is identical with the energy of activation (E_A). The relation E_i/kT [Eq. (2.3.16)] of course decreases with increasing temperature (T). This is illustrated in Fig. 2.13. While 36.8% of all particles have reached the energy level $E_i/kT = 1$, a value twice as high is only reached by 13.5%, and for $E_i/kT = 3$ there are only 5% reaching this level. The percentages drop fast with rising values of E_i/kT.

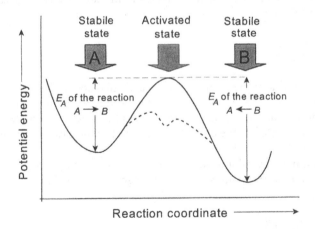

Fig. 2.12. Schematic potential energy diagram of a molecule in the states A and B taken along the reaction coordinate. The *solid curve* represents the reaction without a catalyst; all the explanations correspond to this curve. The *dashed line* indicates how the energy of activation is modified in the presence of a catalyst (enzyme). In this case, the curve with a single maximum, dividing the states A and B is transformed into another with two maxima, between which the minimal energy of an enzyme-substrate complex is positioned

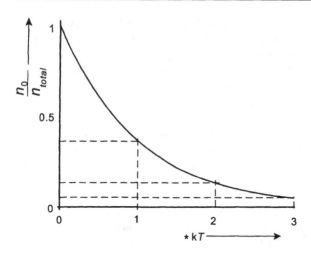

Fig. 2.13. The relative number (n_i/n_{total}) of molecules as a function of the energy in kT-units. This curve corresponds to Eq. (2.3.16)

The theory of absolute reaction rate is based on the assumption that an activated transition state between the two states A and B exists. Its potential energy is identical with the energy maximum in Fig. 2.12. This activated state is, of course, highly unfavorable and unstable from the energetic point of view. At any point in time it will be assumed only by a very small number of molecules which will drop back from the energy maximum into the stable states of an energy minimum to the right and left of this maximum. An equilibrium constant (K^*) can be defined which characterizes the relation between activated and stable states according to the mass action law. The proportionality between the rate constant (k_R) and the number of molecules in the activated state can then be expressed as follows:

$$k_R = q \frac{kT}{h} K^*. \tag{2.3.21}$$

The symbol * indicates the reference to an activated state. The relation between kT and Planck's constant h comes from kinetic quantum theory as derived by Wilson-Sommerfeld. The factor q gives the fraction of activated molecules which leave to the right or left of the energy maximum. For the case of symmetry of the relations, it becomes $q = 1$.

The phenomenological thermodynamics allows calculation of the relation between the molar free Gibbs energy of reaction (ΔG) and the equilibrium constant (K):

$$\Delta G = -RT \ln K. \tag{2.3.22}$$

In the case of a reaction A $\rightarrow$ B, i.e. for a reaction in which the total number of molecules remains constant, and under the above-mentioned condition of $q = 1$, it is possible to calculate the conditions for the activated state (using the symbol *):

$$k_R = \frac{kT}{h} e^{-\frac{\Delta G^*}{RT}}. \qquad (2.3.23)$$

If one further considers the following relation:

$$\Delta G = \Delta H - T\Delta S, \qquad (2.3.24)$$

it follows that:

$$k_R = \frac{kT}{h} e^{\frac{\Delta S^*}{R}} e^{-\frac{\Delta H^*}{RT}}. \qquad (2.3.25)$$

The relation of this formula with the Arrhenius equation [Eq. (2.3.17)] can be derived by modification of the above equation in the following way:

$$\ln k_R = \ln T + \ln\left(\frac{k}{h} e^{\frac{\Delta S^*}{R}}\right) - \frac{\Delta H^*}{RT}, \qquad (2.3.26)$$

and:

$$\frac{d \ln k_R}{dT} = \frac{1}{T} + \frac{\Delta H^*}{RT^2} = \frac{RT + \Delta H^*}{RT^2}. \qquad (2.3.27)$$

Comparing Eq. (2.3.27) with Eq. (2.3.20) the following correlation is obvious:

$$E_A = RT + \Delta H^*. \qquad (2.3.28)$$

For other values of q, e.g. for reactions of synthesis or lysis, this relation can be modified.

In Eq. (2.3.25) the temperature independent exponential part is of particular interest, which contains the *entropy of activation* (S^*). The entropy of activation indicates conformations of the molecules which occur during the reaction. This is important, especially in biochemical reactions in which macromolecules are included. The parameter ΔS^* can become positive as well as negative. This depends on whether the molecule during the reaction gets a larger ($\Delta S^* < 0$) or a lower ($\Delta S^* > 0$) degree of order. On the other hand, the entropy changes of the water molecules during this reaction must be considered too. We will discuss the entropy modifications of water during reactions of hydration later in detail (Sect. 2.4.2; Fig. 2.26).

The acceleration of biochemical reactions by enzymes is based on a change in the reaction coordinate. The energy levels of the starting and the end product of the reaction will remain unaltered by this. However, with the help of an enzyme as a biological catalyst the transition from one state into another is possible with lower barriers of activation energy. This is indicated by the dashed curve in Fig. 2.12. First, an enzyme-substrate complex is formed, which requires less activation energy then that needed to form the activated complex without the enzyme. This enzyme-substrate complex is then broken up again by a small input of activation energy, and substance B will be released. It goes without saying that the reaction A → B cannot take place spontaneously unless the energy level of B is below that of A even if an enzyme is involved.

The concept of the theory of absolute reaction rate is not only applicable to processes of chemical reactions. Processes of diffusion can be described in a similar manner. Diffusion of a substance can be imagined as a reaction consisting entirely of location jumps from one position in the vicinity of a molecule to another. In this situation the reaction coordinate will largely coincide with the direction of diffusion. Investigations of activation energy of diffusion processes may provide the path of diffusion, for example, the mechanism of permeability of a molecule through a membrane.

Let us now discuss the problem of the life span of a single bound, or a whole molecule. Using the Arrhenius equation, it is also possible to calculate the stability of a given molecule. In this case, however, it is not the reaction rate of the molecular transformation which is of interest, but its reciprocal value, namely the mean time during which a bond with a given bonding energy will resist the attack of the energy of thermic noise (kT, resp. RT). In this case the energy of activation of a decay reaction (E_D) is of interest. This mean time of stability (t) is inversely proportional to the rate of transformation. Thus:

$$ t = \tau e^{\frac{E_D}{kT}}. \tag{2.3.29} $$

The time constant τ in this equation, as well as the value A in Eq. (2.3.17) depends on a number of structural properties of the molecule. That is why only values relating to several orders of magnitude are available. For smaller molecules, τ will lie between 10^{-14} and 10^{-13} s. The characteristics of this function are illustrated in Fig. 2.14. Semi-logarithmic plotting (ln t against E_D) will result in a straight line. It can be clearly seen that even small changes of the activation energy of decomposition (E_D) will lead to dramatic changes of the life span (t). The range may stretch from 10^{-12} s to geological periods of time amounting to 10^5 years.

Considering the energies of covalent bonds (Table 2.1, page 12), one understands that they practically cannot be destroyed by thermic impacts. Hydrogen bonds, on the other hand, with bonding energies between 13 and 25 kJ mol^{-1} (see Sect. 2.1.5) have only a short lifetime. When we discuss the structure of water (Sect. 2.4.2), we will show that in these cases oscillations of bonds occur.

In all investigations of the stability of covalent bonds one must of course also consider the possibility of their destruction by energy rich quanta of radiation (ionizing radiation). This we will consider later in connection with the target theory of radiation biophysics (Sect. 4.5.5).

The mean life span of a molecule is a typical variable of statistical thermodynamics. It is only a value of probability which cannot predict the concrete destiny of an individual molecule. The assumption can be made that by biological selection only DNA molecules with monomers of high bonding energy

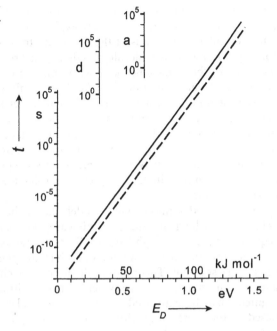

Fig. 2.14. The mean lifetime (t) of a bond depending on the energy of activation of decay (E_D), according to Eq. (2.3.29) (T = 293 K; *solid line*: $\tau = 10^{-13}$; *dashed line*: $\tau = 10^{-14}$ s)

are left. Nevertheless, spontaneous molecular transformations take place which are known as mutations. A mutation results in a new molecule whose stability has not yet been approved by selection. The life span of this molecule can be substantially changed by a slight displacement of its bond energy. This may explain why remutations or mutations of already mutated molecules occur more often than could be expected statistically.

Further reading: Blumenfeld and Tikhonov 1994; Westerhoff and van Dam 1987. The theory of absolute reaction rate: Eyring and Urry in: Waterman and Morowitz 1965

2.3.6
Thermal Molecular Movement

The disadvantage of the method of illustrating structures of organic molecules by using mechanical models composed of spheres and connecting wire axes, as well as of computer simulations of molecular structures, is their suggestion of a more or less static point of view. In reality, however, in liquids as well as liquid-crystalline phases, the molecules are always in a state of vigorous thermal motion. The following three forms of molecular movements are to be considered: vibrations, rotations, and translations.

Vibrations are oscillations in the binding distances between the atoms in a molecule. The term *rotations* means not only the rotation of the whole molecule, but additionally, the spin of individual atoms or atomic groups around the axes of their bonds. Of course, as already explained in Section 2.1.2., in this case the intramolecular interactions of the atoms are to be considered. So, for example, considering a simple molecule of ethane, the rotation around the CH_3—CH_3 bond cannot be homogeneous. Energy minima of rotations occur periodically every 120°. Hence, rotation is "locked" at certain angels of low energy levels.

A covalent bound, rotating around its angle relative to another, consequently describes a cone. The next bond rotates, as well, but it does so quasi on the spot, moving itself in an orbit of that lateral cone surface (Fig. 2.15). In this way chain molecules can assume a stochastic orientation unless other strong attracting or repelling interactions between the monomers are preventing this. The molecular structures resulting in this way can be calculated by the laws of statistical thermodynamics.

The so-called random flight model allows the simplest description of the behavior of a long molecule. In this case, the real molecule is described as a sequence of segments with length l. These segments are connected to each other in such a way that the orientation of each is independent from the orientation of the former segment. This random flight chain is characterized by the length (l) and the number (n) of its segments. The positions of the end points of the segments can be expressed as vectors in a polar coordinate system which is based on the center of gravity of the molecule (Fig. 2.16). The distance between the endpoints of such a molecule (d_{max}) at maximal elongation is:

$$d_{max} = nl. \tag{2.3.30}$$

The real distance of the endpoints of a random flight chain with large enough amounts of n segments, and after a sufficiently long time of observation, however, is distributed according to a Gauss statistic. As shown in Fig. 2.10, the mean of this distance is not identical with its maximum, i.e. the value of maximal probability. The mean value of this distance between the end points of the chain can be calculated by the following equation:

$$d = \sqrt{\langle r^2 \rangle} = l\sqrt{n}. \tag{2.3.31}$$

Fig. 2.15. The rotation cones of the atomic groups in a chain molecule

Fig. 2.16. Projection of a random flight chain, composed of 15 segments with a characteristic length l. ∗ Center of gravity of the molecule, d distance between the two ends of the chain, r_i distance of the i-th point from the center of gravity, R_G radius of gyration, R_S Stock's radius. This model corresponds to an unfolded polypeptide, composed of 150 amino acids, or a double strand DNA. (According to data from G. Damaschun)

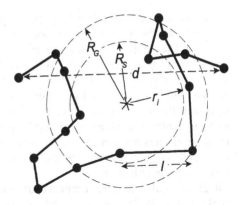

In this case the expression $\langle r^2 \rangle$ is the mean of the squares of the distances as defined as follows:

$$\langle r^2 \rangle = \frac{\sum r_i^2}{n}. \tag{2.3.32}$$

Comparing Eqs. (2.3.30) and (2.3.31) it can be seen that the mean distance (d) between the endpoints of the chain is smaller by $\sqrt{n}$ in relation to its maximum (d_{max}). Considering a chain consisting of 100 segments, the mean distance, therefore, will be only 10% of the maximal length, for chains of 1000 segments, it will be only 3.2%.

The geometric dimension of such a molecule can be described by the mean of the square distance of all atoms of the molecule from the common center of gravity of the molecule. This gives the so-called *radius of gyration* (R_G). For a random flight chain with $n \to \infty$ it gives:

$$R_G = \sqrt{\frac{\langle r^2 \rangle}{6}} = l\sqrt{\frac{n}{6}}. \tag{2.3.33}$$

The radius of gyration, therefore, is proportional to the square root of length of the chain ($\sqrt{n}$). For compact molecules, however, in which the monomers are not arranged in this way, one gets $R_G \sim \sqrt[3]{n}$, and for rod-shaped molecules, $R_G \sim n$.

Furthermore, it is possible to define a so-called *Stockes' radius* (R_S) for macromolecules, which is measurable experimentally. This is an example of a typical effective parameter. It predicts that the molecule moves in the water like a macroscopic sphere with a hydrophilic surface. In this case, Stockes' law can be applied. It gives the frictional force (**F**) of a sphere with radius r, moving in a fluid with viscosity η and with a velocity **v**:

$$\mathbf{F} = 6\pi\eta r\mathbf{v}, \tag{2.3.34}$$

(the bold letters emphasize the vector natures of force and velocity).

The Stockes' radius, therefore, is a measurable parameter which, in the same way as the radius of gyration, does not correspond to any "visible" radius of the molecule. Both are effective parameters which, however, are well defined (see Fig. 2.16). We will come back to this point when considering the hydration radius of ions in Section 2.4.2. The relationship between the experimentally determined parameters R_G and R_S gives important hints about the configuration of a given macromolecule. If the molecule really resembles a random flight chain, one gets: $R_G/R_S = 1.51$. If the molecule has a spherical but compact structure, this relation becomes 0.8. In the case of root-shaped molecules, this parameter depends on the relation of radius to length of it.

Real molecular chains, composed of covalently bound atoms (Fig. 2.15) are usually not as flexible as predicted by the random flight model. Nevertheless, this model is also applicable in these cases. For this, monomeric constituents were connected together as segments, matching the conditions of the random flight model. The stiffer the molecules are, the larger the characteristic length of the segments which must be chosen. In case of a quite flexible molecule of poly-L-glycine, the characteristic segments are formed by approximately four amino acids with a characteristic length of the segments of $l = 1.38$ nm. For the more stiffer molecule of poly-L-alanine, in contrast, twelve amino acids must be put together in one segment with $l = 4.10$ nm.

A repulsive force between the atoms, e.g. by electrical charges of equal sign, results in an additional expansion of coiled molecules. A similar limitation of the random flight model is given by the exclusion principle, i.e. by the fact that two atoms cannot be at the same location at the same time. Therefore, the characteristic length of the segment of a DNA molecule as a strong polyelectrolyte is $l = 90$ nm.

If such a chain molecule is fixed at two points (Fig. 2.17) then attractive or repulsive forces are developed if the distance between the points is larger or smaller, respectively, then the parameter d [Eq. (2.3.31)]. This is the so-called rubber or entropy elasticity. It is caused by the decrease of the entropy of the molecular chain during stretching. In entropy-elastic materials, the points of fixation are built by interconnections of the molecules. This property is the most common type of elasticity of biological materials. We will come back to this in Section 3.6.3.

The third kind of molecular movement mentioned above, *translation*, means a movement in space. This translational movement, driven by thermic hits, is important not only for single molecules, but also for microscopically visible particles. This was first observed in 1827 by the Scottish botanist Robert Brown, observing the stochastic movement of gymnosperm pollen in suspension. This movement later was named *Brownian motion*. The reason for this stochastic translocation of microscopically visible particles are the hits, resulting from thermic translations of the surrounding molecules. The movement results from the vector sum of these hits (Fig. 2.18). If the particles are sufficiently small, then a real vector occurs as the sum. For larger particles, this mean vector force becomes zero.

Fig. 2.17. The molecular mechanism of rubber elasticity. If the distance between the two points of fixation increases, the entropy content of the molecule decreases (schematically illustrated in the elongation of the chain in the lower part of the figure). A force is generated with the tendency to recover the previous situation (the state of maximal entropy). (According to data from G. Damaschun)

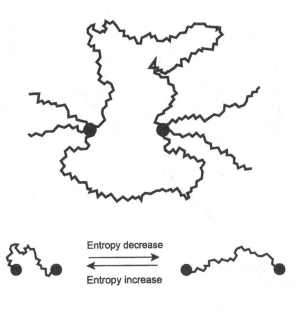

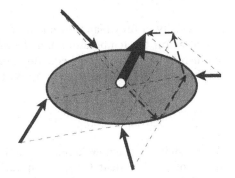

Entropy decrease

Entropy increase

Fig. 2.18. A snapshot indicating the thermic hits by surrounding molecules on a microscopically small particle. The individual hits, drawn as fine vectors, induce by vector addition a resting force (large vector) which effects the Brownian motion of the particle

The spatial translocation of the particle, in fact, is stochastic as well as the molecular translocations themselves. Observing the position of the particle during a sufficiently long time, there will be no translocation at all. As a measure of the Brownian motion, therefore, the mean of the squares of the translocations are used. For this, the position of the particle will be registered at definite time intervals Δt (Fig. 2.19). Subsequently, when the distance x_i is measured, the particle has moved in a projection of an arbitrarily drawn axis. If these values are measured n times, the parameters x_i for $i = 1, \ldots, n$ are obtained. The square of displacement $\langle x^2 \rangle$ can be calculated in analogy to Eq. (2.3.32) in the following way:

$$\langle x^2 \rangle = \frac{\sum x_i^2}{n}.$$

$$(2.3.35)$$

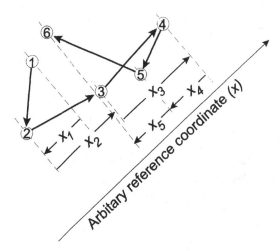

Fig. 2.19. Method to evaluate Brownian motion. At definite intervals of time (Δt) the position of the particle was marked. This two dimensional picture was subsequently projected on an arbitrarily drawn axis, which allows measurement of the distances x_i

This parameter corresponds to the square of the mean path length of the moving particle:

$$\bar{x} = \sqrt{\langle x^2 \rangle}. \tag{2.3.36}$$

The intensity of Brownian movement depends on shape and size of the particle as well as on the temperature and the viscosity of the medium. Einstein and Smoluchowski derived an equation for spherical particles, meeting the conditions of the Stockes' equation [Eq. (2.3.34)]:

$$\langle x^2 \rangle = \frac{RT\Delta t}{3N\pi\eta r} = \frac{kT\Delta t}{3\pi\eta r}. \tag{2.3.37}$$

The Einstein-Smoluchowski Eq. (2.3.37) therefore connected the measurable square of displacement ($\langle x^2 \rangle$) of Eq. (2.3.35) in relation to the radius of the particle (r), the viscosity of the fluid (η) and the temperature (T), this means by the energy of the thermic movement ($RT/N = kT$). Δt is the time interval for which the values x_i were measured.

This equation indicates that the square of the mean translocation length which was achieved by Brownian motion per time interval was inversely proportional to the particle radius (r) and the viscosity (η). A particle with diameter $r = 1$ μm $= 10^{-6}$ m moves in water ($\eta = 0.00089$ Ns m^{-2}) at $T = 298$ K during $\Delta t = 1$ s at a mean distance of:

$$\bar{x} = \sqrt{\frac{1.38 \cdot 10^{-23} \cdot 298 \cdot 1}{3 \cdot \pi \cdot 0.00089 \cdot 10^{-6}}} = 0.7 \cdot 10^{-6} \text{ m}.$$

The size of the particle (r), as well as the mean translocation ($\bar{x}$) are just in the range of resolution of a normal microscope. Measuring $\bar{x}$, and knowing r, it is possible to measure the viscosity (η). This was sometimes used to measure the viscosity of cytoplasm. On the other hand, it is possible to measure the diameter of very small particles by dark field microscopy in fluids with known viscosity, measuring the parameter $\bar{x}$.

The Einstein-Smoluchowski equation is not only important for using these methods. With the help of this equation it is also possible to calculate the time a particle needs to cross a certain distance by diffusion. If, for example, at one point in the cell a vesicle was created and is bound at a definite other place by specific receptors, the mean time this particle needs to cross this distance by stochastic movements can be calculated. Phenomenologically such an event appears to be the result of directed transport. (This, by the way, is an example of a kind of biological Maxwell demon, as discussed in Sect. 2.3.2!).

This translational movement is the basis of all processes of diffusion. Einstein found the connection between the square of displacement ($\langle x^2 \rangle$) and the diffusion coefficient (D) of a molecule in solution:

$$D = \frac{\langle x^2 \rangle}{2\Delta t}. \tag{2.3.38}$$

Introducing Eq. (2.3.37), it gives

$$D = \frac{RT}{6N\pi\eta r} = \frac{kT}{6\pi\eta r}. \tag{2.3.39}$$

We will come back to deriving the Fick's law of diffusion in Section 3.3.1 using the approaches of phenomenological thermodynamics.

This equation allows the conclusion that the diffusion constants of two molecules with similar molecular shapes are inversely proportional to the square root of their molar masses (M):

$$\frac{D_1}{D_2} = \frac{\sqrt{M_2}}{\sqrt{M_1}}. \tag{2.3.40}$$

Further reading: Cantor and Schimmel 1980; Pain 1994

2.4
Molecular and Ionic Interactions as the Basis for the Formation of Biological Structures

In Section 2.3 the relationship between order and life was discussed, and in this context, the molecular basis of biological structures was considered from the perspective of statistical thermodynamics. Emphasis has been placed on the energy of thermal movement and the thermal noise (kT). These stochastic forces of movement are opposed by intermolecular forces which tend to create order and build up molecular and supramolecular structures. Therefore, the expression E_i/kT [see Eqs. (2.3.16), (2.3.17), (2.3.29)] characterizes the relationship between any sort of structure forming energy and the destroying energy of thermic noise.

Now, we will consider in particular those forces which contribute to the formation of molecular and supramolecular equilibrium structures. The forces operating in these so-called self-assembling processes are mostly weak bonds, i.e. bonds with energies of the same order as thermic noise (kT). These bonds, therefore, can be split by thermic hits in a realistic period of time. Examples of such weak bonds, are hydrogen bonds as well as the whole complex of Van der Waals interactions.

2.4.1
Some Foundations of Electrostatics

Because electrical phenomena are of major importance at all levels of biological organization an account will be given of some fundamental parameters, definitions, and implications.

The basis of electrostatic considerations are electric charges. In the CGS-System the electric charges are defined by mechanical parameters. So, for example, the electrostatic unit (eu) is an amount of charge which interacts with an opposite polarized charge of the same size at a distance of 1 cm by an attractive force of 1 dyn. This definition eliminates the need for defining specific electrical units. In the Système International d'Unités (SI), however, the application of mechanical parameters has been abandoned, and the charge is defined by the unit for current – the ampere (A), as coulomb (C) = ampere (A) · second (s). In this book, we will use this now generally accepted international system of units.

In electrochemical as well as in biophysical calculations, the smallest electric charge is the charge of a univalent ion or a corresponding charged group. The *Faraday constant* (F) gives the number of charges per mole of singly charged ions, an amount which is usually called *val*:

$$F = 9.6485 \cdot 10^4 \, C \, val^{-1}.$$

Dividing this molar amount by the Avogadro number (N), gives the absolute value of the charge of a single ion (e), the same value as that of a positron:

$$e = \frac{F}{N} = 1.6021 \cdot 10^{-19} \, C.$$

Using the units of the SI system, a conversion factor must be applied to calculate mechanical parameters resulting from electrical interactions. This conversion factor is the so-called *permittivity of free space* (ε_0):

$$\varepsilon_0 = 8.854 \cdot 10^{-12} \, C \, V^{-1} \, m^{-1}.$$

The Coulomb's law makes it possible to calculate the repulsion force (**F**) of two charges (q_1, q_2) at a distance r from each other:

$$\mathbf{F} = \frac{q_1 q_2}{4\pi\varepsilon_0 \varepsilon r^2}.$$

(2.4.1)

We already used this equation in Section 2.1.3 to calculate the force of interaction of two ions. In contrast to Eq. (2.1.3) which indicates the force of two charges in vacuum, Eq. (2.4.1) additionally contains the *dielectric constant* (ε), which is also called *permittivity*. In contrast to ε_0, which unfortunately is written with the same symbol, the dielectric constant ε is just a number without dimension. It is a factor indicating how much the interaction force between two charges will be diminished, if not a vacuum, but a real substance is in between. The dielectric constant for pure water at 18 °C, for example, amounts to 81.1. For various kinds of glass, it is between 5 and 7, and for arranged layers of lipids about 3.5.

Using the Coulomb's law [Eq. (2.4.1)] one can also calculate the energy of electrostatic binding (E) of two charges in a medium with $\varepsilon > 1$. As already mentioned (Sect. 2.1.3), with the inverted sign, it equals the work required to move both charges from the distance $r = \infty$ to the distance $r = r_i$:

$$E = -W = \int_{\infty}^{r_1} \frac{q_1 q_2}{4\pi\varepsilon_0 \varepsilon r^2} \, dr = -\frac{q_1 q_2}{4\pi\varepsilon_0 \varepsilon r}.$$

(2.4.2)

Any charge in space generates an electric field which is characterized by the gradient of the electric potential. The *electric potential* (ψ) is a state parameter of a particular point in space. It is defined as the work which is required to move a positive charge from infinite distance to the point r. The unit of the potential is the volt (V). In the case of the electric field around a charge q, the electric potential is a radial function of the distance r from this point. According to this definition and Eq. (2.4.2), the potential ψ_i at the distance r_i from the charge q is given by:

$$\psi_1 = \frac{q_1}{4\pi\varepsilon_0 \varepsilon r_1}.$$

(2.4.3)

It is important to stress that the electrical potential is a scalar parameter in space, similar, say, to pressure or temperature. Later we will discuss surface potentials, donnan potentials, diffusion potentials etc. These are all just electrical potentials. Their names only indicate the reason for their occurrence. At any given time, at one and the same point in the space, there can be only one value of an electrical potential, as there can be only one single temperature at one time at one point!

Points of equal potential in space can be interconnected by *equipotential lines*. Perpendicular to these equipotential lines electrostatic forces occur. Correspondingly, an electric field strength (**E**) as a vector parameter can be defined as follows:

$$E = -\text{grad}\,\psi = -\nabla\psi. \tag{2.4.4}$$

The differential operator "grad" as well as the so-called *Nabla operator* ∇ means: derivation of the potential ψ to all three dimensions of space. It transforms the scalar ψ into the vector E.

Figure 2.20 indicates the equipotential lines and the field vectors around a positively charged point. The direction of these vectors corresponds with the vector of the electric field strength (E). According to a convention, these arrows are always directed from the positive, toward the negative pole. Similar illustrations can be seen in Figs. 3.31 and 3.32.

Considering the gradient of the electric field just in one direction (x) of the space, Eq. (2.4.4) becomes:

$$E_x = -\frac{d\psi}{dx}\,i. \tag{2.4.5}$$

In this case i is the unit vector in the direction of the x coordinate.

In Fig. 2.21 the potential and the field strength in the x-direction are depicted in the vicinity of a homogeneous membrane which separates two different electrolyte solutions (phase I and phase II). This demonstrates that points of discontinuity of the field strength can occur, with a reversal of the field direction. In this case, the reason for these discontinuities are surface charges on the phase boundaries. The field vectors and the equipotential lines in this example would indicate a set of parallels, perpendicular to each other. Actually, the biological membrane is not homogeneous, but it is a mosaic of various compounds with different surface charges and various dielectric constants. Even its cross section is not homogeneous. Its electrostatic structure is far more complicated, as we will see in Section 2.5.5, Fig. 2.48.

The unit of field strength, according to its definition [Eqs. (2.4.4) and (2.4.5)] is V m^{-1}. As an example, let us calculate the field strength across a cell membrane with a thickness of 7 nm = $7 \cdot 10^{-9}$ m. If the electrical potential

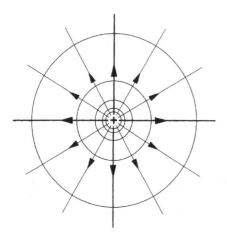

Fig. 2.20. Equipotential lines (———) and field vectors (——→——) around a positively charged point

Fig. 2.21. The electrical potential $[\psi(x)]$ and the electric field strength $[E(x)]$ near a homogeneous membrane which separates two electrolyte solutions with different composition

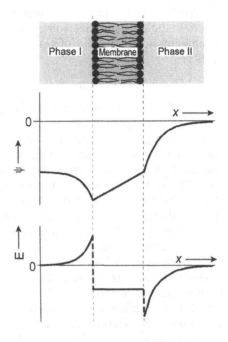

difference between the two surfaces amounts to $\Delta\psi = 50$ mV $= 5 \cdot 10^{-2}$ V, the following field strength results:

$$|E_i| = \frac{5 \cdot 10^{-2}}{7 \cdot 10^{-9}} = 7.14 \cdot 10^6 \text{ Vm}^{-1}.$$

The term $|E_i|$ means the magnitude of the field strength in the x-direction. Later we will discuss the biophysical consequence of this large electric field strength (Sects. 2.5.5 and 3.4.2).

Using Eqs. (2.4.1), (2.4.3), and (2.4.4) one can calculate the force (F), acting on a charge (q) in an electric field (E):

$$\mathbf{F} = \mathbf{E}q. \tag{2.4.6}$$

In addition to charges, dipoles play an important role in electrostatics. A dipole is a structure which contains two equal charges (q) of opposite polarity, positioned at a certain distance (l) from each other. Consequently, an electric dipole , as a whole, is electrically neutral. A dipole moment (μ) is defined as the product of charge (q) and distance (l) between them:

$$\mu = ql. \tag{2.4.7}$$

Dipoles can be formed by ampholytic molecules, this means in molecules simultaneously carrying positive as well as negative charges. They may also result from polarization phenomena of covalent bonds, as described in Section 2.1.2.

According to Eq. (2.4.7), the unit of a dipole is C m or, correspondingly: A s m. In molecular studies, sometimes the Debye-unit (D) is preferred because its amount is in the order of real molecules. The relation between these two units is:

$$1\,D = 3.3 \cdot 10^{-30}\,C\,m.$$

In Table 2.2. for demonstration, dipole moments of several compounds are listed. The dipole moments of chloroform, methanol and water are caused exclusively by polarizations of covalent bonds. In the case of proteins and nucleic acids, the values listed in this Table are to be considered just as orders of magnitude. In these cases charged groups are responsible, whose distance apart, according to the size of these molecules, can be considerable. Single α-helices of proteins have a dipole moment corresponding to a charge $q = e/2$ on both ends of them. Considering a mean length of such an α-helix of $l = 1.3$ nm corresponding to Eq. (2.4.7), a value of $\mu = 1.602 \cdot 10^{-19} \cdot 1.3 \cdot 10^{-9} = 2.08 \cdot 10^{-28}$ C m = 63.1 D results.

Ampholytic molecules exhibit electrostatic neutrality in the vicinity of the isoelectric point, since just at this pH value equal numbers of negative and positive charges are dissociated (for example: $-COO^-$ and $-NH_3^+$ groups, see Sect. 2.4.6). Because of the reversibility of this process, however, a rapid alteration of dissociation and association may take place. A molecule of this type can be considered as a *fluctuating dipole*, the charges of which oscillate in a frequency range of about 10^7 s^{-1}.

If a noncharged molecule is influenced by a strong electric field, shifts of electrons may take place, and an *induced dipole* occurs. The resulting dipole moment depends on the external field strength (E) and on the ability of the molecule to polarize, a property which is described by a factor α:

$$\mu_{ind} = \alpha E. \tag{2.4.8}$$

Such kinds of induction are measurable as *protein electric response signals*.

At the beginning of this chapter, the dielectric constant (ε) had already been introduced. Now we can characterize this parameter in greater detail on the

Table 2.2. Dipole moments of various substances. (After Netter 1959, [1] after Raudino and Mauzerall 1986, [2] see text, [3] after Pethig 1979, [4] after Takashima 1963)

	D	$\cdot 10^{-30}$ C m
Chloroform	1.05	3.47
Methanol	1.68	5.54
Water	1.85	6.11
Urea	8.6	28.4
Glycocoll	15	50
Phosphatidylcholine headgroup[1]	20	66
α-helix ($1 = 1.3$ nm)[2]	63.1	208
Proteins[3]	100–1000	330–3300
DNA from calf thymus[4]	32 000	105 600

basis of dipole properties. The electric field strength (E) of a plate capacitor is proportional to its surface charge density (σ):

$$E = \frac{\sigma}{\varepsilon_0 \varepsilon}. \qquad (2.4.9)$$

The surface charge density expresses the charges per area ($\sigma = q/A$). Introducing a dielectric between the two plates of the capacitor means an increase of the dielectric constant (ε). As can easily be seen in Eq. (2.4.9), an increase of the dielectric constant (ε) has the same effect as a decrease of the surface charge density (σ). This is not only a mathematical relation. In fact, it is caused by the polarization of the dielectrics under the influence of the electric field of the capacitor. Therefore, there is a correlation between the dielectric constant and the dipole moment of a substance.

Further reading: On protein response signal: Keszethelyi and Omos 1989

2.4.2
The Water Structure, Effects of Hydration

The physicochemical properties of water are quite different from those of chemically similar compounds, for example H_2Te, H_2Se or H_2S. Astonishing values could be predicted extrapolating the properties of these substances to H_2O. This would predict a melting point of $-100\,°C$ and boiling point of $-30\,°C$. Water, therefore, would be a gas under the normal conditions. The same deviations between predicted and actually existing values would be found in relation to other physicochemical properties, such as molar heat of evaporation, heat capacity, surface tension, dielectric constant etc.

Another speciality is the so-called *temperature anomaly* of water, which is of particular relevance for the generation and preservation of life. As illustrated in Fig. 2.22, water has its highest density at a temperature of 4 °C. This is the

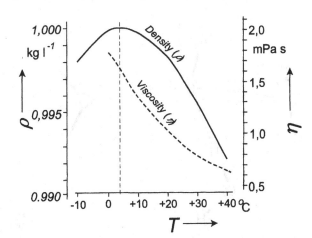

Fig. 2.22. Density (ρ) and viscosity (η) of water as function of temperature (T)

reason why, in contrast to other melting substances, ice floats on the surface, and the temperature of water near the bottom of lakes will not become colder than 4 °C. If water had the same properties as other substances, the cold water would always sink and the lakes would freeze downside up. Life in this lake could not survive the winter.

There are two molecular properties which are responsible for these effects: the dipole moment of water, and the ability of water molecules to build intermolecular hydrogen bonds. In Section 2.1.2 we discussed the electronegativity of elements in covalent bonds and found the following sequence:

$$H < C < N < O < F.$$

Hence, in both H—O-bonds of the water molecule a strong polarization will occur. The common valence electrons in these bonds are strongly attracted by the oxygen atom, which therefore becomes negative in relation to the protons. The positive charges of the two H atoms lead to a repulsion of each other. This increases the angle between them up to a value of 104.5°. The water molecule in all becomes a dipole which is directed towards the bisector of this angle.

This polarization of the H—O-bonds in the water molecule not only has consequences for its electrostatic behavior but is also responsible for its ability to build up hydrogen bonds. They can be built between two water molecules as well as between water and other molecules. This is the reason for the occurrence of the water structure in ice as well as in the fluid water.

The structure of ice has received most attention by investigators. As shown in Fig. 2.23, a tetrahedral structure is built by oxygen atoms which are always

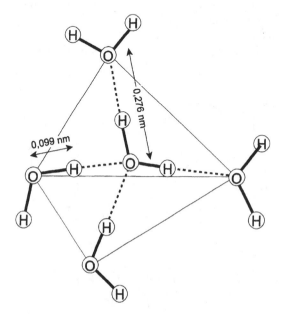

Fig. 2.23. The tetrahedrical structure of water in ice

connected by H-bonds with two hydrogen atoms of other molecules. The continuation of this kind of structure builds up a 3-dimensional network which finally leads to the crystalline structure of ice.

Thermochemical measurements as well as investigations with infrared spectroscopy indicate that these elements of the ice structure will not be destroyed fully if the ice melts. Structural water clusters exist even at temperatures above the melting point of ice. These clusters consist of interconnected water molecules with a structure similar to that of ice. The size of these clusters diminishes at increasing temperatures. Near the melting point, about 90–650 water molecules are clustered together, near the boiling temperature, there are only 25–75 molecules connected in such a cluster (the numbers differ according to different authors). This is the reason for the temperature dependence of the viscosity of water as depicted in Fig. 2.22.

This property of water molecules is enclosed in the mixed cluster model of fluid water as proposed by Nemethy and Scheraga (Fig. 2.24). This model considers that the tetrahedral structure of water, as depicted in Fig. 2.23 is not the only possible geometrical arrangement of the cluster. The strongly oriented crystalline structure of ice is disturbed in floating water by thermic noise. Thus, irregularities and structures on non-tetrahedral geometry occur.

These structures are to be considered as being in permanent movement. An H-bond oscillates with a frequency of about $0.5 \cdot 10^{13}$ Hz. The mean lifetime of a single cluster lasts approximately 10^{-10}–10^{-11} s. Therefore, during the oscillation every H-bond will form 100–1000 times with the same oxygen atom, before it connects to another one. Because of these dynamics, the name *flickering cluster* is used.

In biological systems the water structure is influenced to a great extent by interactions of the water with ions and organic molecules. The interaction of these compounds with the structure of water occurs electrostatically as well as by hydrogen bonds.

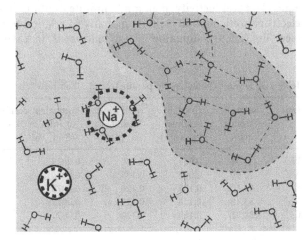

Fig. 2.24. Schematic illustration of the arrangement of water molecules which are partly connected to clusters (*dark region*), and partly occur more or less separately (*light region*). At the same scale, a Na^+- and a K^+-ion are included. The *white circle*, surrounded by the *full line* indicates the crystal radius; the circles with dashed lines indicate the size of the hydration radius

The interaction of ions with water molecules is called *hydration of the first order*. Inorganic ions interact with water molecules exclusively electrostatically. In Table 2.2 (page 56) the dipole moment of water is listed as 1.85 D. The electric field around an ion, therefore, will lead to a more or less strong orientation of water molecules. This orientating force from the electric field of the ion is in competition with influences from other water molecules, and of course with the destructive activity of thermic noise. As we discussed in Section 2.4.1 (Fig. 2.20), the electric field strength decreases with increasing distance from the center of charge. Correspondingly, two areas at different distances from the ion must be considered: near the ion is a region of *primary hydration*. This is a small number of water molecules which are strongly orientated in the electric field of the ion. In Fig. 2.24 this is illustrated in case of the three water molecules near the Na^+-ion. The region of *secondary hydration* follows. At this distance the electric field is always too weak to arrange an orientation of water molecules. On the other hand, it is strong enough to disturb the normal water structure. In Fig. 2.24 this is indicated by the destruction of the darkly marked water cluster near the Na^+-ion. In this, it is like a fault zone without any defined water structure.

To quantify this situation several parameters can be used. They are illustrated in Table 2.3 NMR measurements, and also investigations of coefficients of self diffusion of water, including its activation energy, makes it possible to determine the relative residence time of water molecules in the vicinity of ions. *Self diffusion* means the diffusion of water in water, or in solutions. If t_i stands for the mean time in which the H_2O molecule resists near an ion, and t stands for the time interval during which such a molecule will permanently stand at any point near another water molecule, the relation t_i/t characterizes the degree of demobilization of these molecules near an ion. If $t_i/t > 1$ then obviously an increase of the structural degree of water near the ion occurs, if $t_i/t < 1$ the ion is a structure breaking agent.

An important parameter to characterize the hydration shell around an ion is the *hydration radius*, or *Stockes' radius*. It corresponds to the effective radius which we introduced in Section 2.3.6 for spherical molecules. Measuring the equivalent conductivity of an ion at infinite dilution, it is possible to calculate its electrophoretic mobility, i.e. its mobility in relation to the applied field strength [see: Eq. (2.5.9)]. Equation (2.4.6) allows us to calculate the force which is

Table 2.3. Several characteristic parameters of alkali ions

Ions	Relat. atomic mass M	Crystall radius r_K (nm)	Hydration radius r_H (nm)	Equivalent conductivity at infinite dilution (S m^2 val^{-1})	Relative residence time of a H_2O molecule t_i/t
Li^+	6.94	0.069	0.240	0.00387	2.60
Na^+	22.99	0.098	0.185	0.00501	1.27
K^+	39.10	0.133	0.126	0.00735	0.54
Rb^+	85.48	0.148	0.119	0.00778	
Cs^+	132.91	0.169	0.120	0.00772	0.59

applied to an ion in a given electric field (E). Under conditions of stationary movement, this driving force is fully compensated by the friction which results from the interaction of the moving ion with the surrounding water. Now let us postulate that this frictional force can be calculated by Stockes' law [Eq. (2.3.34)]. This means the ion is taken as a macroscopic sphere with a hydrophilic surface. "Macroscopic" means large enough that water can be considered as a continuum, and not as a molecular structure. In this case, knowing the viscosity (η), the driving force (F), and the mobility (v) of the ion, the Stockes equation makes it possible to calculate an equivalent radius (r) of it as a sphere. This radius is per definition the hydration radius.

This hydration radius is a typical example of an effective parameter. It is exactly defined by its way of measurement and calculation. On the other hand, it does not exist as a visible or at least imaginable border. As indicated in Fig. 2.24, it is in no case identical for example to the thickness of a single layer of water molecules. In fact, the Stockes equation is not the proper law to use to calculate the real movement of an ion in water. The conditions listed above are not realized in this case. Nevertheless this parameter, because of its exact definition, is an important measure for the hydration properties of ions. When using it, for example, to explain membrane permeability for ions, or to discuss other similar mechanisms, it is always necessary to be aware of the conditions of its definition.

As illustrated in Table 2.3, the hydration radius of the alkali ions decreases with increasing atomic mass. In the case of lithium, the hydration radius is much larger than the crystal radius, whereas for potassium it is already smaller. For this, K^+ and Cs^+ are called *structure breaking* ions. The region of secondary hydration, i.e. the fault zone of broken water structure, is larger than that of the electrically orientated molecules. In contrast to this, in small ions, like Li^+ and Na^+, the region of primary hydration is far larger. These are *structure building* ions. This category also includes the two valent cations Mg^{2+} and Ca^{2+} which are not mentioned in the Table. Sometimes, instead of the terms "structure breaking" or "structure building", the expressions "chaotropic" or "cosmotropic" are used.

What is the reason for these differences? Both ions, Na^+ as well as K^+, carry the same single positive charge. The Bohr's radius of these ions, i.e. the effective radius of the outer electron shell which is the same as the crystal radius, is, however, larger for K^+ then for Na^+, because of the additional electron orbital in the case of potassium. With a larger distance from the charged nucleus of the ion, the electric field strength will become lower.

Hydration effects which are not caused directly by electrostatic interactions are called *hydrations of the second order*. In this case hydrogen bonds between organic molecules and water are the reason for the structure of a water layer in the direct vicinity of the surface. In view of these sorts of interaction, two types of molecular surfaces must be considered: proton donator and proton acceptor surfaces. As depicted in Fig. 2.25, this difference is responsible for the orientation of the attached water molecules. This orientation may be transmitted by several layers of water molecules. This kind of hydration can also lead to an interaction between two surfaces. Surfaces with identical orientation of water

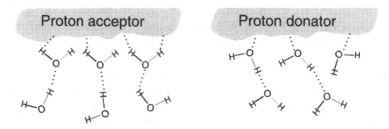

Fig. 2.25. Water orientation near proton-accepting and proton-donating surfaces

molecules repel each other, surfaces with opposite orientation of water show attractions at short distances.

Hydrophobic molecules, in general, behave in relation to the surrounding water like an air bubble. From the phenomenological point of view, water molecules near this surface exhibit a larger amount of energy than molecules in the water bulk. This energy of interfacial molecules in connection with interfacial tension will be described in detail in Section 2.5.1.

There are some peculiarities of the physical properties of water near surfaces in relation to the bulk water. In a layer of 0.1 to 0.2 nm, for example, the effective dielectric constant of water decreases dramatically up to the value of 10 (Takashima et al. 1986). Furthermore, because of the water structure, this region may become anisotropic. This means, for example, that diffusion rates become higher then the average value in the direction parallel to the surface and lower in the direction normal to it (Makarov et al. 1998).

Second order hydration can play a predominant role in conformational changes of molecules and in the formation of supramolecular structures. In these processes not only the entropy of the macromolecules changes, but also the entropy of the surrounding water. Because of the structure-building properties of some molecular surfaces, the entropy of the water near the surface is lower than that in the bulk phase. In the case of conformational changes of macromolecules, or of processes of supramolecular organization, the surface available for interaction with water molecules may decrease. This happens for example if lipid molecules aggregate to form membranes or vesicles. In this case the water between the molecules will be pushed away. Considering these sorts of processes thermodynamically, a higher degree of organization results in the entropy of the macromolecules decreasing, but because of the loss of structure-building surfaces the entropy of water increases. In general, therefore, even if a macromolecular structure of a higher degree of organization results, the entropy of the whole system increases. According to the second principle of thermodynamics, therefore, these kinds of processes may occur spontaneously.

In the same way the spontaneous formation of secondary and tertiary structures of macromolecules can be considered (Fig. 2.26). These kinds of processes are called *entropy driven*.

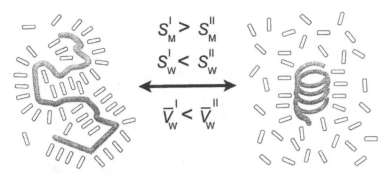

$$S^I_M > S^{II}_M$$
$$S^I_W < S^{II}_W$$
$$\bar{V}^I_W < \bar{V}^{II}_W$$

Fig. 2.26. The orientation of water molecules, schematically drawn as *bars*, near the surfaces of macromolecules. In contrast to the decrease of entropy of the macromolecule during formation of the helix ($S^I_M > S^{II}_M$), the entropy of water during this process increases ($S^I_W < S^{II}_W$). This leads to an increase of the entropy in the whole system. At the same time the partial molar volume of water increases

It was found that the desorption of water from surfaces leads to a certain increase in its partial molar volume (Fig. 2.26). Obviously, the more highly organized water molecules at the surfaces are packed closer together. In this respect, we should remain with the so-called Le Châtelier principle. It predicts that if a constraint is applied to a system in equilibrium, the system adjusts to a new equilibrium that tends to counteract the constraint. An increase of the hydrostatic pressure, in this sense, could be counteracted by the decrease of the partial volume of water in the system. In Fig. 2.26 this would mean an increase of pressure would lead to an increase of hydration, and therefore to a decrease in organization of the macromolecular, or the supramolecular structure. Hence, an increase of hydrostatic pressure should destroy structures which are connected by processes of hydration, i.e. by so-called *hydrophobic bonds*. In fact, extremely high pressures in the order of 10 to 100 MPa are necessary for these reactions. In these cases membranes, microtubules and other supramolecular structures will be destroyed. In some cases high hydrostatic pressures were applied to study physiological properties of membranes. In this context it must be remembered that in the deepest parts of the ocean, at depths of 10 km, a hydrostatic pressure of 100 MPa occurs!

Figure 2.27 illustrates how the shift of the partial volume of water can be measured directly during a dissociation-association reaction. For this, in a vessel with exactly controlled temperature, the macromolecular sample enclosed in a dialysis sac hangs on a very sensitive quartz spring. The surrounding solution of this dialysis sac is exchanged continuously. A modification of the density of the macromolecules will be reflected in a change of the length of the spring. As indicated in the figure, the observable changes are minimal. In this case the change during the pH dependent polymerization of proteins of the tobacco mosaic virus is demonstrated. The time which is necessary to arrive at the equilibrium in this process is quite long.

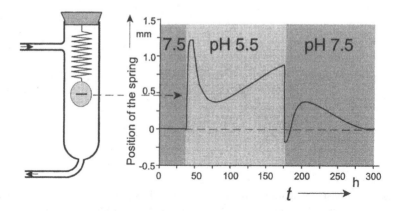

Fig. 2.27. Measurement of the kinetics of a hydration process, using a dialysis sac hanging on a quartz spring. The curve indicates results obtained on the protein of the tobacco mosaic virus. According to data from Lauffer 1975

Hydrophobic bonds, in fact, are not true interatomic connections like covalent bonds. Molecules or parts of them rather were pressed together by a force generated by the entropy increase in the involved water molecules.

Further reading: Benz and Conti 1986; Collins and Washabaugh 1985; Horn 1972; Israelachvili 1994; Lauffer 1975; Makarov et al. 1998; Wiggins 1990

2.4.3
Ions in Aqueous Solutions, the Debye-Hückel Radius

In the previous sections we discussed the electric field, which was induced by a single charged point, and which influences molecules in its vicinity. When considering extremely short distances, calculating for example force and energy of ionic bonds, or the influence on water dipoles nearby, it is possible to consider only this single charged point and to use Coulomb's law [Eq. (2.4.3)] for the calculation. In the case of larger distances from the source of the field, however, the influence of other charges must be considered. In a 100 mM solution of NaCl, for example, the mean distance between the ions only amounts to about 2 nm. This means that under physiological conditions the interactions of ions must be considered as a set of charged points.

These kinds of calculations are possible on the basis of the theory of strong electrolytes, the *Debye-Hückel theory*. This theory considers an ionic cloud around an arbitrarily chosen central ion. This central ion is taken as the center of polar coordinates. It attracts ions with opposite charges (conter-ions) and repels them with identical charges (co-ions). In this way an electric field will build up, generated by the central ion as well as by the ions surrounding it. This field is spherically symmetrical and can be described by the function $\psi(r)$.

At a distance r from the central ion, other ions will be attracted or repelled by the central ion, depending on their charge. We will denote the number of charges of an ion by z_i, for example: $z_{Na} = +1$, $z_{SO_4} = -2$ etc. The electrostatic energy of an ion at a particular point in space with an electrical potential ψ amounts to $z_i e \psi$.

Boltzmann's law of energy distribution [Eq. (2.3.16)] makes it possible to calculate the concentration (c_i) of the ions i at a point with the potential ψ. In this case, the energy of an ion in the electric field ($z_i e \psi$) opposes the energy of thermic noise (kT). This means that an orientation in the common electric field opposes the stochastic distribution of the ions by thermic fluctuations.

$$c_i(\psi) = c_{i0} e^{-\frac{z_i e \psi}{kT}}. \tag{2.4.10}$$

In this equation, c_{i0} is the concentration of the ion i, far away from the influence of the central ion, which means in the bulk solution. This equation, however, will not answer our question. We do not know yet the real amount of the potential ψ at this point, because it depends itself on the concentration of the ions in the ionic cloud. Furthermore, we are not interested in the function $c_i(\psi)$, but would rather know the concentration and potential distribution in space, i.e. the functions $c_i(r)$ and $\psi(r)$. This is possible using the Poisson equation. This is a partial differential equation of the second order, which gives the function $\psi(r)$ in space as a result of a charge density $\rho(r)$:

$$\nabla^2 \psi = -\frac{1}{\varepsilon_0 \varepsilon} \rho. \tag{2.4.11}$$

$\nabla^2 \psi$ means the second derivation of the potential by the coordinates of the space. This Nabla operator has already been described in Section 2.4.1. For ε the dielectric constant of water is used.

The charge density (ρ, in C m^{-3}) can be calculated from the ion concentration c_i of the ions in the cloud by:

$$\rho = \sum_{i=1}^{n} c_i z_i Ne = F \sum_{i=1}^{n} c_i z_i. \tag{2.4.12}$$

Introducing Eq. (2.4.10) in (2.4.12), and both in (2.4.11), one gets the *Poisson-Boltzmann equation*:

$$\nabla^2 \psi = -\frac{F}{\varepsilon_0 \varepsilon} \sum_{i=1}^{n} c_{i0} z_i e^{-\frac{z_i e \psi}{kT}}. \tag{2.4.13}$$

The expression (2.4.13) is a partial differential equation, which can be solved analytically only under simplified conditions. Let us consider the following situation: let the solution contain only a single kind of ions i, let the potential be very low ($z_i e \psi \ll kT$) and simply a spherical symmetric function of the central ion. For this simplified case, the solution of Eq. (2.4.13) becomes:

$$\psi_r = \frac{z_i e}{4\pi\varepsilon_0 \varepsilon r}\, e^{-\kappa r}. \tag{2.4.14}$$

Comparing this equation with Eq. (2.4.3), it shows that the influence of the potential of the central ion will be diminished by the factor $e^{\kappa r}$ ($e^{-\kappa r} = 1/e^{\kappa r}$). The extent of this decrease depends on the concentration of the ions in the solution (see Fig. 2.28). This becomes understandable when considering the definition of the *Debye-Hückel parameter* (κ) in following equation:

$$\kappa = \sqrt{\frac{e^2 N}{\varepsilon_0 \varepsilon\, kT} \sum_{i=1}^{n} c_{i0} z_i^2} = \sqrt{\frac{F^2}{\varepsilon_0 \varepsilon\, RT} \sum_{i=1}^{n} c_{i0} z_i^2} = \sqrt{\frac{2F^2 I}{\varepsilon_0 \varepsilon\, RT}}. \tag{2.4.15}$$

The first expression of this equation can easily be transformed into the second one, using the relation between the Faraday constant (F) and the charge of a single electron (e) as: F = eN, as well as the relation: R = kN. In the last expression of Eq. (2.4.15), the *ionic strength* (I) is introduced.

$$I = \frac{1}{2} \sum_{i=1}^{n} c_{i0} z_i^2. \tag{2.4.16}$$

When calculating the value of κ for real conditions it can be seen that this Debye-Hückel parameter has the measure: m^{-1} (For this of course, the concentration measure: mole per m^3 must be used!). The parameter $1/\kappa$ is therefore a measure of a distance. Accordingly, it is called *Debye-Hückel length* or *Debye Hückel radius* of the ionic cloud. Considering Fig. 2.28 it is obvious that this value $1/\kappa$ again is not a "visible" radius, but just an effective parameter.

In the following example we will demonstrate the calculation of the ionic force of a solution: Let us consider a physiological Ringer's solution with the following composition: 105 mM NaCl, 5 mM KCl, 25 mM Na_2HPO_4, 2 mM $CaCl_2$

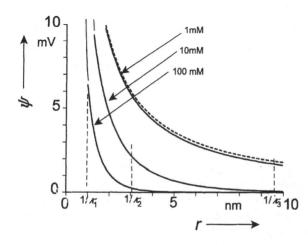

Fig. 2.28. The electrical potential (ψ) as function of the distance (r) from a central ion. The *dashed line* ignores the influence of the ionic cloud according to Eq. (2.4.3); the *solid lines* show the potential $\psi(r)$ which is diminished by the ionic clouds in solutions of various concentrations of a one-one-valent salt ($z = \pm 1$) according to Eq. (2.4.14)

(mM means the usual measure of concentration: mmol l^{-1}). Let us assume a full dissociation of the Na_2HPO_4. Using Eq. (2.4.16), one gets:

$$I = \tfrac{1}{2}(c_{Na}z_{Na}^2 + c_K z_K^2 + c_{Cl}z_{Cl}^2 + c_{HPO_4}z_{HPO_4}^2 + c_{Ca}z_{Ca}^2)$$
$$= \tfrac{1}{2}(0.155 \cdot 1^2 + 0.005 \cdot 1^2 + 0.114 \cdot 1^2 + 0.025 \cdot 2^2 + 0.002 \cdot 2^2)$$
$$= 0.191\,\text{mol}\,l^{-1}.$$

According to Eq. (2.4.15), the Debye-Hückel radius (in nm) at temperature $T = 298$ K can easily be calculated by the following equation, using as a concentration measure mol $\cdot$ l^{-1} = M:

$$\frac{1}{\kappa} = \frac{0.304}{\sqrt{I}} \quad \text{(in nm).} \tag{2.4.17}$$

This means for the case of the above considered solution:

$$\frac{1}{\kappa} = \frac{0.304}{\sqrt{0.191}} = 0.696 \quad \text{(in nm).} \tag{2.4.18}$$

Summarizing, we conclude that with the Poisson-Boltzmann equation [Eq. (2.4.13)] and its integration, the electrical potential can be calculated depending on the distance from a central ion. In this case the screening effect of the ionic cloud is considered. This screening, in fact, increases with increasing ionic strength of the solution. As a measure of the thickness of the ionic cloud, the Debye-Hückel radius ($1/\kappa$) is defined, which, in contrast to the crystal radius or the hydration radius, is not a property of a specific ion but just depends on the ionic strength of the solution. The Debye-Hückel radius is an effective parameter as well as the others (see Sect. 2.4.2).

2.4.4
Intermolecular Interactions

Self-assembly of biological structures is an intricate process consisting of an interplay of a variety of different intermolecular forces. In Section 2.4.2 we explained the role of hydrogen bonds for building up the structure of water, for interactions among ions and between ions and water dipoles, as well as the role of hydrophobic bonds in the process of self organization of macromolecules. In the following, the spectrum of these interactions will be extended to include additional electrostatic and electrodynamic forces. In fact, any kind of interaction between two macromolecules in an aqueous medium is the sum of several forces which differ by their sign (i.e. repulsive or attractive), and by their distance function.

Let us first consider a dipole interacting with an ion. In this case, the energy of electrostatic binding (E_{ID}), can be calculated using Coulomb's law [Eq. (2.4.2)]. It simply results from the sum of all electrostatic components, namely the interaction of the ion (q_1) with one ($-q_2$), as well as with the other ($+q_2$)

charge of the dipole. Considering the simplest case, namely the linear orientation of the dipole in relation to the ion (Fig. 2.29A), it gives:

$$E_{ID} = \left[-\frac{q_1(-q_2)}{4\pi\varepsilon_0\varepsilon\, r} \right] + \left[-\frac{q_1 q_2}{4\pi\varepsilon_0\varepsilon(r+l)} \right] \tag{2.4.19}$$

We will consider simply that the dipole $\mu = q_2 l$ contains two charges with equal size but opposite signs. A simple algebraic rearrangement of Eq. (2.4.19), and the introduction of the dipole moment (μ) results in:

$$E_{ID} = \frac{q_1 q_2 l}{4\pi\varepsilon_0\varepsilon r(r+l)} = \frac{q_1 \mu}{4\pi\varepsilon_0\varepsilon r(r+l)}. \tag{2.4.20}$$

Assuming: $r \gg l$, the Eq. (2.4.20) will be simplified to:

$$E_{ID} \approx \frac{q_1 \mu}{4\pi\varepsilon_0\varepsilon r^2}. \tag{2.4.21}$$

In contrast to the energy of interaction of two ions [Eq. (2.1.5)] as a function of $1/r = r^{-1}$, this equation shows that the interaction energy of an ion with a dipole decreases with r^{-2}.

The mutual interaction between two dipoles (E_{DD}) can be calculated in a similar way. Let us again look at the simplest case of collinear orientation of both dipoles, as depicted in Fig. 2.29 B. Using the same approach, one gets:

$$E_{DD} = -\frac{1}{4\pi\varepsilon_0\varepsilon} \left[\frac{q_1 q_2}{l_1 + r} + \frac{q_1(-q_2)}{l_1 + r + l_2} + \frac{(-q_1)q_2}{r} + \frac{(-q_1)(-q_2)}{r + l_2} \right]. \tag{2.4.22}$$

Again we will rearrange this equation and introduce the dipole moments μ_1 and μ_2:

$$\begin{aligned} E_{DD} &= \frac{q_1 l_1 q_2 l_2}{4\pi\varepsilon_0\varepsilon} \cdot \frac{l_1 + l_2 + 2r}{(l_1 + r)(l_2 + r)(l_1 + l_2 + r)r} \\ &= \frac{\mu_1 \mu_2}{4\pi\varepsilon_0\varepsilon} \cdot \frac{l_1 + l_2 + 2r}{(l_1 + r)(l_2 + r)(l_1 + l_2 + r)r}. \end{aligned} \tag{2.4.23}$$

Using the same approach: $r \gg l$, it gives:

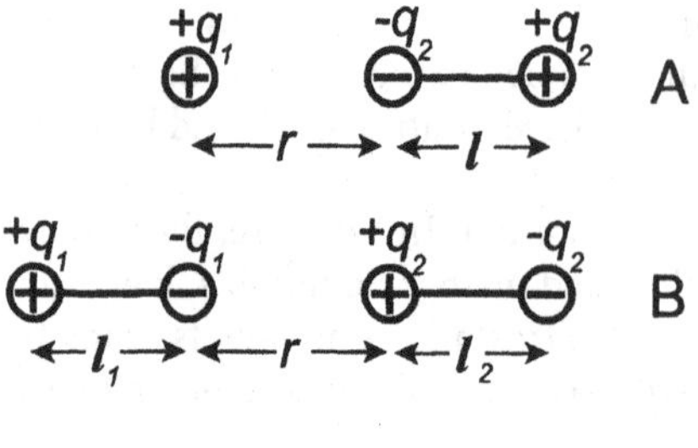

Fig. 2.29. Ion-dipole (A) and dipole-dipole (B) interaction. q – charge, l – length of the dipoles, r – distance

$$E_{DD} \approx \frac{\mu_1 \mu_2}{4\pi\varepsilon_0\varepsilon} \cdot \frac{2r}{r^4} = \frac{2\mu_1\mu_2}{4\pi\varepsilon_0\varepsilon r^3}. \tag{2.4.24}$$

This sequence of possible electrostatic interactions can be extended including interactions with induced dipoles (see Sect. 2.4.1). In this case, a more complicated function of the distance must be expected because the induction of a dipole itself depends on the electric field [Eq. (2.4.8)]. For this case, a function can be calculated, including r^{-4}. If an interaction of two dipoles is considered, inducing each other, a function with r^{-6} will be obtained.

This brings us to the wide field of *van der Waals interactions*. That the expression is in plural indicates that it includes a number of different intermolecular forces. Van der Waals interactions in general are attractive forces between molecular components, whole molecules, or supramolecular particles which are not simply based on electrostatic interactions. Rather they are attributed to electromagnetic interactions, occurring by fluctuations of charges. The reasons for such charge fluctuations in molecules may be different. There can be thermal molecular translations, or oscillations in the thermic level of electrons, as illustrated in Fig. 2.5. Furthermore, fluctuations occur in the electron structure of molecules which can be calculated by quantum mechanics. This is the basis of so-called dispersion forces, or *London-dispersion forces*, sometimes also called *London-van der Waals interactions*. These oscillations occur at frequencies similar to that of visible light.

The understanding of van der Waals forces has been substantially advanced by F. London, H. C. Hamaker and E. M. Lifschitz. Nevertheless, many aspects are unclear even today. The central question concerns the character of this interaction, and its energy as a function of inter-particle distance (r). Furthermore, it still seems unclear whether in these kinds of interactions a molecular specificity can exist which might be related to the frequency of interaction.

The energy of van der Waals interactions is ruled by the *Hamaker constant*, which results from quantum mechanical calculations. The distance relation of this energy depends strongly on particle geometry. This is illustrated in Fig. 2.30. There is a difference as to whether it is an interaction between a spherical and/or a cylindrical particle, and additionally there are different functions $f(r^{-n})$ depending on the total distance between the particles. This means at the same time that simple hyperbolic functions of this kind are only rough approximations.

To understand the relevance of van der Waals forces, it is necessary to consider them in context with electrostatic interactions. Here we touch a central concept of colloid chemistry, especially the concept which is connected with the names of B. V. Derjaguin, L. D. Landau, E. J. W. Verwey, and J. Th. G. Overbeek. It is therefore called the *DLVO-theory* of colloidal interactions. The basic ideas of this approach, considering the Debye-Hückel theory of the ionic clouds, as well as the properties of van der Waals, can be listed as follows:

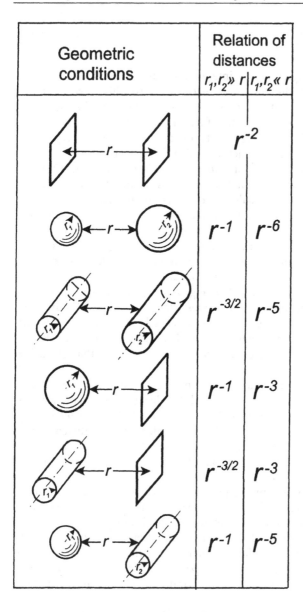

Geometric conditions	Relation of distances $r_1,r_2 \gg r$	$r_1,r_2 \ll r$
	r^{-2}	
	r^{-1}	r^{-6}
	$r^{-3/2}$	r^{-5}
	r^{-1}	r^{-3}
	$r^{-3/2}$	r^{-3}
	r^{-1}	r^{-5}

Fig. 2.30. The distance function $[f(r^{-n})]$ of the van der Waals energy of interaction between particles of various shape. (Parameters after Israelachvili 1974)

- The interaction force between two particles results from the sum of the particular forces, each one decreasing with increasing distances in different ways.
- In contrast to generally attractive van der Waals forces, electrostatic interactions can also become repulsive.
- The intensity of electrostatic interactions, in contrast to van der Waals forces, depends to a high degree on the environmental conditions.

To illustrate the consequences of these postulates, we must imagine that the distance function of the interaction energy $[E(r)]$ results as a sum of various functions, schematically drawn in Fig. 2.31. There are so-called *short distance interactions*, decreasing with r^{-5}, or r^{-6}, and *long distance interactions*, the curves of which do not decrease as fast (r^{-1}). Some of these curves can be looked at upside down, representing repulsive forces, which consequently must be subtracted by the others. Additionally, the dependence of the electrostatic forces on the ionic strength of the solution is to be considered, as illustrated in Fig. 2.28, as well as the various distance functions of the van der Waals forces, as depicted in Fig. 2.30. Hence, the shape of the resulting sum can be quite complicated.

Figure 2.32 illustrates this situation for a more or less simple case of the interaction of two particles in a solution. In this case the interaction is supposed to be the sum of an electrostatic repulsion (r^{-3}) and two kinds of van der Waals attraction forces, one a long- (r^{-1}), another a short-distance interaction (r^{-6}). Considering the sophisticated function of the electrical potential near a charge in electrolyte solutions, as depicted in Fig. 2.28, and furthermore the circumstance that the function of van der Waals interactions itself depends on distance (Fig. 2.30), it is understandable that the real conditions of interactions are much more complicated. Additionally, stearic barriers have to be considered when particles are in closer contact as well as changes in the hydration property and many others.

Nevertheless, the curves of Fig. 2.32 demonstrate at least qualitatively some basic properties of macromolecular interactions. It shows that two particles may attract each other at larger distances, even if they carry charges of the same sign, if additionally a long ranging van der Waals attraction occurs. This is demonstrated in Fig. 2.32 at distance $r > r_{min}$. The point r_{min} marks the position of minimum interaction energy. The attractive force ($\mathbf{F} = dE/dr$) at this point becomes zero. If this distance is decreased a repulsion force

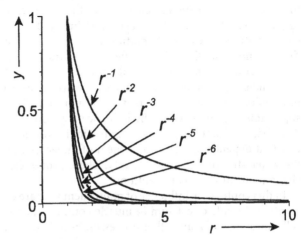

Fig. 2.31. Functions: $y = r^{-n}$ to illustrate the character of short- and long-distance interactions

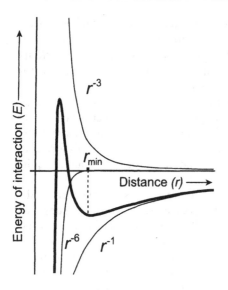

Fig. 2.32. The total energy of interaction (E) (*thick line*) between two particles as a sum of three functions (*fine lines*): one sort of electrostatic repulsion (r^{-3}), and two kinds of van der Waals attractions (r^{-1} and r^{-6}). The point r_{min} indicates the distance of the secondary minimum

appears. The particles, therefore, will rest at this point, which is called the distance of the *second minimum*. If in any way this repulsion force were to be overcome, attractive interactions again would take place and the particles would aggregate at the distance of the *primary minimum*. In Fig. 2.32 this primary minimum is not shown. This would predict an inclusion of further components of interactions at very short distances. So, for example, the influence of Bohr's radii must be considered in this case, since the extremely strong repulsion forces arising from electron shell interactions come into play.

As already mentioned, the electrostatic interaction strongly depends on the ionic conditions of the medium. This is understandable in view of Eq. (2.4.14) (Sect. 2.4.3), as well as Fig. 2.28. By increasing the ionic strength, for example, the energy barrier in the region $r < r_{min}$ will become diminished. Finally, thermic hits could be sufficient to let the particles cross this barrier and to agglutinate. This effect has been experimentally verified. Increasing the salt concentration of a colloid solution leads to the agglutination of the particles. This is known as the *salt-out effect*, in contrast to the *salt-in* process, leading to a re-stabilization of a colloidal system.

In molecular biology this interplay between the various forces involved in intermolecular interactions is very important. This is one way to control supramolecular processes. The influence is possible either by the control of the dielectric constant, or by changing the ionic concentration. The local dielectric constant depends on the arrangement of molecules, of their polarization and of the water structure in their vicinity. These processes, are mainly controlled by changing local ionic compositions.

In this context the influence of multivalent ions are of particular interest. Let us mention the role of calcium in muscle contraction, nerve excitation, exocytosis, cell adhesion, and many other processes. In colloid chemistry the Schulze-Hardyrule is known, an empirically found relation which later was also verified theoretically. This rule predicts that the concentration of a salt, which would be necessary to overcome the energy barrier between the second and the first minimum, is inversely proportional to the 6th degree of its valency. This means that the critical concentration produced by a univalent ion relates to this concentration of a bivalent one, like 1^{-6} to 2^{-6}, which means: 1 to 1/64. Consequently, 2 mM Ca^{++} would be as effective, as 128 mM K^+ or Na^+. This is a relation which reminds us of the real concentrations of in vivo conditions. It must however, be considered that this would not explain the difference in the behavior of Ca^{++} in contrast to Mg^{++}. Other reasons must also be taken into consideration.

Further reading: For van der Waals interactions in molecular biology: Jehle 1969; Israelachvili 1994

2.4.5
Structure Formation of Biomacromolecules

In the preceding sections, problems relating to the information content of biomacromolecules were discussed, the idea of random coil formation by molecular chains as a state of maximum entropy was introduced, as well as the role of water in the entropy balance of macromolecular systems, and the broad spectrum of intermolecular interactions. A knowledge of these aspects is an important pre-requisite to understanding the dynamic structure of proteins and nucleic acids in cells.

Investigations of the 3-dimensional structure of biological macromolecules, their genetic coding, and especially the processes of self assembling of these structures, are the main problems of molecular biophysics. There is no doubt that the total information on the stearic structure of proteins is stored in its primary structure, i.e. in the sequence of the amino acids. The question arises as to how one can decode the 3-dimensional structure of a protein, reading its amino acid sequence and furthermore, how the process of self assembly, which leads spontaneously and reversibly to this genetically predicted protein structure, really works?

The proteins are composed of α-L-amino acids which are connected to each other, forming long chains. In these connections the carboxyl groups of the α-S-Atoms are condensed with the amino group of the following monomer, splitting up a molecule of water. In Fig. 2.33.A the distances and the angles of the bonds of the basic structure of a polypeptide are illustrated. The symbols R mean the side chains characterizing the 20 amino acids appearing in living nature.

These molecular chains have a high degree of flexibility. During Brownian movement, interconnections of various parts of these chains are possible. This occurs mostly by hydrogen bonds connecting atoms of two peptide bonds, or atoms of a peptide bond with those of a side chain of an amino acid, or finally of two side chains together. Covalent bonds between side chains are also possible. This is mostly done by disulfide bridges between the SH-groups of cysteins. Electrostatic and hydrophobic interactions also take place.

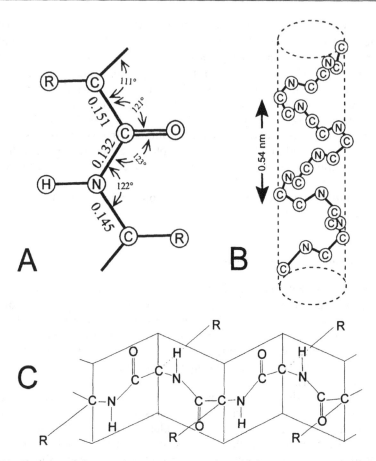

Fig. 2.33. Elements of the protein structure: A angles and distances in a peptide chain; B parameters of an α-helix (for more clarity, only the basic atoms of the polypeptide chain are depicted); C β-sheet

Let us start to consider possible mechanisms of self assembly, hypothesizing that the molecule will move stochastically, reversibly connecting and disconnecting bridges with monomers of its own chain. This proceeds as long as the polypeptide chain does not become a structure of minimal conformational energy, which will be identical with the active structure of the natural protein. This hypothesis can easily be reduced to absurdity, considering the time which would be necessary for this process. Let us take a protein with 300 amino acids, each of which could have 8 rotational positions. In this case $8^{300} = 10^{270}$ positions of the whole molecule are possible. Even in a case where, by reason of stearic hindrance, this amount were to be limited, and even if the conformational changes were extremely fast, the time needed to arrive at the correct structure with this procedure would be astronomically long. Finally, it would be questionable whether the absolute minimum was really found, and whether this really represents the proper structure of the protein.

In fact, however, the number of possible configurations are much more strongly limited. There are always two types of secondary structures which occur, and which are predicted by the interaction of neighboring monomers in the sequence of the chain. These are the α-helix, and the β-sheet. The α-helix is a right-hand spiral, the parameters of which are illustrated in Fig. 2.33B. This structure mostly occurs in proteins which are embedded in the lipid leaflet of membranes (see also Fig. 2.38). In these hydrophobic surroundings they are obviously more stable then in aqueous solutions. In such solutions complexes are sometimes built consisting of several helices, which stabilize each other.

In β-sheets the groups of 6 atoms (C_α-CONH-C_α; see Fig. 2.33A) are positioned in a plane. Schematically such an arrangement of planar monomers is depicted in Fig. 2.33C. Up to nine monomers are usually arranged in such a β-sheet. Mostly, these planar elements are not simply folded in a plane, but the folded elements themselves form a right-hand spiral. In contrast to the α-helix where the side changes (R) are orientated perpendicular to the helical axis (not depicted in Fig. 2.33B!), in the case of the β-sheet they are orientated perpendicular to the plane of the monomer (Fig. 2.33C). The stability of these sheets increases by parallel or by antiparallel arrangements of several sheets which are connected together by hydrogen bonds.

Beside these two types of arrangements of polypeptides, others are sometimes possible, as for example α-helices with another pitch, a different amount of amino acids per turn (3_{10}-helix, π-helix), or even left-hand turned structures.

Apparently there are specific sequences of amino acids which tend to generate helical structures, and others forming β-sheets or unstructured regions connecting them. There is always a search for correlations between amino acid sequences and their secondary structures, to derive rules predicting the occurrence of them. Collagen, as a typical structure protein for example, forms a triplet-helix as a ternary structure. This is already marked in the primary structure by a repeating sequence (Gly-X-Y)$_n$. In this case Gly stands for glycine, whereas X and Y demand some other amino acids. Mostly these are proline and hydroxyproline. This repeated occurrence of glycine in each third position stabilizes the triplet-helix by hydrogen bonds. In some cases amino acids with the ability to form disulfide bridges build a sort of starting point for continuing the formation of super helices. In this case, as in the formation of α-helices, the process resembles a zip fastener. In a similar way myosin forms a left-hand double super-helix. The turn of this helix is predicted by the arrangement of amino acids in the corresponding α-helices of the secondary structure.

Recently a number of enzymes were found significantly accelerating the process of structure formation of proteins, for example the group of *chaperons*. They mostly help to avoid wrong ways of folding, binding reversibly to specific sequences of the primary structure of the polypeptide. For this reason it is not correct to use the term structure forming enzymes for them. In contrast, protein-disulfide-isomerases and poly-cis/trans-isomerases directly promote the process of folding. This is because all peptide bonds are primarily synthesized in trans-form, but 7% actually occur in cis-form. This trans-cis transformation in general occurs quite slowly, but will be accelerated significantly by these enzymes.

A correctly folded protein, i.e. a native protein, can become reversibly denaturated by shifts of pH, temperature or by modifications of the ionic conditions in the solution. In this way it becomes a so-called *molten globe*. The way back again toward the native protein is marked by various intermediates, where transient H-bonds or even disulfide bonds are established, which will be released again on the way to the next intermediate state. Therefore, the path of this reaction has apparently been determined, but is not yet fully understood.

The difference in free molar Gibbs energy (ΔG) between the native, versus the denatured status of the proteins, is sometimes lower then 50 kJ mol^{-1}. Considering that about 100 connections between amino acids of neighboring chains must be loosened during this process, their binding energies must be lower then the energy of thermic motion ($RT = 2.5$ kJ mol^{-1} for $T = 300$ K). The stability of the protein structure, therefore, can be explained only as a phenomenon of cooperativity. Additionally, this estimation illustrates the dynamics of the molecule. Obviously, in the native protein small conformational transitions must continuously occur, caused by thermic fluctuations.

These thermic fluctuations of protein structure are, in fact, prerequisites for their function. Such modifications, for example, can lead to activation or inactivation of enzymes. The enzyme activity can therefore be controlled in this way. Similar modifications can be caused by ligands, by changes of the local ionic conditions, or even under the influence of electric fields. These field effects occur for example during the control of Na- and K-channels in nerve cells, causing the action potential (see Sect. 3.4.3).

These molecular fluctuations may modify the shape, the volume, as well as the energetic state of the molecules. Analysis of these fluctuations makes it possible to calculate the average square of displacement of the energy ($\langle \Delta U^2 \rangle$)as well as that of the volume of the molecule ($\langle \Delta V^2 \rangle$):

$$\langle \Delta U^2 \rangle = kT^2 m C_V, \qquad (2.4.25)$$

$$\langle \Delta V^2 \rangle = kTV\beta_T. \qquad (2.4.26)$$

Hence, these parameters depend on the energy of thermic noise (kT), on the molecular mass (m), on the molecular volume (V), on the isochoric heat capacity (C_V), as well as on the isothermic volume compressibility (β_T). Considering a globular protein with a molecular mass $M = 25\,000$, it results:

$$m = \frac{M}{N} = 4.15 \cdot 10^{-20}\,\mathrm{g} = 4.15 \cdot 10^{-23}\,\mathrm{kg}.$$

Let us consider the following case: $V = 3.2 \cdot 10^{-26}$ m^3, $C_V = 1.34$ kJ kg^{-1}, $T = 298$ K, as well as $\beta_T = 2 \cdot 10^{-11}$ Pa^{-1}. From Eq. (2.4.25) results

$\Delta U = 2.61 \cdot 10^{-19}$ J per molecule. Multiplying this by the Avogadro number ($N = 6.022 \cdot 10^{23}$ mol^{-1}), would mean 157 kJ mol^{-1} in a case where all molecules would fluctuate synchronously. Already this is enough energy to denaturate the protein. The deviations in the volume of this molecule result from Eq. (2.4.26) for: $\Delta V = 5.1 \cdot 10^{-29}$ m^3. This is about 0.16% of the total volume of the molecule. Nevertheless, these deviations would be sufficient to create transient local hollow spaces in the molecule, or to allow the penetration of ions in preformed pores. Hence, this kind of statistical fluctuation should be included in the consideration of protein functions.

It must be added that we considered here the proteins as structures in equilibrium. This corresponds to the situation of isolated molecules. During biological processes however, the proteins are usually considered as far from equilibrium. Therefore, many other structural modifications occur, mostly concerning special functional groups of the molecule. This is a field of further research.

Further reading: Cantor and Schimmel 1980; Creighton 1992; Cooper 1976

2.4.6
Ampholytes in Solution, the Acid-Base Equilibrium

We already learned that charges in biomolecules significantly influence the formation of their molecular and supramolecular structures. This aspect will be underlined in the next section when we discuss properties of interfaces and biological membranes. Furthermore, in Section 3.2.3 we will discuss the role of fixed charges in the cytoplasm to explain the occurrence of Donnan potentials. These aspects demand knowledge about the pH dependence of these sorts of charges.

An acid, per definition, is a proton donor, i.e. a compound which is dissociating hydrogen ions from particular groups. A base, on the other hand, is a proton acceptor and thus is able to associate it. According to this definition, water is a weak acid, dissociating with the following scheme:

$$H_2O \leftrightharpoons H^+ + OH^-.$$

This equilibrium leads to the definition of the ionic product of water and, finally, to the pH-value:

$$pH = -\log c_H \qquad (2.4.27)$$

Strictly of course, it is not the concentration (c_H), but rather the activity of the protons that must be considered, but in view of the extremely small concentration of these protons, this makes no difference.

In the same way, the dissociation of any other acid can be written:

$$HA \leftrightharpoons H^+ + A^-$$

The equilibrium constant of this reaction may be written as:

$$K = \frac{c_A c_H}{c_{HA}} \tag{2.4.28}$$

In the same way, as in the case of pH, it is possible to define a pK-value:

$$pK = -\log K = -\left(\log c_H + \log \frac{c_A}{c_{HA}}\right) \tag{2.4.29}$$

or, using Eq. (2.4.27):

$$pK = pH - \log \frac{c_A}{c_{HA}} \tag{2.4.30}$$

This is the *Henderson-Hasselbalch equation* which makes it possible to understand the buffer properties of various substances and their titration curves.

Substances which contain acid as well as basic groups are called *ampholytes*. As a first example, there are the amino acids. The buffer capacity, as well as the dissociation property of these substances can be investigated by titration experiments. Furthermore, this allows calculation of the dynamics of electrical charges of these molecules. In Fig. 2.34 the charge of glycine is demonstrated as a function of the pH in the solution. For this, the mean number of elementary charges (z) per molecule is used. In the preceding sections we learned that dissociation and association are stochastic processes. Hence these amounts of z can be considered as an average in time or as a mean value of a large number of molecules.

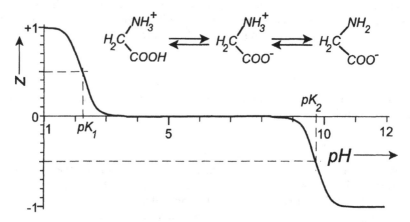

Fig. 2.34. The average number of charges (z) of glycine plotted against the pH in the solution. (pK$_1$ = 2.35, pK$_2$ = 9.78)

Glycine is a cation at extremely low pH, and in contrast an anion at extremely high pH. In the pH region in between, both groups of this molecule are dissociated, demonstrating the zwitterionic character of this molecule. In this case it has a considerable dipole moment and a maximal number of charges but on average it is electroneutral. This curve allows us to determine two pK-values which always correspond to the pH of the medium, where 50% of the charges are dissociated (in Fig. 2.34 corresponding to $z = +1/2$, and $z = -1/2$), as well as an isoelectric point which is positioned exactly in the middle between these two pH-values.

The pH dependent charge of ampholytes of course also strongly influences the structure of the surrounding water. This is the reason why the effective volume of these molecules was often modified by the pH of the medium (see Sect. 2.4.2, Fig. 2.27).

Ampholytes are typical constituents of buffer solutions. Buffers are substances which are able to bind protons, or release them, in this way stabilizing the pH of the solution. A buffer capacity can be defined which corresponds to the slope of the titration curve. It is maximal near the pK-values.

Whereas the Henderson-Hasselbalch equation can easily be applied for binary ampholytes with two pK-values at considerable distance from each other, this becomes impossible in the case of polyampholytes, such as proteins with a large number of different amino acids. This is illustrated in Fig. 2.35, showing the titration curve of hemoglobin. This curve can be considered as a sum of many sigmoidal curves with different properties. Near the isoelectric point, this curve can be fitted by the simple function:

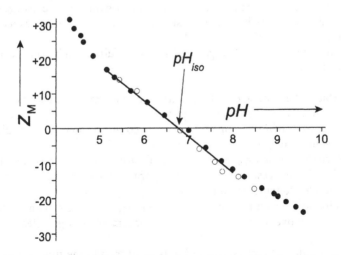

Fig. 2.35. The mean number of charges of horse hemoglobin as a function of the pH. ∘ – points for oxidized hemoglobin, deviating from others. The line between pH 5 and pH 8 corresponds to Eq. (2.4.31) with a buffer capacity of $z_{M0} = 10.2$ val mol^{-1}, and an isoelectric point pH$_{iso} = 6.68$. (According to measurements of German and Wyman 1937)

$$Z_M = -Z_{M0}(\text{pH} - \text{pH}_{iso}) \hspace{4cm} (2.4.31)$$

(z_{M0} - buffer capacity, pH_{iso} - isoelectric point).

The titration curve, of course, indicates only the effective charges of molecules as a function of the pH in the solution. In the case of polymer molecules, the interaction of their monomers must be considered, depending on their mutual distance in the molecule. Furthermore, near the place of dissociation, a local pH may exist, differing from that of the bulk solution significantly. This is caused by fixed charges and therefore can be predicted by local electric potentials in the sense of the Boltzmann or Nernst equations [Eq. (3.2.43)]. In this case, the term *intrinsic* pK is used, meaning the true pK of specific dissociable groups in a polymer, considering the real local proton activity. This, in fact, is possible to measure. Using nuclear magnetic spin resonance (NMR) for example, the dissociation of specific carboxyl groups can be studied, traced by the isotope ^{13}C. Investigations of polypeptides indicated that considerable differences may exist between the effective and the intrinsic pK of some groups even if the influences of neighboring amino acids are quite low.

Further reading: On electrostatic properties of proteins: Matthew 1985; Pethig and Kell 1987

2.5
Interfacial Phenomena and Membranes

After explaining various kinds of molecular interactions in the previous section, we will now come to the next level of biological organization, considering self assembly and molecular dynamics of supramolecular structures, especially of membranes.

The cell membrane is a well-organized structure fulfilling a broad spectrum of physiological functions:

- As a surface, it forms a dynamic matrix for enzymatic reactions, receptor processes, and immunological recognition
- As a barrier of diffusion it controls the ionic composition of the cytoplasm by highly specific transporters
- As an electrically isolating leaflet it contains a mosaic of various passive and active electric devices, controlling membrane potential as well as near-membrane electrodynamic conditions
- As a mechanical structure it guarantees the integrity of the cell and influences its shape and movement as well as the displacement of organelles

We will discuss these biophysical mechanisms of membrane functions in many sections of this book in detail. Here, we will consider the molecular dynamics and the processes of self organization. Unfortunately, in some cases it is necessary to deal with some topics which will not be explained in detail until later.

2.5.1
Surface and Interfacial Tensions

Each molecule of a homogeneous isotropic phase interacts energetically with all its surrounding neighbors. If these forces are averaged for a sufficiently long period of time, all the moments of interaction are abolished. Considering, however, a molecule not in the bulk phase, but at the phase boundary, or moreover, at the liquid-air surface, the situation is quite different (Fig. 2.36). In this case the forces of interaction from one side differ from those the other side. A molecule of a liquid for example at the liquid-gas surface feels a stronger attraction to its neighbor molecules in the liquid than to the molecules of the gas phase. The energy of the surface molecules is higher than that of the molecules in the bulk phase. This is the reason why water droplets in air spontaneously form a sphere as the shape of minimal relative surface and therefore minimal surface energy.

To increase the surface of a liquid against air or against a gas which is saturated by the vapor of this liquid, it is necessary to heave more molecules to this higher level of energy. The energy which is required to enlarge the surface of a liquid phase in this way by 1 m^2 is named *specific surface energy*, which from the physical point of view, is the same as the *surface tension* γ (J $\cdot$ m^{-2} = N $\cdot$ m $\cdot$ m^{-2} = N $\cdot$ m^{-1}).

Water as a polar liquid with strong intermolecular interactions (see Sect. 2.4.2) has a high surface tension which decreases with increasing temperature. At 25 °C it amounts to 0.0728 N m^{-1}. The surface tension of organic liquids, for example of benzene, is only 0.0282 N m^{-1}, and of ethanol only 0.0223 N m^{-1} at the same temperature.

For biological reactions the surface tension at the liquid-air interface is interesting in two ways: on one hand it is important for organisms living directly at the surface of a lake or the sea. These are microorganisms and small algae, forming the so-called neuston, as well as various higher plants, and finally

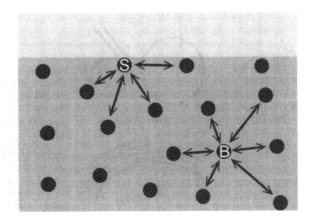

Fig. 2.36. Interaction forces of molecules at surface (S) and in the bulk solution (B)

various insects using hydrophobic properties of their legs to run on the water surface. On the other hand surface tension plays an important role in the biomechanics of the lung, where a highly solvated surface is in contact with the air.

To explain this situation the behavior of bubbles or vesicles formed by a liquid must be considered. In the same way that surface tension tends to decrease the surface of a droplet, bringing it into a spherical shape, the surface tension of a liquid-formed vesicle tends to decrease its inner volume, generating pressure against the included air (Fig. 2.37A). The pressure difference $\Delta p = p_i - p_e$ can be calculated by the Laplace equation:

$$\Delta p = \frac{2\gamma}{r}. \tag{2.5.1}$$

This means that the internal pressure of a vesicle is proportional to the surface tension of the liquid (γ) which forms it, and inversely proportional to its radius (r). The alveoli of the lung are vesicles of various diameters, connected each other and changing their diameter periodically. As indicated in Fig. 2.37B, according to Eq. (2.5.1), the pressure in the smaller vesicles would be higher than in the larger vesicles. This would lead to a collapse of the smaller vesicle in favor of a larger one if they were connected to each other. To avoid this, the surface tension of the water-air interface in the alveoli of the lung is strongly reduced by a surface-active material known as *pulmonary surfactant*. This layer consists primarily of phospholipids. The periodic change of the alveolar volume during inhalation additionally requires the property of this layer to be expanded quickly and reversibly. For this, a reservoir is formed, reversibly folding and unfolding these layers (Fig. 2.37C). This is a complicated process stabilized by special surfactant-associated proteins.

In most cases it is not the *surface tension* of the liquid-air interface, but rather *interfacial tensions* between two liquid phases which are important, or, in other words, the specific energy of interfacial molecules. In the ideal case the

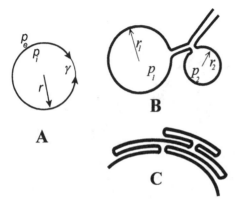

Fig. 2.37. A A vesicle with a surface tension γ tends to decrease the internal volume and generates an internal pressure $\Delta p = p_i - p_e$; B Two vesicles, built by a liquid of the same surface tension generate different internal pressures which are receprocally proportional to their radius. In case of communication, the smaller vesicle should collapse in favor of the larger one. C The change in the surface of the surfactant layer by folding

interfacial tension of two liquids is equal to the difference between their surface tensions.

The attempt to measure the surface tension of living cells and to use this parameter to explain various physiological properties such as phagocytosis, exocytosis, cell division, amoeboid movement, cell shape etc. is quite old. In 1898 L. Rhumbler demonstrated that a drop of chloroform in water could "phagocytose" a thin, shellac-covered glass fibre in a manner similar to that of an amoeba ingesting a filamentous alga. In the same way cell shapes in a tissue were used to try to explain a result of surface tensions, comparing the tissue to a lather. D'Arcy Thompson, in his classical book *On Growth and Form*, first published in 1917 discussed this problem in detail.

Those comparisons, however, as visually impressive as they may be, are not very helpful in explaining these biological processes. In the case of a liquid droplet, surface molecules can always be drawn back into the bulk, or extruded again in the surface layer very quickly. This process is reversible; the mechanically deformed drop will re-assume its spherical shape as soon as the external mechanical disturbance is discontinued. In contrast, a living cell is surrounded by a membrane. An increase of the membrane area by inclination of additional molecules from the cytoplasm is a process which is much slower.

Further reading: Historical papers: Rhumbler 1898; Thompson 1966; for the surfactant of the lung: Batenburg and Haagsman 1998; Galla et al. 1998

2.5.2
Self Assembly and the Molecular Structure of Membranes

The spontaneous orientations of molecules in phase boundaries are of particular interest for questions of self-organization and stability of biological membranes. Let us first consider the behavior of molecules in the boundary between two phases, one being hydrophillic (water), the other hydrophobic (for example oil). A protein tends to orient its polar groups as much as possible toward the aqueous phase, and its nonpolar, i.e. hydrophobic groups toward oil (Fig. 2.38A). This leads to an orientation of the molecule at the phase boundary. If the polar groups of the molecule are distributed homogeneously over the whole molecule, then it could become unfolded by orientation in such a boundary. The protein, now being a fibril will spread over this interface.

The most important constituents of biological membranes are phospholipids. This is a group of compounds which consists of two fatty acids and a phosphate group, all attached to a polyhydric alcohol such as glycerol. The phosphate group is usually esterified with a nitrogen-containing alcohol. The structure and formula of a simple phospholipid is depicted schematically in Fig. 2.39. Two long hydrophobic fatty acid chains and a single phosphate group esterified with an ethanolamine residue are linked to a central glycerol molecule. The phosphate and the amino-group of the ethanolamine residue represents the hydrophilic part of the molecule.

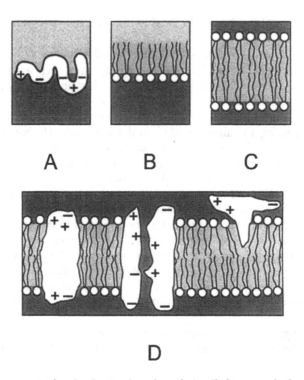

Fig. 2.38. Arrangement of molecules in phase boundaries (*light area* – hydrophobic phase, *dark area* – hydrophilic phase): A orientation of a protein with hydrophobic and hydrophilic regions; B orientation of a monolayer of phospholipids (this corresponds also to the lipid layer at a water-air surface); C a lipid bilayer in a homogeneous aqueous phase. In the region of the fatty-acid chains, orientated against each other, a hydrophobic micro-phase is formed; D orientation of proteins in such a phase, where the hydrophilic parts of the molecules are orientated towards water and the polar head groups of the lipids. In this way the formation of hydrophilic membrane pores is also possible

The large variety of biological lipids is realized by the diversity of the fatty acids as well as by their different head groups. In the case of phosphatidyl-choline, the head group is polar but does not carry net charges under physiological conditions. The negative charge of the phosphate is compensated by a positive charge of the amino group. Phosphatidyl-serine, in contrast to this, contains an additional negative charge, namely that of a carboxyl group. This is the reason why it significantly influences the surface charges of biological membranes.

Phospholipids strongly orientate themselves at interfaces. In this case the hydrophobic chains of fatty acids are orientated parallel to each other and perpendicular to the interface (2.38B). In this way, monomolecular layers, so-called *monolayers*, are formed.

In case of homogeneous aqueous phases, phospholipids reach a state of minimal surface tension if they orientate their hydrophobic portions toward one

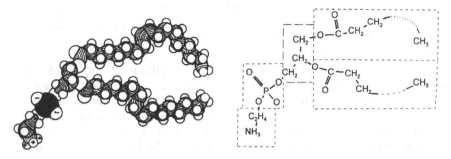

Fig. 2.39. Model and formula of a simple phospholipid (phosphatidylethanole amine with olein- and palmitin acid-rests. (After Haggis 1965 redrawn)

Fig. 2.40. Self-assembly of lipids in aqueous solutions. **A** Micelle; **B** Unilaminar liposome; **C** Sector of a multilaminar liposome. The *dotted areas* always indicate local micro phases of water

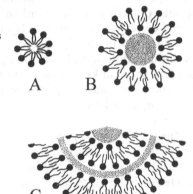

another (Fig. 2.40A). There is a large diversity in lipid aggregates in aqueous solutions. Using special methods, for example sonications of lipid suspensions with ultrasound, it is possible to produce structures with lipid double layers (*bilayers*) and vesicles, containing aqueous solutions (*liposomes*). There are two types of liposomes: unilamellar liposomes covered only by a single bilayer (Fig. 2.40B), and multilamellar liposomes (Fig. 2.40C) containing numerous lipid bilayers, separated from each other by a thin water layer.

It is also possible to prepare planar lipid membranes in aqueous phases (Fig. 2.38C). If such a membrane was formed, its fatty acid chains would have built a hydrophobic micro phase. In this microphase other hydrophobic molecules can be included and orientated, as illustrated for the case of intrinsic membrane proteins in Fig. 2.38D.

In this context it must be mentioned that surface tension is, at least partly, the phenomenological expression of processes which on the molecular level are the result of the dynamics of water structure and hydrophobic bonds. All the lipid structures mentioned here can be destroyed by extremely high pressure in the sense of the Le Châtelier principle, as discussed in Section 2.4.2.

During recent decades ingenious methods have been developed to produce bimolecular membranes with high structural and chemical specificity. These are the so-called BLMs which is an abbreviation for black lipid membrane, or bimolecular lipid membrane.

Experiments with such membranes have significantly promoted our knowledge on self-assembly of biological membranes, their molecular dynamics and processes of membrane transport. Standard experiments are the measurement of electrical conductivity of these membranes, using chambers connected by a hole, which is covered by a BLM. With such equipment, transport properties of these membranes can also be investigated. For this, specific transport proteins are inserted into these membranes to study their specific properties in vitro.

Measurement of electric conductivity of cell suspensions indicated a long time ago that the cell must be surrounded by an electrically isolating layer with a specific capacity of about 10 mF m^{-2} (in detail, see Sects. 2.5.5 and 3.5.3). This would correspond to a double layer of lipids with a thickness of 8 nm. Singer and Nicolson in 1972 proposed a *fluid-mosaic model* of the cell membrane. Meanwhile, this model has been substantially developed, and today it forms the fundament of our knowledge about the molecular structure and the dynamics of biological membranes. According to this model the membrane consists of a highly organized dynamic mosaic of proteins, embedded in a double layer of phospholipids.

The proteins are organized in the membrane according to their hydrophobic (i.e. apolar), and hydrophilic (polar) regions (see Fig. 3.38D). Many proteins proceed to the external surface of the membrane as *glycoproteins*. They are located with their C-terminal end in the cytoplasm. Their N-terminal end, which is modified by various carbohydrates, extends over a long distance into the outer environment of the cell. These carbohydrates in some cases represent the receptor-sides of the proteins and are the location of immunological reactions. Preferentially, at the end of the protein chains, monomers of the *N*-acetylneuraminic acid (also called sialic acid) are located, which carry a dissociable carboxyl group. In this way, these are the most important carriers of fixed negative surface charges of the cell (see Fig. 2.48). Depending on the type of cell, there are between 1 and 10 groups of sialic acid per 1 nm^2 membrane area. Hence, the glycoproteins form a loose external layer of the cell, the so-called *glycocalyx* or surface coat. The charges, located at the ends of these filaments modify the thickness of this structure electrostatically (for detail, see Sect. 2.5.5).

An important role for the lateral membrane structure is played by a network of cytoplasmatic proteins as a part of the cytoskeleton with connections to the inner side of the membrane. It is best investigated in human erythrocytes, where it consists mostly of spectrin. This is a heterodimeric protein (molecular masses: 260 and 220 kDa) which forms rods of approximately 100 nm length, aggregating head to head, to form tetramers of the double length. These spectrin rods are connected to each other by the globular actin molecules (molecular mass: 100 kDa). Other proteins, the ankyrins, connect this network with the intrinsic membrane proteins. The cytoskeleton is a dynamic structure which controls the lateral distribution of the proteins in the membrane.

The bilayer of membrane lipids forms a two-dimensional phase with amphiphilic properties. This can be thought of as a fluid matrix in which the proteins are embedded. The lipid composition of the inner and outer membrane leaflet is highly specific. This concerns the specificity of their fatty acids, as well as their polar head groups (see Fig. 2.39). The fluidity of the membrane, i.e. the lateral movement of its constituents is strongly determined by the length of the aliphatic fatty acids chains, as well as by their degree of saturation (see also Sect. 3.6.3). The property of the polar head groups determines their mutual distance in the membrane surface, i.e. their density of packing. The heterogeneity of the lipid composition in the membrane in some cases may lead to formation of specific lipid domains. These are lipid clusters which can be considered as sort of lateral microphases. Specific lipid domains also surround the proteins. Furthermore, the lipids are distributed asymmetrically between the two leaflets of the membrane. The negatively charged phosphatidylserine, for example, in animal cells is located nearly exclusively at the inner leaflet of the membrane, in contrast to phosphatilcholine, which is concentrated mostly in its outer leaflet. This asymmetric distribution of lipids is controlled by specific lipid transport systems.

Further reading: Bloom et al. 1991; Hianik and Passechnik 1995

2.5.3
Mechanical Properties of Biological Membranes

The mechanical properties of the biological membrane are very important for understanding a number of cell physiological functions such as cell movement, cell division, vesiculation, and membrane fusion. In the same way, the mechanical properties of tissues, as well as the streaming properties of the blood, which we will discuss in Section 3.7, are controlled by the viscoelastic properties of cell membranes.

It must be stressed that physical parameters, for example viscosity, elasticity, strain etc. are defined for homogeneous phases and materials (see Sect. 3.6). Hence, they are not proper parameters of highly organized supramolecular structures like biological membranes. In some cases it may be convenient to use these parameters, but then they must be considered as *effective parameters*, which we explained in the introduction to Section 2, and which we have already used in many other cases.

The effective membrane viscosity, for example, is a parameter which was measured using specific effects, as for example the rotation or translocation of a spin-marker in the membrane. This is possible using special fluorescence methods or electron spin resonance techniques (ESR). Considering the membrane as a homogeneous phase, and the marker molecule as a macroscopic body with a particular shape, one can calculate the viscosity, applying usual phenomenological equations for movement. Knowing the method of measurement, as well as the applied equations, this measurement can be reproduced any time, and the measured parameter is exactly defined. The problem which arises from using an equation, which is not adequate for the movement of a molecule in a heterogeneous phase, is not relevant because it is included in the definition. On the other hand, however, it means that by measuring these effective viscosities by different methods, one gets different parameters but not a viscosity according to the original physical definition. This is the difference between molecular, and the usual macroscopic kinds of measurement, where viscosity is independent of the measurement method.

Applying different markers, the parameters measured in this way depend on the kind of marker molecule (protein, lipid), and on the regions of the molecule where the marker was bound; for example, different locations in the fatty acid of a lipid. These differences indicate the inhomogeneity in the plane of the membrane, as well as the anisotropy of the mechanical properties of the biological membrane in general.

The lipids in biological membranes show a high lateral mobility because they can easily exchange their positions. The mobility of proteins, in contrast, is strongly limited by their fixation on the cytoskeleton. Local enrichments of proteins to particular locations in the membrane can lead to inhomogeneities of its bending property.

Lipids may also exchange between the two leaflets of the membrane. This is a so-called flip-flop process. The rate of this exchange strongly depends on the type of lipids. Half life times of between minutes up to many hours are known. We already mentioned that there are special mechanisms of lipid transport, catalyzing this process.

Summarizing all these aspects, the following mechanical properties for the biological membrane are to be expected: it is easily deformable by shearing forces (Fig. 2.41C), independently of the position of the intrinsic proteins which occasionally are fixed by the cytoskeleton (Fig. 2.41D). In contrast to this shearing deformation which occurs without an increase area, it is nearly impossible to extend the area of the membrane. The stress (σ') which must be applied to extend a membrane area A by the amount ΔA can be calculated by:

$$\sigma' = Y' \frac{\Delta A}{A}.$$

(2.5.2)

Y' is the specific modulus of elasticity, a coefficient which is related to the thickness of the membrane. The symbol Y stands for Youngs modulus, another name for the same parameter (see Sect. 3.6.3). For the membrane of erythrocytes $Y' = 0.45$ N m^{-1}, for lymphocytes 0.64 N m^{-1}. To find a relationship between membrane properties and properties of macroscopic materials, like those in Table 3.2 (page 213), Y' must be divided by the thickness of the membrane (d), which is approximately $8 \cdot 10^{-9}$ m. In this case, for erythrocytes the result is:

$$Y = Y'/d = 5.6 \cdot 10^7 \, \mathrm{N\,m^{-2}} = 56 \, \mathrm{Mpa}$$

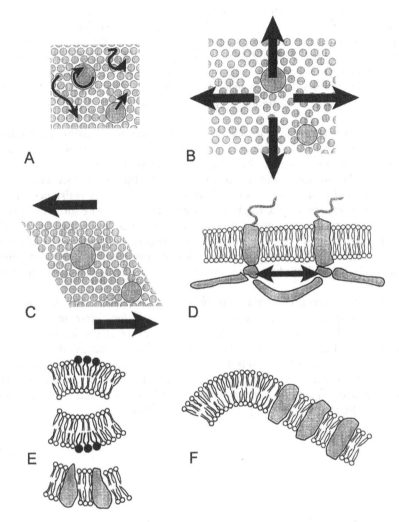

Fig. 2.41. Possible kinds of molecular displacement and mechanical deformations of biological membranes. A Translation and rotation of membrane components; B planar extension of the membrane area; C shear deformation; D translocation of membrane proteins by the spectrin network; E bending of the membrane, induced by asymmetric lipids or proteins, or by inclination of additional molecules; F extrusion of membrane components with lower flexibility from highly bended areas

Relating this value to the parameters of Table 3.2 one can see that it already tends toward the values of steel-elastic materials. The membrane will rupture if it is expanded by more than 1–2% of its area. This is understandable considering that any increase of the membrane area leads to an increase of the distance between the lipid head groups.

The membrane therefore resembles a material with a high degree of flexibility in plan, but with minimal ability for area extension. Hence, it has properties very unlike a rubber sheet, which can easily be expanded on account of its thickness. Living cells, therefore, can swell only by spherulation, by unfolding of the membrane, or by inclusion of new membrane material. Each swelling of the cell, which extends the surface of a sphere by more than 1–2% of its area or an unfolding beyond its capacity, will lead to immediate cytolysis.

As we will see later, using the elastic modulus, it is also possible to calculate forces, which oppose the bending of a material (Sect. 3.6.4). Applying this to a biological membrane, again, the inhomogeneity of the membrane must be considered. According to the heterogeneity of the lipid composition and the differences in the distribution of the proteins, the bending resistance of the membrane differs from place to place. It is possible to calculate forces, acting to extrude small areas of enlarged stiffness, like protein patches, out from strongly bent regions (Fig. 2.41F). In shapes with regions of extremely different membrane bending, for example in the case of crenated erythrocytes, (so-called echinocytes), it could be shown that the number of proteins is lower in strongly bent regions.

In some cases membranes are actively bent by themselves. This is done by a bilayer couple mechanism, if one of the membrane leaflets increases its area by inclusion of additional molecules. Another possibility is the introduction of conically shaped molecules (Fig. 2.41E). A factor (f) characterizes this property. It relates the area of the head group (A_H) to the mean area (A_F) which is required for the region of fatty acids (Fig. 2.42). A_F can be calculated by the mean length of the fatty acids (l) and the effective volume of the region of fatty acids (V_F) by: $A_F = V_F/l$. The shape factor f, therefore, is defined as:

$$f = \frac{V_F}{lA_H}.$$

(2.5.3)

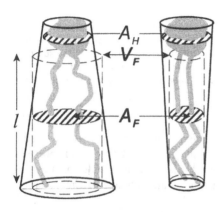

Fig. 2.42. The effective shape of a phospholipid (explanations in text)

In contrast to the area of the head group (A_H), which is more or less constant, the effective volume of the fatty acid chains (V_F) strongly depends on their phase conditions, i.e. on their degree of order. There are dynamic variations possible as well as phase transitions by temperature modifications. Other parameters of the molecule depend directly on their composition. So, for example, lysophospholipids, containing only one single fatty acid chain, are to a large degree conical with a dominating area A_H. In this case, $f < 1/3$. In contrast, phosphaticylethanolamine, usually containing a large amount of unsaturated fatty acids, gives $f > 1$, with dominating A_F. Mostly lipids are more or less cylindrical, with $1/2 < f < 1$.

Further reading: Bereiter-Hahn et al. 1987; Bloom et al. 1991; Hianik and Passechnik 1995; Mohandas and Evans 1994; Schmid-Schönbein et al. 1986; Silver 1985

2.5.4
Electrical Double Layers and Electrokinetic Phenomena

An interface carrying fixed charges induces in its vicinity an electric field, and accordingly, modifies the local ionic conditions, analogous to a point charge, as described in Section 2.4.3. In the simplest case, a double layer occurs with the fixed charges on one side, and the electrostatically attracted mobile conterions on the opposite side. This concept was originally put forward by H. von Helmholtz. It only applies for extremely high concentrations of mobile ions in the solution. In general, thermic movement repels a great part of the ions from their positions near the fixed charges. This leads to the formation of a *diffuse double layer*, the electrical potential of which declines exponentially with the distance from the surface.

Figure 2.43 illustrates the generalized model of Stern, which includes the Helmholtz double layer as well as its diffuse part. To avoid overloading of the picture, only the fixed charges (in this case proposed as negative), and the compensating mobile conterions are depicted. Of course, there are also mobile anions present, and there is a great number of both cations and anions that should be included to give a true picture. This picture indicates an enlarged concentration of cations near the surface for the case of negative surface charges, and the electroneutrality of the total system, including the bulk solution.

Despite the thermic noise, some of the mobile ions directly oppose the fixed charges, forming the so-called Helmholtz layer. Between the surface and the end of the Helmholtz layer, the potential is proposed to decline linearly from ψ_0 to ψ_H. Further, the potential declines according to the model of the diffuse double layer which was calculated by Gouy and Chapman.

The theory of the diffuse double layer again is based on the Poisson-Boltzmann equation, which we explained in relation to the Debye-Hückel theory of ion clouds (Sect. 2.4.3). This equation makes it possible to calculate the distribution of mobile charges in a given electric field:

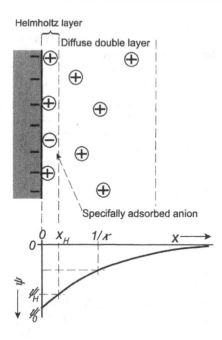

Helmholtz layer

Fig. 2.43. Schematic illustration of an electrical double layer. *Above* fixed charges and mobile counter-ions near a surface; *beneath* the function of the electrical potential, according to the model of Stern. The charge of the surface in this particular case is enforced additionally by a chemically adsorbed negatively charged ion. ψ_0 – surface potential, ψ_H – potential at the end of the Helmholtz layer, x_H – thickness of the Helmholtz layer, $1/\kappa$ – Debye-Hückel length as a measure of the effective thickness of the diffuse double layer

$$\nabla^2 \psi = -\frac{F}{\varepsilon_0 \varepsilon} \sum_{i=1}^{n} c_{i0} z_i e^{-\frac{z_i e \psi}{kT}}. \tag{2.5.4}$$

In this case the amount of fixed charges is compensated by the excess of charges in the whole double layer.

Applying this equation for one-one-valent electrolytes, like KCl or NaCl, it reduces to:

$$\nabla^2 \psi = -\frac{F c_0}{\varepsilon_0 \varepsilon} \left(e^{-\frac{e\psi}{kT}} - e^{+\frac{e\psi}{kT}} \right) = \frac{2F c_0}{\varepsilon_0 \varepsilon} \sinh \frac{e\psi}{kT}. \tag{2.5.5}$$

In this case, a rearrangement was made, using the function

$$\sinh x = (e^x - e^{-x})/2.$$

To solve the Poisson-Boltzmann equation, i.e. to calculate the function $\psi(x)$, one needs to define two initial conditions. For our case this means:

$$\psi(\infty) = 0,$$
$$\psi(0) = \psi_H,$$

whereas the end of the Helmholtz layer (ψ_H) is taken as the starting point of the function [$\psi(0)$]. The algebraic integration of this equation again, is possible only after linearization. For $x > x_H$ one gets the following function $\psi(x)$:

$$\psi(x) = \psi_H e^{-\kappa x}. \tag{2.5.6}$$

In this equation, κ is again the Debye-Hückel parameter, which was introduced in Section 2.4.3 [Eq. (2.4.15)]. The Debye-Hückel length $(1/\kappa)$ is the distance, over which the potential ψ_H is dropped by the factor e:

$$\frac{\psi(1/\kappa)}{\psi_H} = \frac{1}{e} = 0.368. \tag{2.5.7}$$

The value of this parameter, as a function of the ionic strength (I) is given in Eqs. (2.4.15–2.4.17).

What does "linearization" of Eq. (2.5.5) mean, and what are the limitations of this simplification? In fact, the basis for such approaches is the expansion of these functions in series, using the following equations:

$$e^x = 1 + x + \frac{x^2}{2} + \frac{x^3}{3} + \frac{x^4}{4} + \cdots, \tag{2.5.8}$$

and:

$$\sinh x = x + \frac{x^3}{3} + \frac{x^5}{5} + \frac{x^7}{7} + \cdots. \tag{2.5.9}$$

For the case:

$$\frac{ze\psi}{kT} \ll 1, \tag{2.5.10}$$

one can apply these equations, using only the first terms of the series:

$$e^{-\frac{ze\psi}{kT}} \approx 1 - \frac{ze\psi}{kT}, \text{ and: } \sinh\frac{e\psi}{kT} \approx \frac{e\psi}{kT}. \tag{2.5.11}$$

Let us investigate the scope of this simplification. For $T = 298$ K one can calculate:

$$z\psi\frac{e}{kT} = z\psi\frac{1.60218 \cdot 10^{-19}}{1.3807 \cdot 10^{-23} \cdot 298} = 38.94z\psi. \tag{2.5.12}$$

Using, for example, a potential $\psi = 0.01$ V and a charge number $z = 1$, one gets: $e^{-0.3894} = 0.6774$, and according to Eq. (2.5.8): $1 - 0.3894 = 0.6106$. Correspondingly: sinh $0.3894 = 0.3993$. Both calculations, really, indicate small differences, which will become larger if the electrical potential exceeds the value of 10 mV. This is also important for the application of this linearization in the previous Section 2.4.3, as well as in the following considerations.

Beside the calculation of the diffusion layer it is necessary to know the relation between the potentials ψ_0 and ψ_H. In general, the difference between these two parameters will become smaller as the ionic strength of the solution drops. In solutions of extremely low ionic strength, this difference can be neglected.

In fact, besides the screening of fixed charges by mobile ions, other processes must be considered, such as dipole orientations, interactions with water molecules or other processes of adsorption, based on Van-der-Waals interactions. In this way, it is even possible that against electrostatic repulsion, coions were absorbed at the interface, increasing its charge density. In this case the Helmholtz-potential (ψ_H) can become even larger than the surface potential (ψ_0).

We will discuss the relevance of electrical double layers for functions of biological membranes in detail in Section 2.5.5. The most important consequences are local ionic concentrations, including local pH values (Fig. 2.44). This can be calculated according to the Nernst equilibrium [Eq. (3.2.34), Sect. 3.2.2]. The relevance of these local conditions is understandable, considering the function of membrane-bound enzymes.

A number of electro-mechanical interactions occur which are caused by these electrical double layers near charged surfaces, and which are summarized by the term: *electrokinetic phenomena*. On the one hand, an externally applied electric field may induce mechanical effects, like movement or streaming, on the other

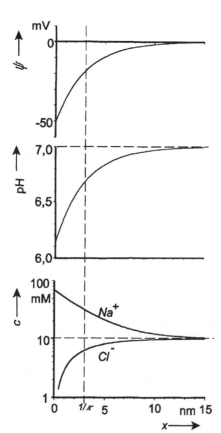

Fig. 2.44. Electrical potential (ψ), pH, as well as Na^+-, and Cl^--concentrations (c) as a function of the distance (x), near a negatively charged surface in a diffuse double layer. The curves consider the following parameters: $\psi_0 = -50$ mV, $pH_\infty = 7.0$, $c_\infty = 10$ mM, $T = 298$ K (the index ∞ means: at $x \to \infty$). In this case the Debye-Hückel length ($1/\kappa$) amounts 3 nm

hand, a mechanical movement may lead to the occurrence of electrical potential differences. In Fig. 2.45 these phenomena are depicted schematically.

Electrophoresis means the movement of charged molecules and particles in an externally applied electric field. The displacement of cells and other particles can be measured by microscopic observations. Their mobility (*b*) is defined as the relation between their velocity (**v**) and the electric field strength (**E**):

$$b = \frac{\mathbf{v}}{\mathbf{E}}. \qquad (2.5.13)$$

The particles move in the opposite direction from their double layers. Because of the increasing frictional force, a stationary velocity, will be reached.

The classical theory of cell electrophoresis is based on the model which was proposed by Smoluchowski, considering the movement of large spherical particles. "Large" in this sense means that the diameter of the particle, and the

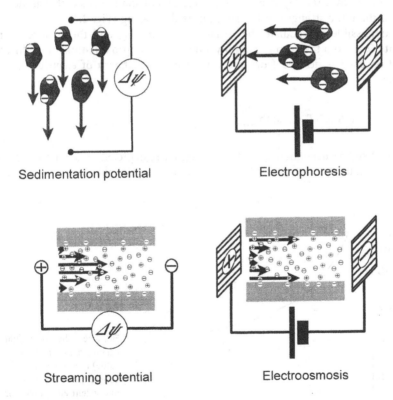

Fig. 2.45. Schematic representation of electrokinetic phenomena of particles (above), and surface-near regions in tubes or capillaries (beneath); *left* the occurrence of electrical potentials as a result of mechanical movement (grey arrows); *right* the induction of movement (above) or streaming (beneath) as a result of an applied electric field (black arrows)

corresponding radius of bending, must be large in relation to the Debye-Hückel parameter ($1/\kappa$). Smoluchovski postulated that a shearing plane exists, caused by the electrokinetically induced streaming. At this plane, the electroneutrality of the system is disturbed. The electrical potential at this distance from the surface is called *electrokinetic potential*, or ζ-potential. It can be calculated by the measured electrophoretic mobility (b) using the following equation:

$$\zeta = \frac{\eta \mathbf{v}}{\varepsilon_0 \varepsilon \mathbf{E}} = \frac{\eta b}{\varepsilon_0 \varepsilon}, \tag{2.5.14}$$

where η is the viscosity of the medium. Surprisingly, the radius (r) of the particle is not included in this equation. Sometimes, this equation contains an additional factor, to take into account stronger deviations from the shape of a sphere.

Of course, the above introduced ζ-potential is not identical with the surface potential (ψ_0). With some approximation, however, it can be considered as equal to the Helmholtz-potential (ψ_H; see Fig. 2.43). In the same way, as discussed above, the ζ-potential increases, by constant surface charge density (σ_0), if the ionic strength in the solution is decreasing (Fig. 2.46).

It is possible to calculate the surface potential (ψ_0) and the surface charge density (σ_0) from the ζ-potential only with certain approximations. For this, the equation used is derived from the Gouy-Chapman theory of electrical double layers:

$$\sigma_0 = \sqrt{8\varepsilon_0 \varepsilon R T} \sqrt{I} \sinh \frac{F\zeta}{2RT}. \tag{2.5.15}$$

In Fig. 2.46 the dependence of the ζ-potential is depicted as a function of the ionic strength of a one-one-valent electrolyte, which in this case is equal to its

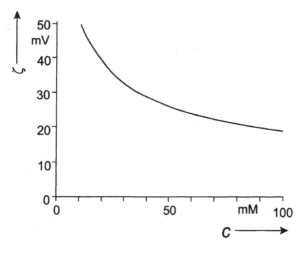

Fig. 2.46. The ζ-potential (depicted as a negative value) as a function of the concentration of a one-one-valent electrolyte for a surface with a constant surface charge density ($\sigma_0 = 0.014$ C m^{-2}) according to Eq. (2.5.15)

concentration. In this case a surface charged density of $\sigma_0 = 0.014$ C m^{-2} was chosen which corresponds to that of a human erythrocyte.

Equation (2.5.15) can be simplified by expanding it as series, corresponding to Eq. (2.5.9). If the Debye-Hückel parameter (κ) was used according to Eq. (2.4.15), we have

$$\sigma_0 = \zeta\kappa\varepsilon\varepsilon_0, \tag{2.5.16}$$

which, when combined with Eq. (2.5.14), results in

$$\sigma_0 = \eta\kappa b. \tag{2.5.17}$$

Introducing this kind of linearization we found that this simplification can be accepted as a reasonable approximation only for potentials ≤10 mV. As indicated, however, in Fig. 2.46, this limit is exceeded for most cases of biological membranes. Furthermore, all equations were derived considering the fixed charges as located on a planar boundary. The thickness of the Helmholtz layer, which amounts only to a few tenths of a nanometer explains what the word "planar" means. The roughness in the order of a molecular diameter is always enough to exceed this assumption.

To measure electrophoresis, equipment is used which allows observation of the movement of cells in a homogeneous dc-field microscopically (Fig. 2.47). Usually, for electrophoretic measurements, a field strength in the order of 1 kV m^{-1} is used. The chamber needs a sufficient temperature control because, according to Eq. (2.5.14), the electrophoretic movement depends on the viscosity of the solution. Furthermore, it must be considered that an electro-osmotic flow occurs, caused by the surface charges of the glass, which are mostly negative. Hence, when the chamber is closed this flux must return through the

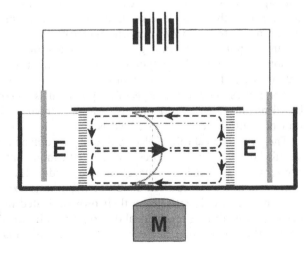

Fig. 2.47. Schematic illustration of equipment to measure cell electrophoresis; M microscope; E chamber for electrodes, which are connected to the measuring chamber by a conducting medium. The *arrows* indicate the direction of the electro-osmotic flow.
·········· – the flow profile in the chamber – · – · – · – the neutral planes

center of the chamber. Between these two flows of opposite direction, neutral planes must exist, where the streaming velocity becomes zero, and where the measurement of the correct electrophoretic mobility of cells is possible. These planes are located at a distance 21.1% of the total thickness of the chamber, on both sides of it (see Fig. 2.47).

Since the development of the method of cell electrophoresis many cells and organisms have been investigated. Under physiological conditions, nearly all biological objects indicate a ζ-potential between -10 and -70 mV. Only at pH < 4, in some cases, do these potentials become positive. This also holds for bacteria, yeast and plant cells, which are surrounded by a cell wall. The cell wall is a thick (in relation to the membrane) porous ion exchange layer with fixed negative charges.

In contrast to cell electrophoresis, which is used routinely to measure surface charges, the reversible effect, i.e. the *sedimentation potential*, some times also called the *Dorn-effect*, seems to be not so important. In this case, a small electrical potential difference occurs in suspensions of quickly sedimenting charged particles.

An electrical potential difference which occurs if a solution is streaming along a charged surface, or inside a capillary with surface charges, is called *streaming potential*. The reason for this is the superposition of the streaming profile and the profile of mobile charges in an electrical double layer. In detail, the theory of streaming potential is quite complex. In the center of a capillary, for example, the streaming velocity is maximum, and the concentration of the coions of the surface charges minimum. Therefore, these coions will flow faster than the counter-ions near the surface. This leads to a certain charge differentiation at the ends of the capillary. The resulting electric field, however, enforces the movement of counter-ions and lowers the movement of these coions. The effect, therefore, is lowered by this reaction. In capillaries with an extremely small diameter, a Donnan-effect must additionally be considered. This means that the amount of surface charge always shifts the total electrolyte composition in this fine capillary toward a Donnan equilibrium (see Sect. 3.2.3). The total amount of counter-ions, therefore, is larger than that of the coions. This can even lead to a streaming potential with opposite polarity. It is possible to measure the streaming potential in biological tissue if a flow of solutions is induced by osmotic gradients. The potential difference measured in such experiments is only in the order of a few millivolts.

The reversible effect of streaming potential is *electro-osmosis*. In this case, based on the same considerations, near a charged surface, or in a capillary, a flow of solution occurs if a dc-field is applied.

The physiological role of electrokinetic phenomena is not very clear. It cannot be ruled out that electrophoresis of charged components in the membrane, or of proteins in intercellular spaces occurs under the influence of potentials generated in vivo (see Sect. 3.5.2). A hypothesis exists that microorganisms can move electrophoretically in their own generated field. Streaming potentials occur in bone during mechanical deformations. In this case, it is induced by the flow in the small canals of the bone, the so-called canaliculi (see Sect. 3.5.2).

Further reading: Theory of electrical double layers: Donath and Voigt 1986; Glaser 1996; Voigt and Donath 1989; cell electrophoresis: Bauer 1994; streaming potential in bones: Guzelsu and Walsh 1993; self-electrophoresis of microorganisms: Lammert et al. 1996

2.5.5
The Electrostatic Structure of the Membrane

After explaining some structural and mechanical properties of biological membranes in Sections 2.5.2 and 2.5.3, and after having discussed ionic properties near fixed charges, we will now introduce the membrane as an electrical structure. This seems to be a crucial point for many biophysical considerations, and it will pervade this whole book. Electrical properties of the biological membrane will be mentioned in the context of many topics. Whereas we introduce here, the electrostatic properties of the membrane, we will continue this discussion later in Sections 3.4.2 and 3.5.2, including electrodynamic processes, such as generation of electrical currents by processes coupled with the cell metabolism. In Section 3.5.3 the explanation of passive membrane properties will be continued, including its behavior in electromagnetic fields. This knowledge will be applied in Section 3.5.4 to introduce techniques of cell manipulations with electric fields. Finally, in Section 4.4 we will use this model to discuss the influence of weak electric and electromagnetic fields on cells and organisms.

In contrast to the external medium and to the cell plasma, the cell membrane has a high electrical resistance and a low dielectric constant. This allows us to consider it as an extremely thin hydrophobic electrically isolating interface between two aqueous phases, behaving like a capacitor with certain capacity (C), and resistance (R). From this point of view, it is possible to consider the passive electric behavior of the cell as an RC-element (see Figs. 3.25 and 3.38).

The specific capacity (C_{sp}) can be calculated from the membrane thickness (Δx) and the dielectric constant (ε):

$$C_{sp} = \frac{\varepsilon_0 \varepsilon}{\Delta x}. \tag{2.5.18}$$

The specific capacity of the cell membrane is relatively constant, because both parameters, the dielectric constant (ε), as well as the membrane thickness (Δx) cannot vary significantly. For the normal cell membrane it amounts to approximately 10 mF m^{-2} (for methods of measuring this parameter, see Sects. 3.5.3 and 3.5.4). Equation (2.5.18) allows calculation of the mean dielectric constant of the membrane, assuming: $\Delta x = 8 \cdot 10^{-9}$ m:

$$\varepsilon = \frac{C_{sp}\Delta x}{\varepsilon_0} = \frac{10^{-2} \cdot 8 \cdot 10^{-9}}{8.854 \cdot 10^{-12}} = 9.0.$$

This value is surprisingly large, because the dielectric constant of lipids amounts to only $\varepsilon = 3.5$. The reason for this difference is the heterogeneity of the membrane, in particular, its protein content.

The specific capacity of the biological membrane is a very important parameter, indicating the relation between the amount of charge (σ, in C m^{-2}) which is required to generate a membrane potential difference ($\Delta\psi$, in V):

$$C_{sp} = \frac{\sigma}{\Delta\psi}. \tag{2.5.19}$$

In Section 3.4.2 we will use this relation to explain the role of the membrane as an electrochemical accumulator.

In Section 2.5.2 we described the membrane as a dynamic structure containing fixed charges in a 2-dimensional pattern of the membrane area and in a charge profile perpendicular to this. This particular profile along an x-coordinate, cutting the membrane area perpendicularly, is depicted in Fig. 2.48 schematically. This picture does not contain the mobile ions in the double layers, and the ionic distribution which is responsible for generation of the transmembrane potential ($\Delta\psi$). We will discuss the circumstances leading to the occurrence of this transmembrane potential as a Donnan potential, as a diffusion potential, or finally as a result of electrogenic transporters later in Section 3 in detail.

The fixed surface charges of the outer part of the cell membrane are the result of dissociations of carboxyl groups of neuraminic acids (also called sialic acids),

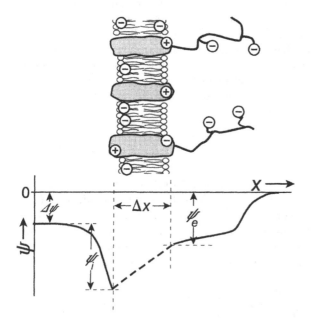

Fig. 2.48. The fixed charges of a cell membrane and the corresponding electrical potential (ψ). The transmembrane potential ($\Delta\psi$) is indicated as the difference between the potentials of the internal and external phases. The parameters ψ_i and ψ_e indicate the internal and external surface potentials; Δx means the thickness of the membrane. Inside the membrane, the function $\psi(x)$ depends on the positions where the membrane is cut. Therefore, the function at this point is marked only by a simple linear connection between the surface potentials

positioned at the ends of the glycoproteins. The isoelectric point of these neuraminic acids is near pH 2.6. For most membranes the negative charges are between −0.01 and −0.02 C m^{-2}. Human erythrocytes carry on their surfaces, which are about 140 μm^2, approximately 10^7 dissociated neuraminic acid groups.

Additional carriers of electric charges are the polar groups of several phospholipids. The negatively charged phosphatidylserine is of particular importance. In the membrane of human erythrocytes it is located exclusively in the inner leaflet of the membrane. In this case, its surface charge amounts to approximately −0.09 C m^{-2}.

As depicted in Fig. 2.38, proteins are located in the membrane in such a way that their polar groups are orientated towards the water phase. Hence, they also contribute to the surface charge of the membrane by charges of both sign. These charges of membrane proteins are very important for controlling various protein functions, such as ion transport through the membrane.

These considerations indicate that the models of electrical double layers, resulting from charges of a planar surface (Fig. 2.43) are just rough approximations of the real situation on the cell surface. The glycocalyx forms an external layer outside the membrane. The fixed charges of it are to be considered as charges in a space, not on a surface. It must therefore be introduced in the Poisson-Boltzmann equation [see Eq. (2.5.4)] as space density (ρ, in C m^{-3}), instead of the surface charge density (σ, in C m^{-2}):

$$\nabla^2 \psi = -\frac{1}{\varepsilon_0 \varepsilon} \left(\rho + F \sum_{i=1}^{n} c_{i0} z_i e^{-\frac{z_i e \psi}{kT}} \right). \tag{2.5.20}$$

Even for the simplest functions, $\rho(x)$, an integration of Eq. (2.5.20) is possible only by iteration. The resulting function $\psi(x)$ (Fig. 2.48), of course differs from that for a classical double layer (Fig. 2.43). The effective range of the electrical potential in this case is predicted firstly by the thickness of the glycocalyx, and only indirectly by the Debye-Hückel length ($1/\kappa$). This thickness, however, is itself determined by electrostatic interactions of its charges. In solutions of high ionic strengths, these charges are more or less shielded, and the glycoproteins are closely attached to the membrane. In solutions with lower ionic strength, however, the fixed charges repulse each other, the glycoprotein filaments are therefore packed more loosely, and the glycocalyx becomes thicker. Hence, the function $\rho(x)$, i.e. the distribution of the charges of the neuraminic acids, itself depends on $\psi(x)$. The surface structure of the cell is therefore controlled by the cell itself, as well as by the conditions of its surroundings.

Not only external, but also internal membrane surfaces carry fixed charges. These are mostly the charges of phosphatidylserine, and additionally, a number of surface-associated proteins, which are not depicted in Fig. 2.48. In contrast to the external surface charges of the membrane, which can be measured by cell electrophoresis, the charges of the cytoplasmatic side cannot be measured easily.

As well as the charge profile in a direction perpendicular to the membrane surface, there is a lateral mosaic of charges, which is partly controlled by the cytoskeleton. This charge distribution has been investigated by electron microscopy, and recently by atomic force microscopy, using electrostatically bound particles of heavy metals. It is also possible to follow the movement of charged proteins, marked by fluorescence labels in strong external fields. The investigation of the dynamics of this lateral electrostatic structure of biological membranes is, however, only at the beginning.

Although we will discuss the role of the membrane electric field, as well as the application of external fields on cell membranes in detail later in Sections 3.4, 3.5, and 4.4.4, we will nevertheless mention here the biological relevance of this electrostatic structure, and its role in several cell physiological processes. In general, on the one hand, the profile of the electrical potential directly creates a profile of specific ion conditions. On the other hand, the electric field strength, as the derivative of the potential, affects some processes directly.

Inside the membrane, and in its vicinity, electric field strengths in the order of 10^7 V m^{-1} occur. In Fig. 2.48 the course of the potential inside the membrane is drawn only with a dotted line as a linear connection between the potentials of both membrane surfaces ψ_i, and ψ_e, because the real function $\psi(x)$ in this region strongly depends on the location of the cut in the membrane. This function would be quite different if the x-axis were to cut a protein, or if it were located inside the lipid phase. The local charge densities, as well as the local dielectric constants would differ strongly. Furthermore, it must be remembered that the function $\psi(x)$ will change dramatically if the transmembrane potential $\Delta\psi$ depolarizes during the action potential in a time course of about 1 ms (see Fig. 3.27). Hence, we come to the conclusion that inside and near the membrane, an extremely strong electric field occurs which is not only a 1-dimensional function $\psi(x)$, as depicted in Fig. 2.48, but which must be considered as a 3-dimensional, time dependent field.

It is clear that this field must influence all polar, or polarizable molecules in this region. A number of transport proteins in the membrane are influenced by this field. Nerve cells are the classical example for a feedback circle, generating the action potential: the alteration of the membrane permeability, induced by any kind of excitation, changes the diffusion potential $\Delta\psi$, and therefore the resulting modification of the field strength in the membrane changes the permeability of the Na$^+$-, and subsequently K$^+$-channels, which again influences the membrane potential. We will explain these reactions in detail in Section 3.4.3. Potential sensitive transporters are distributed in all cells.

Proteins and lipids are influenced by the field strength inside the membrane. The field influences the orientation of the polar head groups of the lipids, and modifies in this way the effective area which they occupy in the membrane. We discussed the consequence of this modification in Section 2.5.2. The field may also induce an intrinsic membrane pressure, which may alter the phase transition in the hydrophobic region of the membrane. Such processes may be the basis of mechanisms of electro-mechanical coupling.

Furthermore, the function of the electrical potential near the membrane generates particular micro-conditions near charged groups. Depending on the ionic strength, and therefore on the Debye-Hückel length ($1/\kappa$), a lateral mosaic of local concentrations and local pH values can occur. In Fig. 2.44 the local ionic concentrations and local pH values are illustrated as a function of the distance from a charged surface. It must be stressed that the field changes the concentration of multivalent ions, for example Ca^{++}, to a larger degree than univalent ions like K^+ or Na^+. In this context, the role of local charges must be considered for the control of membrane-bound enzymes and transporters.

Further reading: Cevc 1990; Glaser 1996; McLaughlin 1989; surface mapping with atomic force microscopy: Heinz and Hoh 1999

CHAPTER 3

Energetics and Dynamics of Biological Systems   3

The functions of biological systems are based on a highly organized molecular structure. So, in Section 2, we considered firstly the physical principles ruling molecular and supramolecular structures. This was done using approaches of quantum mechanics and statistical thermodynamics.

Now we will come to various functions of cells, tissues, and organs. For this, of course, molecular considerations form an important fundament, but at the same time, we enter the field of so-called continuum physics. Here, we will use parameters such as concentration, volume, viscosity, dielectric constants, conductivity etc., which of course are based on molecular properties of matter, but which have been introduced and defined simply by phenomenology. This means that these parameters are defined for sufficiently large homogeneous phases. This, however, does not meet the conditions of biological structures. In Section 2 we were constantly meeting this problem, considering for example mechanical or electrical properties of biological membranes. Even in the following considerations we must always be aware of this situation. The step from a molecular to phenomenological approach, nevertheless, must be taken not only in the physics of simple materials, but also in biophysics, if one considers larger systems like cells and organisms.

We will come back to this point in general in the context of the electrical structure of organisms, when discussing levels of biological organization in Section 3.5.1 (Fig. 3.30).

3.1
Some Fundamental Concepts of Thermodynamics

One of the first important treatises on the problems of thermodynamics was the famous monograph by Sadi Carnot, entitled *Réflexions sur la puissance motrice du feu et sur les machines propres a développer cette puissance* which was published in 1824. In this book, thermodynamics was introduced as the theory of heat engines. Julius Robert Mayer (1814–1878) was a physician and scientist to whom we owe the first numerically defined correlation between heat and work (1842). He can, therefore, be considered as the discoverer of the first law of thermodynamics. He had already discussed the relation between functioning of the human body and the heat engine as a result of his own physiological observations. Meanwhile, thermodynamics have become the theoretical fundament of all kinds of energy transformation, and consequently, of all kinds of movement.

Thermodynamic considerations of biological processes require an extension of the classical thermodynamics of equilibria towards the direction of thermodynamics of irreversible processes, i.e. towards nonequilibrium thermodynamics. This extension was taken in two steps: firstly, only small deviations away from equilibrium are taken into consideration. In this case, linear relations between forces and rates can be assumed. In contrast to these linear approaches, the thermodynamics of non-linear processes can calculate systems far from equilibrium, where steep gradients of potentials exist. In this case, so-called *dissipative structures* appear, which are stationary states with completely new qualities.

It seems important to stress here that although the far-from-equilibrium condition of an organism represents an absolute precondition of life, nevertheless, subsystems in the organism exist, which can be properly calculated using equilibrium thermodynamics, or thermodynamics of linear approaches. This means that biophysics must span the whole scale of thermodynamic approaches.

3.1.1
Systems, Parameters and State Functions

In Section 2.3.3, the term "system" was introduced with the explanation of the term "structure". We defined the system as an aggregate of elements with certain interrelations. Furthermore, we explained that systems with interrelations which are not simply relations, but interactions, are called dynamic systems. These are the kinds of systems where thermodynamic approaches can be applied.

The question of what kind of model we should use, what we should consider as a system, and what are its elements, depends exclusively on the particular problem, and the corresponding point of view. An element may be a system itself, extending the problem. An organism, for example, as an element in an ecological system can become a system itself, if we ask a physiological question such as what are the interactions between its organs. The organ can be considered a system of cells, the cell as a system of organelles, and so on.

A dynamic system can be analyzed in different ways. In contrast to system theory, which calculates the kinetic interplay of individual elements, thermodynamics considers a system simply as a continuum which stands in a defined interrelation with its environment. The limit of this continuum does not have to be a wall or a membrane. It can also be a process that changes the quality of the subject of study. Chemical reactions as well as processes of proliferation and evolution are examples of this.

In thermodynamics, systems are classified as follows according to the nature of their boundary against their environment:

– The *isolated system*: this is an idealized system that does not exchange any kind of energy or matter with its environment.

- The *closed system*: this system can exchange all kinds of energy with its environment, but not matter.
- The *open system*: it can exchange both energy and matter with its environment.

The closed system can be influenced by its environment, and can cause changes in its environment. However, it cannot be involved in an exchange of matter.

The state of a system can be described by a number of *state variables*. These are either extensive or intensive parameters. *Intensive* parameters are non additive. They are independent of the size of the system (e.g. temperature, concentration, pressure, density). *Extensive* parameters on the other hand, are additive when two systems are combined (e.g. mass, volume).

Changes in a system are often characterized by differentials of its state variables. A *differential* describes a very small change of a dependent variable (dy), if in a function $y = f(x)$, a small change in the variable (dx) occurs. It can be calculated from the product of the first derivative of the function $f(x)$, multiplied by dx:

$$dy = f'(x)dx. \tag{3.1.1}$$

Most thermodynamic equations are functions with several variables. Hence, the derivatives can be obtained with respect to one variable if the others are kept constant. This procedure is called *partial differentiation*. It has a special notation with the parameters that are to be kept constant put as subscript to the symbols in parentheses. The following example is quoted out of context, to demonstrate this.

$$\left(\frac{\partial G}{\partial n_i} \right)_{p,T,n_j} = \mu_i \quad \text{for: } j \neq i. \tag{3.1.2}$$

The partial derivative of the Gibbs free energy G with respect to the molar number of substance i, when pressure (p), temperature (T), and the molar number of all the other substances (n_j) are kept constant, gives per definition, the chemical potential (μ_i) of the substance i. In general, this is the same procedure as is used when a function with several dependent variables is represented graphically in a 2-dimensional plot against one selected variable, keeping all other variables constant.

Small changes in a state function with several variables can be represented by a so-called *total differential*. For this, all partial differentials of this function must be summarized. These partial differentials are calculated as shown in Eq. (3.1.1), using, however, partial derivatives. The following equation, for example, would apply to the Gibbs free energy [$G(p, T, n_i)$]:

$$dG = \left(\frac{\partial G}{\partial p} \right)_{T,n_j} dp + \left(\frac{\partial G}{\partial T} \right)_{p,n_j} dT + \sum_{i=1}^{m} \left(\frac{\partial G}{\partial n_i} \right)_{p,T,n_j} dn_i. \tag{3.1.3}$$

The mathematical definition of the total differential is of very great physical importance to thermodynamics. This will be indicated by the following chain of reversible logical deductions:

dG is a total differential	↔	G is a state function	↔	G depends only on the state of the system, and not on the way in which that state was achieved

For this reason it is important that this property of a function is able to be mathematically proven.

A differential equation is not always written in the easily followed way shown in Eq. (3.1.3). Often it is presented as a Pfaffian differential equation:

$$dn = L\,dx + M\,dy + N\,dz. \tag{3.1.4}$$

The capital letters here represent any variable. There is a mathematical indicator to determine whether dn is a total differential. This is the so-called *Cauchy condition*, stressing that dn is a total differential when:

$$\frac{\partial L}{\partial y} \overset{!}{=} \frac{\partial M}{\partial x}; \quad \frac{\partial M}{\partial z} \overset{!}{=} \frac{\partial N}{\partial y}; \quad \frac{\partial L}{\partial z} \overset{!}{=} \frac{\partial N}{\partial x}. \tag{3.1.5}$$

If this is applied to Eq. (3.1.3), this means:

$$\frac{\partial \left(\frac{\partial G}{\partial p}\right)}{\partial T} = \frac{\partial \left(\frac{\partial G}{\partial T}\right)}{\partial p} = \frac{\partial^2 G}{\partial T\,\partial p}. \tag{3.1.6}$$

Additionally, it follows from Eq. (3.1.6) that for such functions if they are differentiated several times, the sequence of differentiations is unimportant. We will use this property in a later derivation [Eq. (3.2.6)].

Total differentials do not only result from energetic parameters. This formalism of course can be applied to any state function. Volume changes in mixed phases for example, can be described by the following total differential equation:

$$dV = \bar{V}_1\,dn_1 + \bar{V}_2\,dn_2 + \bar{V}_3\,dn_3 + \cdots + \bar{V}_m\,dn_m, \tag{3.1.7}$$

where dV is the change in volume that occurs when the molar number of one or more components of the system is changed. Equation (3.1.7) allows us to define the partial mole volume of a given substance i:

$$\bar{V}_i = \left(\frac{\partial V}{\partial n_i}\right)_{n_i} \quad \text{for: } j \neq i. \tag{3.1.8}$$

The partial molar volume has the inverse unit to the concentration, namely: $m^3 \, mol^{-1}$.

3.1.2
Gibbs Fundamental Equation

The scientific basis of thermodynamics is its three principles which are founded on experimentally verifiable, empirical facts. Upon this solid foundation a framework of definitions and relations has been built up which enables far-reaching postulations on all kinds of energy transformations.

The principle of conservation of energy, the so-called *first law of thermodynamics* states that there must exist a physical parameter having the property of a state function, which includes the consequences discussed in Section 3.1.1. Work (W), as a physical parameter does not comply with this condition. General experience shows that a change of a system from state A to state B can be achieved through many ways that differ greatly from one another in the amount of work that is required. Therefore, work cannot be a state function which could be used to characterize the energy state of a system independently of the way in which it was achieved. In specific cases such as the work differential of an ideal gas, this can be proved using the Cauchy condition [Eq. (3.1.5)].

Let a parameter called *internal energy* (U), now be defined as a state function which, as such, has a total differential dU. Furthermore, let the internal energy of a system be increased by dU if a certain amount of heat (dQ) is introduced into the system, and/or if certain work (dW) is done on the system.

$$dU = dQ + dW. \tag{3.1.9}$$

This equation contains the essence of the first principle of thermodynamics. Both the differentials dQ, as well as dW, are reflections of a change of energy. However, according to the second principle of thermodynamics, heat (Q), as a physical parameter, differs from all other forms of energy because it possesses a specific property: any form of energy can be completely transformed into heat, but heat itself can only partly be transformed into work. Entropy (S) makes its appearance in phenomenological thermodynamics as a kind of measure of the quality of heat. Entropy is defined as follows for a quasi-reversible process:

$$dS = \frac{dQ_{rev}}{T}. \tag{3.1.10}$$

This allows us to find an expression for dQ which can be introduced in Eq. (3.1.9).

Let us now consider the differential of work (dW) in more detail. It can be expressed as a sum of products of sort of work coordinates, extensive parameters indicating a kind of measure of the performed work, and work coefficients, intensive parameters, reflecting the effort required for this. For example, the work (dW_p) which is done when a gas is compressed by a pressure p resulting in a volume alteration dV will be:

$$dW_p = -p\,dV, \tag{3.1.11}$$

where p represents the work coefficient, and dV the work coordinate. The sign of this equation depends on the definition of the work differential (see also Sect. 2.1.3). A positive dW means that there is an increase in the work done in favor of the system. In this case work is achieved through the compression, i.e. a negative differential of the volume.

Work can be done in many different ways. An expansion of a material, for example, means elongation (dl) in response to the application of a force $(\mathbf{F})$:

$$dW_l = \mathbf{F}\,dl. \tag{3.1.12}$$

A cell can do work by transporting a certain number of atoms or molecules (dn) against a concentration gradient. At this point, the chemical potential (μ) must be introduced as the work coefficient. We will come back to this parameter in detail later [Eq. (3.1.33)]. In this case the work differential can be written as follows.

$$dW_n = \mu\,dn. \tag{3.1.13}$$

Finally, let us, from among the many other possible examples, consider charge transport. If a particular amount of charge (dq) is transported against an electric potential (ψ), then the electrical work done will be:

$$dW_q = \psi\,dq. \tag{3.1.14}$$

Equations (3.1.11) to (3.1.14) can be combined:

$$dW = -p\,dV + \mathbf{F}\,dl + \mu\,dn + \psi\,dq. \tag{3.1.15}$$

Considering that usually in the system a number of m substances are transported, then this equation can be expanded as follows:

$$dW = -p\,dV + \mathbf{F}\,dl + \sum_{i=1}^{m} \mu_i\,dn_i + \psi\,dq. \tag{3.1.16}$$

This is the more detailed form of the work differential which, together with Eq. (3.1.10), can be introduced in Eq. (3.1.9):

$$dU = T\,dS - p\,dV + \mathbf{F}\,dl + \sum_{i=1}^{m} \mu_i\,dn_i + \psi\,dq. \tag{3.1.17}$$

Equation (3.1.17) is a differential form of *Gibbs fundamental equation*. Of course it can be expanded by adding more kinds of work differentials.

Alternatively, this equation will be automatically reduced if certain changes become irrelevant. For example, suppose a defined transformation within a system is not accompanied by a mechanical strain. Then l remains constant, and consequently, $dl = 0$. As a consequence, the corresponding term disappears from the equation.

In Eq. (3.1.17) the Gibbs fundamental equation appears in the form of a Pfaffian differential. Such expressions can be integrated under certain conditions, which apply in this case. This gives:

$$U = TS - pV + Fl + \sum_{i=1}^{m} \mu_i n_i + \psi q. \tag{3.1.18}$$

It must be noted that the transition from Eq. (3.1.17) to Eq. (3.1.18) does not mean a simple elimination of the differential operators; it is the result of a proper integration which is not described here!

Using the rule given in Eq. (3.1.3), this process of integration can be reversed. It gives:

$$dU = T\,dS + S\,dT - p\,dV - V\,dp + F\,dl + l\,dF$$
$$+ \sum_{i=1}^{m} \mu_i\,dn_i + \sum_{i=1}^{m} n_i\,d\mu_i + \psi\,dq + q\,d\psi. \tag{3.1.19}$$

A comparison of this result with the initial Eq. (3.1.17) shows that the following condition must be satisfied:

$$S\,dT - V\,dp + l\,dF + \sum_{i=1}^{m} n_i\,d\mu_i + q\,d\psi \overset{!}{=} 0. \tag{3.1.20}$$

This is the so-called Gibbs-Duham equation. It is useful for some calculations because it allows reduction of the degree of freedom of a system by one variable.

It has proved useful to define not only the internal energy (U), but also three further energy functions. In some books the introduction of these parameters is explained in a physical way, by discussing processes of vapor compression etc., but it seems to be simpler just to accept the definitions of these parameters, and subsequently substantiate their usefulness.

The definitions are:

enthalpy,	$H = U + pV;$	(3.1.21)
Helmholtz free energy,	$F = U - TS;$	(3.1.22)
Gibbs free energy,	$G = H - TS.$	(3.1.23)

The Gibbs fundamental equation [Eq. (3.1.17)] can now easily be written down for these new defined functions. Let us first transform Eq. (3.1.21) into a total differential, according to Eq. (3.1.3). According to the definition [Eq. (3.1.21)], the enthalpy (H) is a function of: U, p, and V. This gives:

$$dH = \left(\frac{\partial H}{\partial U}\right)_{p,V} dU + \left(\frac{\partial H}{\partial p}\right)_{U,V} dp + \left(\frac{\partial H}{\partial V}\right)_{U,p} dV. \tag{3.1.24}$$

From Eq. (3.1.21) follows directly:

$$\left(\frac{\partial H}{\partial U}\right)_{p,V} = 1; \quad \left(\frac{\partial H}{\partial p}\right)_{U,V} = V; \quad \left(\frac{\partial H}{\partial V}\right)_{U,p} = p, \tag{3.1.25}$$

which, when combined with Eq. (3.1.24), results in

$$dH = dU + V\,dp + p\,dV. \tag{3.1.26}$$

Combining with this, Eq. (3.1.17) gives the Gibbs fundamental equation for dH:

$$dH = T\,dS + V\,dp + \mathbf{F}\,dl + \sum_{i=1}^{m} \mu_i\,dn_i + \psi\,dq. \tag{3.1.27}$$

In the same way it is possible to derive from Eq. (3.1.22) the relations:

$$dF = -S\,dT - p\,dV + \mathbf{F}\,dl + \sum_{i=1}^{m} \mu_i\,dn_i + \psi\,dq, \tag{3.1.28}$$

and:

$$dG = -S\,dT + V\,dp + \mathbf{F}\,dl + \sum_{i=1}^{m} \mu_i\,dn_i + \psi\,dq. \tag{3.1.29}$$

The choice of whether Eqs. (3.1.17), (3.1.27), (3.1.28), or (3.1.29) should be used to calculate a particular system depends on the external conditions and the question which is being asked. Investigating a system under isobaric conditions (p = const, i.e. dp = 0), it is useful to apply the equation for dH [Eq. (3.1.27)], or for dG [Eq. (3.1.29)], because in this case the term volume expansion work ($V\,dp$) vanishes. This corresponds to the situation in most biological investigations. Therefore, we will use mostly enthalpy (H) and Gibbs free energy (G) in all further biophysical calculations instead of the inner energy (U) and the Helmholtz free energy (F).

If the conditions are isothermal (dT = 0), as well as isobar (dp = 0), then in Eq. (3.1.29) the terms connected with heat and volume work will vanish. Hence, dG expresses directly the deviation of the energy content, as a result of work which was done. The gradient of free Gibbs energy therefore indicates the direction of a spontaneous process in the same way as a gradient of potential energy indicates the direction of a rolling sphere on an uneven surface.

The differential dH, on the other hand, is a measure of the thermal characteristics of isobar processes.

All these forms of the Gibbs fundamental function [Eqs. (3.1.27), (3.1.28), (3.1.29)], as well as Eq. (3.1.7) for the partial volume, can be integrated according to Eq. (3.1.17) for dU.

For questions of equilibrium thermodynamics, only the situation at the beginning and at the end of a reaction is of interest. For this, differences between parameters of these states are defined. So, for example, an entropy change of a reaction $\Delta_R S$ is defined as follows:

$$\Delta_R S = S_{\text{product}} - S_{\text{substrate}}. \tag{3.1.30}$$

In a corresponding way, parameters like $\Delta_R U$, $\Delta_R H$, $\Delta_R F$, and $\Delta_R G$ are defined. In contrast to entropy, however, no absolute values exist for energetic parameters. They always need a defined reference value. So, for example, *standard energies of formation* ($\Delta_F U$, $\Delta_F H$, $\Delta_F F$, and $\Delta_F G$) are defined as energetic changes that occur when a substance is formed from its elements under standard conditions ($T = 297$ K), or better: *would occur*, because in most cases a direct synthesis of the substance from its elements is impossible. Analogous with Eq. (3.1.30) it results in:

$$\Delta_R G = \Delta_F G_{\text{product}} - \Delta_F G_{\text{substrate}}. \tag{3.1.31}$$

The chemical potential (μ_i) of the substance i is particularly important for the following calculations. It can be easily defined using Eqs. (3.1.17), (3.1.27), (3.1.28), or (3.1.29):

$$\mu_i = \left(\frac{\partial U}{\partial n_i}\right)_{S,V,l,n_j,q} = \left(\frac{\partial H}{\partial n_i}\right)_{S,p,l,n_j,q} = \left(\frac{\partial F}{\partial n_i}\right)_{T,V,l,n_j,q} = \left(\frac{\partial G}{\partial n_i}\right)_{T,p,l,n_j,q}.$$

$$\tag{3.1.32}$$

The chemical potential, therefore, is a partial expression having the dimensions J mol^{-1}. The chemical potential of the substance i (μ_i) can be calculated from the concentration (c_i), or better from the chemical activity (a_i) of this substance using:

$$\mu_i = \mu_i^0 + RT \ln a_i. \tag{3.1.33}$$

In this equation a_i is just a number, i.e. the number of moles per liter, as the chemical activity is measured. The chemical activity is a kind of effective concentration. Its relation to the concentration is given by:

$$a_i = f_i c_i. \tag{3.1.34}$$

In this equation, f_i is the *coefficient of activity*. In ideal solutions, $f_i = 1$, that is, the activity of the dissolved substance is equal to its concentration c_i. Usually f_i decreases as the concentration in the solution increases (see Fig. 3.1). For dissociating salts, f represents an average activity coefficient for the ions. For example, the ions in a 100 mM solution of NaCl show an average activity coefficient of 0.8. The chemical activity of this solution therefore is equal to an ideal solution with a concentration of only 80 mM. In contrast to the coefficient of activity, which is a dimensionless number, the chemical activity has the same units as the concentration.

In some cases it may be useful to employ the mole fraction as a measure of concentration. The *mole fraction* of a substance i is defined as the number of moles of that substance (n_i), divided by the total number of moles of all substances present:

$$x_i = \frac{n_i}{\sum_{i=1}^{m} n_i}, \tag{3.1.35}$$

where:

$$\sum_{i=1}^{m} x_i = 1. \tag{3.1.36}$$

According to Eq. (3.1.34), the mole fraction of a substance can also be expressed as the mole fraction activity (a_{xi}).

The standard potential (μ_i^0) can be easily defined by means of Eq. (3.1.33). It follows from this equation that when $a_i = 1$, $\mu_i = \mu_i^0$. The standard potential

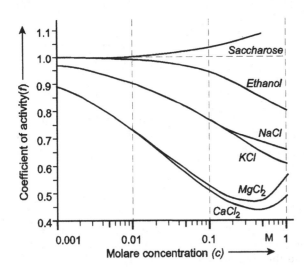

Fig. 3.1. Coefficients of activity (f_i) of various substances as functions of their concentrations (c_i) in aqueous solutions under standard conditions ($T = 297$ K)

therefore is the chemical potential of a substance i in a solution, with an activity of 1 M, if this measure of concentration is applied. In Eq. (3.1.33), using the mol fraction (x_i) of a substance, or its mol fraction activity (a_{xi}) as concentration measure, the chemical standard potential (μ_i^0) is determined by the chemical potential of the pure substance ($a_{ix} = 1$).

A further extension of the Gibbs fundamental equation concerns the term $\psi \, dq$. The charge of a single charged ion is determined by the Faraday constant (F). The charge on n moles of a z-fold charged ion is obtained as follows:

$$q = zn\text{F}. \tag{3.1.37}$$

This is a function with one independent variable (n). Therefore it is easily transformed into a differential according to Eq. (3.1.1):

$$dq = \left(\frac{dq}{dn}\right) dn = z\text{F} \, dn. \tag{3.1.38}$$

If more than one ion is in the solution, the charges can be summarized:

$$dq = \sum_{i=1}^{m} z_i \text{F} \, dn_i. \tag{3.1.39}$$

Introducing this expression into the Gibbs fundamental equation, the equation then gets two terms with the differential dn_i. It is obviously useful to combine these terms. First, consider these two terms of the Gibbs equation in isolation from the other terms of this equation:

$$\psi \sum_{i=1}^{m} z_i \text{F} \, dn_i + \sum_{i=1}^{m} \mu_i \, dn_i = \sum_{i=1}^{m} (\mu_i + z_i \text{F} \psi) dn_i = \sum_{i=1}^{m} \tilde{\mu}_i \, dn_i. \tag{3.1.40}$$

Here, the expression inside the brackets taken together is described as the electrochemical potential ($\tilde{\mu}_i$) of the substance i.

$$\tilde{\mu}_i = \mu_i + z_i \text{F} \psi. \tag{3.1.41}$$

The electrochemical potential is the basis for most electrochemical calculations, and thus forms an important starting point for further considerations.

3.1.3
Force and Motion

After introduction of the energetic state functions in the previous section we will now consider their application to determine forces leading to any sort of motion.

A sphere is rolling downhill. It is moving spontaneously from a position with a higher potential energy to one with a lower potential. The direction of its movement follows a force vector (**X**) and is, consequently, determined by the negative gradient of the energy U:

$$\mathbf{X} = -\text{grad}\, U. \tag{3.1.42}$$

If consideration of the energy gradient is confined to the direction of the x-coordinate, this equation can be simplified to give:

$$\mathbf{X}_x = -\frac{dU}{dx}\mathbf{i}, \tag{3.1.43}$$

where **i** is simply a unit vector, i.e. a vector with the amount of 1, and an arrow, directing toward the x-coordinate. For dU, any appropriate energy state function should be substituted as shown in Section 3.1.2. Let us consider for example the force acting on a charge (q) in an electric field in the x-direction $[\mathbf{E} = (-d\psi/dx) \cdot \mathbf{i}]$. Substituting for dU in Eq. (3.1.43), the expression, according Eq. (3.1.17), gives:

$$\mathbf{X}_q = -\frac{dU}{dx}\mathbf{i} = -\frac{d\psi}{dx}\mathbf{i}q = \mathbf{E}q. \tag{3.1.44}$$

This assumes that there is no other gradient in the x-direction, i.e. that neither p, nor T, nor μ are functions of x. The introduction of the field strength (**E**) is in accordance with the definition given in Eq. (2.4.5). Equation (4.1.44) is identical to Eq. (2.4.6), which was derived in another way.

Equation (3.1.44) cannot be applied to practical calculations of ion transport because only the transport of a charge is considered. In contrast to the movement of an electron, the transport of ions always means that there is an additional change in concentration. For transport of a noncharged substance (i), the negative gradient of its chemical potential (μ_i) is the driving force:

$$\mathbf{X}_i = -\text{grad}\, \mu_i. \tag{3.1.45}$$

Ions, in contrast, are driven by their gradient of the electrochemical potential $(\tilde{\mu})$ which, according to Eq. (3.1.41), includes the electric potential. Applying the differential operator, the electric potential (ψ) transforms into an electric field strength (**E**):

$$\mathbf{X}_i = -\text{grad}\, \tilde{\mu}_i = -(\text{grad}\, \mu_i - z_i\mathbf{F}\mathbf{E}). \tag{3.1.46}$$

There are many kinds of movement in biological systems that are the concern of biophysics. Their range covers electron transfer, structural changes of molecules, chemical reactions, fluxes of molecules and ions, streaming of liquids in the body, and finally, mechanical movements of limbs and whole bodies. Fluxes occupy a central position in the study of movements in biology, and therefore these will be considered here for the sake of simplicity, as a sort of generalized movement.

The term flux (J_i) stands for the amount of a substance (i) that passes perpendicularly through a unit of surface per unit time. This definition shows that the flux in general is a vector. Often, flux through a membrane is considered, the direction of which is always predicted.

The relation between Flux J_i of the component i, and its velocity v_i is:

$$J_i = c_i v_i. \tag{3.1.47}$$

A system which is traversed by a flux, and where no substance is added or removed to this flow, is called a *conservative* one. In this case the following conditions apply:

$$\text{div} J_i = 0. \tag{3.1.48}$$

The differential operator div ("divergence") can be replaced by the Nabla operator (∇), as explained in Section 2.4.1. In contrast to the operator 'grad', which, applied to a scalar, and results in a vector, the 'div' operator is to be applied to a vector, producing a scalar. The Nabla operator (∇) is applicable to differentiate vectors as well as scalars.

If there is a point in such a system which is traversed by a flux, where $\text{div} J_i > 0$, then there will be a source adding substance to the flux. If $\text{div} J_i < 0$, there will be a removal of some of the substance like a sink. Figure 3.2 illustrates this situation using an example where the flow is simply in the x-direction.

This formalism is applied, describing for example fluxes through a biological tissue. For the transport of ions, this system is conservative, because no

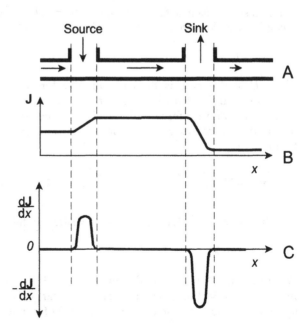

Fig. 3.2. A Graphic representation of a linear flow with a source, and a sink; B the flux (J) as a function of x; C the change of the flow (dJ/dx) as a function of x

accumulation or depletion of ions occurs. In contrast, for the transport of oxygen the condition div $J_o < 0$ holds, because the tissue uses oxygen for its own respiration.

If a constant force acts on a body, then the latter accelerates. However, as the velocity of this body increases, the friction is likely to increase too. When both the driving force and the frictional force become the same amount, then the body will move with a constant velocity. This is a special case of a stationary state, the so called *stationary motion*.

It is a fundamental experience in physics that there are regions, characterized by their relations between force and movement, having quite different qualities. If, for example, a comparatively minor force acts on a body, then the body will attain a velocity that is proportional to the force. This is a region where a linear analysis of irreversible thermodynamics is applicable. If the same body is more forcibly moved, then the frictional force will increase in a stronger way, and a non-linear approach is necessary.

This concept, illustrated here by a mechanical example, has a more general application. A linear approach to a general equation of motion can be formulated as follows:

$$J_i = L_i X_i, \tag{3.1.49}$$

where L_i is a coefficient which is a kind of a generalized conductance. In the same way, the following equation can be written:

$$X_i = R_i J_i, \tag{3.1.50}$$

In this case a resistance factor is applied: $R_i = 1/L_i$. Ohms law is a special form of this general expression:

$$U = RI, \tag{3.1.51}$$

where U in this case is the electromotive force, depending on the potential gradient (grad ψ), and I is the electric current.

In a similar way force and velocity can be linearly related. For this it is necessary to introduce a mobility factor (ω), and its corresponding coefficient of friction (f):

$$\mathbf{v} = \omega \mathbf{X} = \frac{\mathbf{X}}{f}. \tag{3.1.52}$$

The flux equation (3.1.47) can now be written as follows:

$$J_i = c_i \omega_i X_i \tag{3.1.53}$$

We introduced the Stockes' equation (2.3.34), when discussing molecular movement in Section 2.3.6. This is also a particular expression of this linear approach. It indicates the frictional force ($\mathbf{F}$) of a sphere with radius (r), moving in a liquid with viscosity (η) at velocity (**v**). Under stationary conditions, this friction force is equal to the driving force:

$$\mathbf{F} = 6\pi r \eta \mathbf{v}. \tag{3.1.54}$$

In view of Eq. (3.1.52), the following analogy is obvious:

$$f = \frac{1}{\omega} = 6\pi r \eta. \tag{3.1.55}$$

The definitive mechanical behavior of bodies moving in water will be discussed in detail in Section 3.7. Here it is only being used as an example for the consideration of general principles.

Figure 3.3 shows the frictional coefficient (f) for a sphere with a diameter of 1 cm in water as a function of its velocity (**v**). The frictional coefficient only remains constant up to a velocity of about 1 mm s^{-1}, corresponding to Eq. (3.1.55). As the velocity increases above this, the frictional factor deviates, at first slightly, and then greatly from the horizontal line. This illustrates the transition of the system from linear to non-linear behavior. This is the transition to far-equilibrium conditions, where non-linear thermodynamic approaches must be applied.

Let us explain the genesis of such a non-linear relation by a simple example. Let a phenomenological coefficient be a sum of a constant term (L_i), and a variable term L_i', which is proportional to the flux (J_i). In this case, the linear approach of Eq. (4.1.49) transforms into:

$$\mathbf{J}_i = (L_i + L_i' \mathbf{J}_i)\mathbf{X}_i. \tag{3.1.56}$$

Solving this equation for J_i, it gives:

$$\mathbf{J}_i = \frac{L_i \mathbf{X}_i}{1 - L_i' \mathbf{X}_i}. \tag{3.1.57}$$

Hence we have a non-linear function $J_i(\mathbf{X}_i)$.

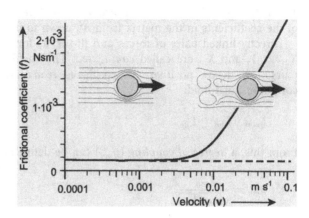

Fig. 3.3. The coefficient of friction ($f = $ F/v) of a sphere with a diameter of 0.01 m in water as a function of its velocity (**v**)

The qualitative consequences of these non-linear approaches will be considered further in the next section. Let us at present remain in the field of linear thermodynamics.

If different kinds of motions and forces occur simultaneously in a system, they will influence each other. Let us consider again the flux (J_i) as a generalized kind of motion. The nonequilibrium thermodynamics in its scope of linear approaches allows us to write down a set of phenomenological equations forming a flux matrix. It includes the idea whereby all forces and fluxes in the system are coupled with each other. The simple Eq. (3.1.49), therefore, will be expanded to the following set of equations:

$$J_1 = L_{11}X_1 + L_{12}X_2 + L_{13}X_3 + \cdots + L_{1n}X_n$$
$$J_2 = L_{21}X_1 + L_{22}X_2 + L_{23}X_3 + \cdots + L_{2n}X_n$$
$$J_3 = L_{31}X_1 + L_{32}X_2 + L_{33}X_3 + \cdots + L_{3n}X_n \tag{3.1.58}$$
$$\cdots$$
$$J_n = L_{n1}X_1 + L_{n2}X_2 + L_{n3}X_3 + \cdots + L_{nn}X_n.$$

In these equations the vector notation of fluxes and forces is still retained, regardless of whether, in rare cases (which we will discuss in the following) they may really be scalars.

The parameters L_{mn} are phenomenological coefficients, also called *coupling coefficients, cross coefficients*, or *Onsager coefficients*. (This equation was first used by L. Onsager in 1931). This general set of equations in reality reduces, because a flux J_m is coupled with a force X_m only when $L_{mn} \neq 0$.

Equation (3.1.58) shows that n forces with their corresponding fluxes require a set of equations with n^2 coupling coefficients. Onsager, however, was able to show that this matrix is symmetric. It means that near the equilibrium the following relation holds:

$$L_{mn} = L_{nm} \quad \text{for: } n \neq m. \tag{3.1.59}$$

This is *Onsager's law on the reciprocal relation*. It leads to a significant reduction of the coefficients in the matrix from n^2, down to the number of $n(n + 1)/2$.

Directly linked pairs of forces and fluxes, as for example J_1 and X_1, J_2 and $X_2, \ldots, J_n$ and X_n, are called *conjugated*. The coefficients (L_{nn}), linking these pairs, are always positive. If two fluxes really are coupled, the following condition must hold:

$$L_{mm} \cdot L_{nn} \geq L_{mn}^2. \tag{3.1.60}$$

From this, a *degree of coupling* (q_{mn}) can be defined:

$$q_{mn} = \frac{L_{mn}}{\sqrt{L_{mm}L_{nn}}}. \tag{3.1.61}$$

This degree of coupling can vary as follows: $1 \geq q_{mn} \geq 0$. When $q_{mn} = 0$, the fluxes are completely independent of each other, when $q_{mn} = 1$, there is maximal coupling.

As mentioned before in Eq. (3.1.58) all fluxes and forces are printed in bold letters as vector parameters. At the same time we mentioned that we will consider here fluxes in a very general sense, symbolizing all kinds of motion. This means that true fluxes of matter, which really are vectors going in a particular direction, as well as, for example, chemical reactions can be considered. The flux, as mentioned in this equation, therefore, can also mean a production of a substance, or a disappearance of it, by a chemical reaction. In this case, however, the flux does not remain a vector, but becomes a scalar. How can we include these scalar fluxes into a matrix together with vectors, without violating mathematical rules?

Let us consider a simple set of flux equations, including transport of a substance ($\mathbf{J}_i$) and a chemical reaction, the rate of which we will denote by the scalar flux J_R. Formal mathematics appears to allow only the following possibility:

$$J_R = L_{RR}X_R + \mathbf{L}_{Ri}\mathbf{X}_i$$
$$\mathbf{J}_i = \mathbf{L}_{iR}X_R + L_{ii}\mathbf{X}_i. \qquad (3.1.62)$$

In the first equation, the product of two vectors ($\mathbf{L}_{Ri}\mathbf{X}_i$) gives a scalar, as well as the product of scalars ($L_{RR}X_R$), and so this equation is a sum of scalars resulting in a scalar flux (J_R). In contrast, in the second equation all the terms of the sum are vectors. In both equations, therefore, the mathematical requirement for homogeneity has been satisfied.

But: what is the meaning of a vectorial coupling coefficient? Introducing this parameter, we declared it as a sort of conductivity. What does a conductivity vector mean? In fact, in so-called *anisotropic systems* vectorial coefficients can appear. In an isotropic system, for example, in an aqueous solution, the mobility of an ion in all directions is equal. The parameter L, therefore is a scalar. Considering, however, the same ion in a pore of a membrane, its movement is possible only in a predetermined direction, and its conductivity consequently becomes a vector.

These circumstances are considered in the so-called *Curie-Prigogine principle*. It states that the direct coupling of scalar and vectorial fluxes is possible only in anisotropic systems. This principle is important in biophysical considerations of active ion transport. In this case a hydrolysis of ATP is coupled to the transport of an ion against an electrochemical gradient.

3.1.4
Entropy and Stability

The second principle of thermodynamics states that an isolated system moves spontaneously towards a maximum in its entropy. When this state is achieved, then it is in thermodynamic equilibrium. In the same way, the decrease of the free energy down to a minimum can be considered as the way towards the equilibrium in the sense of the second principle.

Any movement as a result of energy transformation leads to an increase in the entropy of the system or its environment. The term *entropy production* ($\sigma = dS/dt$) has been introduced to characterize this process. The entropy production is always positive, but can approach zero asymptotically. The condition: $\sigma = 0$ would mean an idealized reversible process. Thermodynamically, a process is defined as being reversible if it can be repeated an arbitrary number of times without requiring the supply of additional energy.

To prevent misunderstanding, the different meanings of the term 'reversible' in physics, chemistry, and biology must be pointed out. In physics, the term 'reversible' is used according to the thermodynamic definition, i.e. connected with the condition: $\sigma = 0$. When a chemist speaks about a 'reversible reaction', or a 'reversible electrode', he or she only means processes that in principle could run in both directions, independently of the required energy. Finally, biologists say that a change of a biological system is 'reversible' when it is able to reverse an induced change so that no irreparable damage is caused (for example: reversible inhibition of metabolic processes).

Considering the total entropy balance of a system, it must be pointed out that, in contrast to a closed system, where as a result of energy transformations only an entropy increase is possible, in an open system an entropy decrease is even possible. This can be done if substances are incorporated which have a low entropy content, in exchange for entropy-rich substances which are being extruded. To characterize this process, an entropy flux (J_S) is formulated which penetrates the whole system. Hence, the total entropy balance of the system can be written as follows:

$$\frac{dS}{dt} = -\nabla J_S + \sigma. \tag{3.1.63}$$

The overall change of entropy in an open system (dS/dt), therefore, results as a sum of the always positive entropy production (σ), and the divergence of the entropy flux ($\nabla J_S \equiv \operatorname{div} J_S$), penetrating the system. In reference to the definition in Section 3.1.3 (see Fig. 3.2), the system in fact is not conservative in relation to this entropy flux. Depending on the relation of the two terms in the sum [Eq. (3.1.63)], the total energy change (dS/dt) can become positive as well as negative. The control of the term ∇J_S can be considered as the work of a Maxwell demon, as described in Section 2.3.2 (Fig. 2.9).

For further considerations it may be useful to introduce a thermodynamically based classification of the various kinds of stationary states. We define a *stationary state* as a state in which the structure and parameters are time independent. The reasons leading to this quality can be quite different. The water level of a lake, for example, can be time independent, i.e. constant, either because there is no inflow into the lake, and no outflow, or because inflow and outflow are equal to each other. These two kinds of stationary states can be distinguished by their entropy production. In the first case no energy is required to maintain this state, therefore there is no entropy production, the system is in *thermodynamic equilibrium* ($\sigma \overset{!}{=} 0$). In contrast, the lake with exactly the same in- and outflow is in a *steady state*. This is a stationary state with entropy production ($\sigma > 0$). The thermodynamic definition of the steady state is the only possible one. A steady state cannot be defined by its kinetic properties.

Let us illustrate this statement by an example: using radio-nuclides it is possible to demonstrate that human erythrocytes exchange chloride as well as potassium ions with their environment. With this method, it is possible to measure directly the corresponding exchange rates. This kinetic method of analysis may give the impression that both ions, ^-Cl, as well as ^+K, are in a steady state because in both cases the unidirectional fluxes, outside $\rightarrow$ in and inside $\rightarrow$ out, are equal. This, however, is an incorrect conclusion. The chloride ions in fact are distributed passively between the external medium and the cytoplasm, according to their thermodynamic equilibrium. The observed exchange of radioactive chloride results from their self diffusion by thermal motion. This process is not accomplished by entropy production because no energy is converted. It is like a stochastic exchange of water molecules between the two vessels in Fig. 3.4A. In contrast to this, potassium is pumped actively into the cell by use of metabolic energy against its electrochemical gradient, and diffuses passively back, driven by this gradient (Fig. 3.4B). Both are true fluxes in the thermodynamic sense, producing entropy. Potassium, therefore, in contrast to chloride, really is in a stationary state. This example indicates that the above described kinetic experiment of compartment analyses is unsuitable for distinguishing between an equilibrium state and a steady state. We will come back to this problem later in detail (Sects. 3.4.1 and 5.2.1).

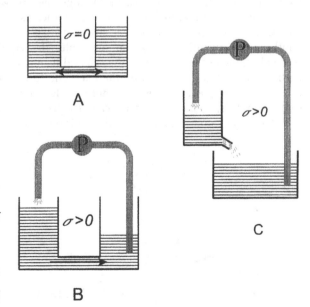

Fig. 3.4. Stationary states in hydraulic models: A thermodynamic equilibrium ($\sigma = 0$); B a steady state system ($\sigma > 0$), which becomes an equilibrium (A), if the pump (P) is stopped; C a steady state system ($\sigma > 0$), where the stop of the pump (P) would lead to a complete outflow of the liquid from the upper vessel

As we will see later (Sect. 3.3.3), the steady state of potassium can be transformed into an equilibrium state if the pump is inhibited. In this case, a Donnan equilibrium will be established which means an equilibration of the electrochemical potentials. The steady state of sodium and potassium in the cell therefore resembles the case B in Fig. 3.5, changing into the equilibrium state (Fig. 3.5A), when the active transport is inhibited. In contrast to this, various substances do not show the possibility of an equilibrium distribution. If the influx is stopped in these cases, the substance disappears completely (Fig. 3.4C).

An important property of all stationary states is their kind of stability. Let us illustrate this again by a mechanical example, using the behavior of a sphere on a surface (Fig. 3.5). The requirement for a stationary state in this case simply means an equilibrium and means the kind of position of the sphere on any small, but horizontal part of the surface. In the case of an *indifferent* state, the surface has a large horizontal region. In this case the energy of the sphere will not be changed by a change of its position. In case of a *stable* state, every change of the position leads to an increase of the energy of the sphere, and generates a force driving the sphere back to its original state. In contrast, an *unstable state* is characterized by a situation where even small changes of the position release forces that cause the system to be deflected even more. Additionally, sometimes so-called metastable states are considered. As *metastable*, a stable state can be considered which is delimitated from another one by a small barrier which can be easily overcome.

Figure 3.6 indicates all possible kinds of stationary states. First of all the presence, or the absence of entropy production indicates whether the given stationary state is a thermodynamic equilibrium ($\sigma = 0$), or whether it is a steady state ($\sigma > 0$). Furthermore, in the case of thermodynamic equilibrium, global and local equilibria must be distinguished between. In case of a *global equilibrium*, the function of free energy indicates only one minimum. This means that no alteration, how ever strong it may be, can bring the system into another equilibrium state. As an example, the equilibrium distribution of particular kinds of ions between the cell and its environment can be mentioned. In the case of *local equilibrium*, the energetic function indicates two or more minima which are separated by more or less large energy barriers. As an example of this, reaction isotherms can be considered, as illustrated in Figs. 2.12 and 2.32. The stability of such local equilibria is as high as the energy barrier

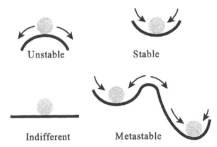

Unstable   Stable

Indifferent   Metastable   **Fig. 3.5.** Various kinds of stability conditions

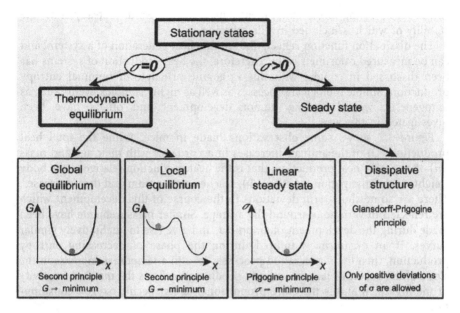

Fig. 3.6. Possible kinds of stationary states and conditions of their stability

between these minima is. If a barrier between two minima is so small that thermal fluctuations can lead to quick transitions, the state is called metastable. This, for example, is the typical situation for enzyme-substrate complexes. For the schemes in Fig. 3.6 it must be considered that in reality G is not simply a function of a single reaction coordinate (x), but rather a hyperplane in a n-dimensional space.

To consider the two kinds of stationary states ($\sigma > 0$) from Fig. 3.6 in detail, we must learn something more about entropy production. Entropy production is not only an indicator for distinguishing between the equilibrium state and steady state, but is also important for some stability considerations of the latter.

In concordance with its definition, entropy production has the measure: $J K^{-1} s^{-1}$. In some cases, additionally, this parameter is related to the mass in kg or to the molar mass.

Multiplying entropy production by temperature, one gets *Rayleights dissipation function* (Φ), which can be calculated from fluxes and forces as follows:

$$\Phi = \sigma T = \sum_{i=1}^{m} J_i X_i. \tag{3.1.64}$$

This equation holds for the region of linear, as well as of non-linear flux-force relations. Particularly for the linear region, I. Prigogine was able to show that systems tend to develop towards a reduced entropy production. This is the *Prigogine principle of minimal entropy production*. Systems which are not far from thermodynamic equilibrium, and which are kept in imbalance by continuously acting forces consequently may move towards a steady state, the stability of which is included in this criterion.

The dissipation function reflects the specific heat generation of a system, and can be measured calorimetrically. Therefore, the heat production of systems has been discussed in context with this Prigogine principle of minimal entropy production. Simple biological processes as well as highly complex events such as ontogenesis, wound healing, tumor development and others have been investigated in this way.

Figure 3.7 shows some observations made in mice. While the total heat production ($\dot{Q}$) of the animals increases in accordance with their age and mass (m), the specific heat production, that is the heat production relative to the body weight, as the dissipation function (Φ), reaches a maximum and then decreases. There are some short-term deviations in the course of this development which are reflected in variations around the average. Similar measurements have been made during the development of amphibia, and give rise to qualitatively similar curves. If an organism is injured during the phase of decreasing entropy production, then the regeneration processes lead to a temporary increase in the slope of the curve. In fact the entropy production reflects the metabolic activity of the organism and is therefore proportional to its specific oxygen consumption.

Despite these results, whereby after passing through a phase of rapid development, the system moves towards a stationary state of minimal entropy production, it is still controversial as to whether the Prigogine principle can be applied to large systems which include a great number of different subsystems, particularly those which are far from equilibrium.

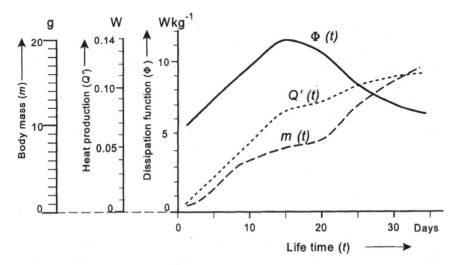

Fig. 3.7. Dissipation function (Φ), heat production ($\dot{Q}$), and mass (m) of mice as a function of their life span from the time of birth, up to the 35th day of life. (Data from Lamprecht and Zotin 1978)

If a system deviates from the region of linear approaches, then the Prigogine principle is no longer valid. In contrast to steady states in the scope of linear thermodynamic approaches which are always stable and do not show any kinds of metastability, systems in the region of non-linear approaches show more complicated behavior.

Considering non-linear systems we must at first generalize the term 'stationary'. We already mentioned in context with the Stockes law [Eq. (3.1.54)] the so-called stationary movement as a movement with constant velocity where the frictional force is equal to the driving force. This sort of movement is also a kind of stationary, i.e. time independent state. The term 'stationary state' can also be applied to states that are not at rest, but show repetitive periodic movements in a stationary manner. For example, cardiac function can be called 'stationary', if the frequency and amplitude of the heart beat does not change during the period of observation.

In the case of a linear steady state, fluctuations of the system parameters produce only positive deviations of entropy production, bringing the system back to the stationary state, which is therefore stable in any case. In contrast to this, in the region of non-linear approaches, far from equilibrium, fluctuations of a stationary state can also lead to negative deviations of the entropy production, leading to a destabilization of the system. The system may jump from one into another stationary state. The steady states in the region of non-linear approaches therefore are mostly metastable (see Fig. 3.5). Their stability condition requires the occurrence of only positive deviation of the entropy production. This is the so-called *Glansdorff-Prigogine principle*.

The transition from linear to non-linear approaches of thermodynamics is not only associated with an increase in the coefficient of friction, or with the occurrence of several stationary states, but also with the spontaneous development of so-called *dissipative structures*. The general concept of the diversity of structures has already been discussed in Section 2.3.3. We postulated that there are basically two different types of structures: equilibrium structures ($\sigma = 0$), and dissipative structures ($\sigma > 0$). In contrast to equilibrium structures which are always structures in space, dissipative structures can also be structures in time, or in a space-time continuum.

In order to illustrate this, Fig. 3.3 shows the function $f(v)$ of a sphere, and the pattern of flow close to it. The laminar flow, which occurs at a low velocity, can be compared with the turbulent flow, which appears when the non-linear region of the function is attained. When the flux-force relation changes from the laminar to the non laminar region, then the unstructured laminar flow becomes unstable and vortices appear which, in the terminology of thermodynamics, are dissipative structures (for details of streaming behavior, see Sect. 3.7.1).

There is extensive literature on the theory of dissipative structures and on their occurrence in nature. Most of these dissipative structures are periodic structures in space, such as cloud patterns, flow phenomena in liquids with an applied temperature gradient, so-called Benard cells, plasma waves in electron beam tubes etc. In addition, there are many time structures, including for example all kinds of sound production, from the electronic organ to vibrating strings, and to the wind instruments.

In biological systems, in spite of many speculations, dissipative structures in space have not been unequivocally demonstrated. On the other hand, time patterns showing this property have been found to occur quite often. Examples of this include oscillations in metabolism, periodical changes in the membrane potential of certain cells, for example in the cells of the cardiac pacemaker in the sino-auricular node, and finally more complex, oscillatory movements in ecosystems. Sometimes, such oscillations in local concentrations of certain substances become 'frozen', so that structures arise which do not require entropy producing processes to sustain them, but which originally had been built as dissipative structures. One can conceive, for example, the genesis of the first viable bio-macromolecule occurring in this way. Similar processes could form the basis of morphogenesis (for further detail, see Sects. 5.3.2 and 5.3.3).

Let us come now to a further aspect of stationary states. If a system is considered as stationary, this is always in relation to a defined period of time. An organism could be said to be in a stationary state for a number of hours, or for days. That is, its volume, the content of certain substances, its shape, temperature etc, are constant within defined tolerance for this period of time. Nevertheless, the organism in fact is ageing or growing. If a longer time period is considered, then the changes in these parameters will exceed the defined tolerance limits.

A biological system consists of sub-systems, each of which is associated with a different time constant. For example, the rate of aging in man is slow, compared with the rate of mitosis of a hemopoetic cell. Thus, the conditions in bone marrow can be regarded as stationary during the course of several cycles of mitosis in spite of the ageing processes affecting the organism as a whole. In vivo, the life span of a human erythrocyte is about 100 days. The ionic composition of juvenile erythrocytes differs somewhat from that of the mature cells. If one is interested in the ionic regulation of these cells, then, because of the temporal characteristics of such processes, experiments lasting a few hours will be sufficient to provide the required information. Within such short time periods the ionic concentration can be regarded as stationary. Figure 3.8 shows some characteristic times on a logarithmic scale that will serve to extend this list of examples.

These considerations hint at the existence of a *time hierarchy* of stationary states, related to their time constants which range over several orders of magnitude. The kinetic consequences of this circumstance will be considered in Section 3.2.3 and especially 5.2.2. The following concept, however, is important for our further thermodynamic considerations.

The living organism as a whole, when considered within a limited period of time, is in a stationary state with entropy production, i.e. in a steady state. It is made up of a great number of subsystems which are ordered in a defined time hierarchy. The steady state of the system as a whole does not imply that all of the sub-systems are also in a steady state. A large proportion of them, particularly those with a short time constant, are in thermodynamic equilibrium. If the

Fig. 3.8. Characteristic time constants (s – seconds, h – hours, a – years) of various processes in a logarithmic scale, to illustrate the time hierarchy in biological systems

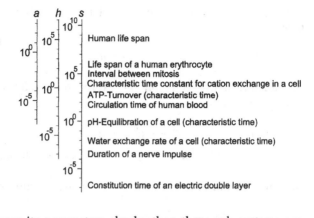

system as a whole changes its parameters slowly, then these sub-systems are capable of following such changes quickly, so that they almost completely adapt within their characteristic time, and thus are always in a stationary state. This is sometimes called a *quasi-stationary*, or *quasi-equilibrium* state.

The following example will illustrate this: the water content of a tissue depends on the ionic composition of its cells. Sodium and potassium ions are being actively transported against passive fluxes, giving rise to a steady state. In this way the active transport, and the corresponding passive fluxes regulate the osmotic properties of the cells. The characteristic time of the water flux is much shorter than that of the cations (see Fig. 3.8). As a result, the water in the interior of the cells is always in osmotic equilibrium with the surrounding medium. In Section 3.2.3 we will discuss this example in detail in context with the Donnan-osmotic quasi-equilibrium.

3.1.5
Thermodynamic Basis of Biochemical Reactions

In the previous formulation of the Gibbs equation [Eqs. (3.1.17), (3.1.27)–(3.1.29)], the chemical reaction was not explicitly enclosed. In fact, a chemical reaction can be analyzed by referring to the chemical potential of its components. The reaction:

$$v_a A + v_b B \rightarrow v_c C + v_d D \tag{3.1.65}$$

can be considered as a replacement of the substances A and B, by the substances C and D.

Considering this process for isobaric ($dp = 0$) and isothermic ($dT = 0$) conditions, where only concentration changes take place ($dn_i \neq 0$, $dq = 0$, $dl = 0$), the equation for Gibbs free energy [Eq. (3.1.29)] reduces to:

$$dG = \sum_{i=1}^{m} \mu_i \, dn_i. \tag{3.1.66}$$

The change in the Gibbs free energy ($\Delta_R G$) according to the definition (3.1.31) then gives:

$$\Delta_R G = v_c \mu_c + v_d \mu_d - v_a \mu_a - v_b \mu_b. \tag{3.1.67}$$

Introducing in this equation the expression for the chemical potential [Eq. (3.1.33)], it follows:

$$\begin{aligned} \Delta_R G = {} & v_c \mu_c^0 + v_d \mu_d^0 - v_a \mu_a^0 - v_b \mu_b^0 \\ & + RT(v_c \ln a_c + v_d \ln a_d - v_a \ln a_a - v_b \ln a_b). \end{aligned} \tag{3.1.68}$$

Now, we can introduce a molar free standard reaction energy:

$$\Delta_R G^0 = v_c \mu_c^0 + v_d \mu_d^0 - v_a \mu_a^0 - v_b \mu_b^0. \tag{3.1.69}$$

Introducing this in Eq. (3.1.68), and combining the logarithmic terms, we obtain *Van't Hoffs equation*:

$$\Delta_R G = \Delta_R G^0 + RT \ln \frac{a_c^{v_c} a_d^{v_d}}{a_a^{v_a} a_b^{v_b}}. \tag{3.1.70}$$

In the case of thermodynamic equilibrium: $\Delta_R G = 0$. If the symbol a_i^0 stands for activities of the system in equilibrium, we can introduce the equilibrium constant (K_p) for isobaric processes as follows:

$$K_p \equiv \frac{a_c^{0 v_c} a_d^{0 v_d}}{a_a^{0 v_a} a_b^{0 v_b}} = e^{-\frac{\Delta_R G^0}{RT}}. \tag{3.1.71}$$

Introducing this expression into Eq. (3.1.70) it gives:

$$\Delta_R G^0 = -RT \ln K_p. \tag{3.1.72}$$

The molar free standard Gibbs energy of a chemical reaction ($\Delta_R G^0$) can be calculated from the standard energies of formation ($\Delta_F G^0$) obtained from tables, using Eq. (3.1.31). Here, we use the property of Gibbs energy as a state function, as discussed in the introduction (Sect. 3.1.1), whereas its value is independent of the way in which the state was achieved.

If a reaction is not in equilibrium, then the direction it will take can be determined by calculating $\Delta_R G$ [Eq. (3.1.70)] from the given activities of the components. Spontaneously, the reaction can run in the direction which is indicated in Eq. (3.1.65) only in the case where: $\Delta_R G < 0$, giving a reduction of the free energy during this process.

So, we have considered chemical reactions simply as a changes in concentrations of their compounds, not differentiating whether the substance was introduced to the system by a transport process, or by a chemical reaction.

This, in fact, is correct and in most cases useful. In order to analyze systems in which chemical reactions take place in addition to real transport, this approach can lead to misunderstandings. In these cases it is useful to introduce a special term into the Gibbs equation, called the *degree of advancement* of a chemical reaction ($d\xi$):

$$d\xi = \frac{1}{v_i} dn_i. \tag{3.1.73}$$

A positive value of $d\xi$ represents a particular step in the reaction from left to right, whereas $dn_i > 0$ means an increase, and $dn_i < 0$ a decrease of the molar amount of the substance. In order to keep to these sign definitions, the stoichiometric coefficients (v_i) of the initial substrates must become negative, and those of the end products, positive.

Taking for example the reaction:

$$2\,H_2 + O_2 \Rightarrow 2\,H_2O,$$

the following applies:

$$d\xi = -\tfrac{1}{2}dn_{H_2} = -dn_{O_2} = \tfrac{1}{2}dn_{H_2O}. \tag{3.1.74}$$

Now, a work coefficient according to our discussion in Section 3.1.2, the *chemical affinity* (A) is introduced:

$$A = -\sum_{i=1}^{m} v_i\mu_i, \tag{3.1.75}$$

where $i = 1 \ldots m$ mean the components of the given reaction. Using Eq. (3.1.71), the following relation is obtained:

$$\sum_{i=1}^{m} \mu_i\, dn_i = -A\, d\xi. \tag{3.1.76}$$

Both expressions can be present in the Gibbs equation simultaneously if transport processes and chemical reactions are considered together. This may be important when calculating processes of active ion transport through membranes.

Affinity, from its definition, proves to be an energy difference and, as such, the driving force of a chemical reaction. Consequently, it can be directly inserted into the flux matrix [Eq. (3.1.58)] instead of a generalized force (**X**).

In contrast to $\Delta_R G$, which determines the direction of a chemical reaction, the molar reaction enthalpy $\Delta_R H$ indicates the heat of a reaction. This means:

$$\Delta_R H > 0 \quad \text{– endothermic reaction,}$$
$$\Delta_R H < 0 \quad \text{– exothermic reaction.}$$

According to the definition (2.1.23), for isothermal systems:

$$\Delta_R G = \Delta_R H - T\Delta_R S. \tag{3.1.77}$$

We already pointed out that if a reaction proceeds spontaneously then $\Delta_R G$ must apply. According to Eq. (3.1.77), in the case of an endothermic reaction this will only occur when:

$$\Delta_R H < T\Delta_R S. \tag{3.1.78}$$

The direction of such a reaction is therefore determined by the rise in entropy of the system. These sorts of processes, therefore, are called *entropy driven reactions*. The classical example of such a reaction is the melting of ice. In Section 2.4.2 we discussed various biomolecular reactions of this type in the context of water structure.

Equation (3.1.77) at the same time enables the determination of the temperature at which the two phases of a system would be in equilibrium. This would be the melting point (T_M) of ice, or the temperature at which collective phase transitions take place in macromolecular systems.

If, at equilibrium $\Delta_R G = 0$, then it follows from Eq. (3.1.77):

$$T_M = \frac{\Delta_R H}{\Delta_R S}. \tag{3.1.79}$$

In this way, the measurement of the thermal characteristics of a phase change enables the concomitant change in entropy to be determined.

Further reading: There is a very extensive literature on the thermodynamics of irreversible processes. For basic problems and approaches in calculating fluxes, see: Katchalsky and Curran (1965); Prigogine (1967); Schnakenberg (1981). For the problems of thermodynamics and evolution see: Glansdorf and Prigogine (1971); Haken (1983); Lamprecht and Zotin 1985. For entropy driven processes: Lauffer (1975)

3.2
The Aqueous and Ionic Equilibrium of the Living Cell

From the point of view of thermodynamics, a living cell can be considered as a system in non-equilibrium. This non-equilibrium state is maintained by permanent processes of energy transformation. In spite of the state of the whole system, however, some of its subsystems nevertheless may be in thermodynamic equilibrium. Such equilibria, or quasi-equilibria have already been discussed in Section 3.1.4, in connection with the time hierarchy of biological systems. In this section we will direct our attention to these equilibrium states, especially equilibrium distributions of charged and uncharged components.

3.2.1
Osmotic Pressure

The concept of osmotic pressure is very important for understanding a number of cell physiological processes. Unfortunately, osmotic pressure is often confused with hydrostatic pressure, especially with the turgor pressure of plants. In this section it will be shown that osmotic pressure is just a property of a solution or a suspension and that although it can be the origin of the hydrostatic pressure in the cell, it is not at all identical to it.

The effect of osmotic pressure can be demonstrated best by a *Pfeffer cell* (see Fig. 3.20). This consists of a glass bell with a vertical tube or a manometer at the top. The mouth of the bell is closed by a semipermeable membrane. The bell is filled with a solution and submerged in a vessel with pure water. The membrane allows the water, but not the molecules of the solute, to pass through. In his experiments in 1877, the botanist W. Pfeffer used a sheet of porous pottery to close the bell. The pores of this "membrane" were covered by a precipitated layer of cupric(II)-hexacyanoferrate(II). In this experiment the water penetrates the membrane, passes into the glass bell, and increases the internal hydrostatic pressure. Eventually an equilibrium will be reached when the internal pressure balances the osmotic force which drives the water into the Pfeffer cell. If the membrane is truly semipermeable, and only if pure solvent is outside the glass bell, then, and only then, does the hydrostatic pressure difference between the internal and the external phase in this equilibrium equal the osmotic pressure of the solution inside the Pfeffer cell.

In order to analyze this situation let us assume two phases (*I* and *II*) separated from each other by a membrane (see Fig. 3.9). Let the membrane be semipermeable and thus allow the solvent (w), but not the solute (s), to pass through. Because of solvent flux through the membrane, the volumes of both phases and their concentrations will change. This will also alter the chemical potentials (μ_i) of the components (i).

The concentration dependence of the chemical potential [see Eq. (3.1.33)], using the mole fraction (x_i) as a measure of concentration is:

$$\mu_i = \mu_{ix}^0 + RT \ln x_i. \tag{3.2.1}$$

To achieve thermodynamic equilibrium under isothermal conditions, the chemical potential of the exchangeable components of both phases *I* and *II*

Fig. 3.9. Derivation of osmotic pressure: phases *I* and *II* are separated from each other by a semipermeable membrane. Only the solvent (w), but not the solute (s), can penetrate this membrane. The hydrostatic pressure of both sides of the membrane is different

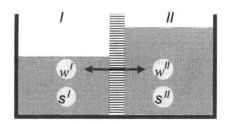

must be equal. In the example considered here, this will only apply to the solvent (w). This requires

$$\mu_w^I \overset{!}{=} \mu_w^{II},\tag{3.2.2}$$

which means that

$$\mu_{wx}^{0I} + RT \ln x_w^I = \mu_{wx}^{0II} + RT \ln x_w^{II},\tag{3.2.3}$$

or

$$RT \ln \frac{x_w^I}{x_w^{II}} = \mu_{wx}^{0II} - \mu_{wx}^{0I}.\tag{3.2.4}$$

Now, let us first direct our attention to the difference in the standard chemical potentials on the right-hand side of Eq. (3.2.4). Because of the pressure difference induced by the flow of the solvent, $\mu_{wx}^{0I} \neq \mu_{wx}^{0II}$. The pressure dependence of the standard chemical potentials therefore must be studied in more detail.

The following definition of the standard chemical potential can be derived from Eq. (3.1.32):

$$\mu_i^0 = \left(\frac{\partial G^0}{\partial n_i}\right)_{T,p,n_j}.\tag{3.2.5}$$

The pressure dependence of μ_i^0 can therefore be expressed as

$$\frac{\partial \mu_i^0}{\partial p} = \frac{\partial \left(\frac{\partial G^0}{\partial n_i}\right)}{\partial p} = \frac{\partial \left(\frac{\partial G^0}{\partial p}\right)}{\partial n_i}.\tag{3.2.6}$$

The way in which the sequence of the derivation steps in a total differential can be changed was discussed in Section 3.1.1 [see also Eq. (3.1.6)].

From the equation of Gibbs free energy [Eq. (3.1.29)] we obtain:

$$dG^0 = V\,dp \quad \text{for: } dT = 0, \quad dl = 0, \quad dq = 0, \quad dn = 0,\tag{3.2.7}$$

and

$$\frac{\partial G_i^0}{\partial p} = V_i.\tag{3.2.8}$$

If this is substituted into Eq. (3.2.6), then

$$\frac{\partial \mu_i^0}{\partial p} = \frac{\partial V_i}{\partial n_i} = \bar{V}_i.\tag{3.2.9}$$

We are already acquainted with the variable $\bar{V}$ which is the partial molar volume of the substance i [Eq. (3.1.8)]. Now, the differential $d\mu_i^0$ can be calculated. This is a small change in μ_i^0 that occurs when there is a small change in the pressure dp. Following the general rule (3.1.1), we obtain:

$$d\mu_i^0 = \left(\frac{\partial \mu_i^0}{\partial p}\right) dp = \bar{V}_i\, dp. \tag{3.2.10}$$

To obtain an expression for the difference in the standard potentials due to alterations in pressure, the equation must be integrated between the corresponding limits:

$$\int_{\mu_i^{0I}}^{\mu_i^{0II}} d\mu_i^0 = \int_{p^I}^{p^{II}} \bar{V}_i\, dp, \tag{3.2.11}$$

$$\mu_i^{0II} - \mu_i^{0I} = \bar{V}_i\left(p^{II} - p^I\right). \tag{3.2.12}$$

(This manner of integration only applies when $\bar{V} = const$, that is, when the partial molar volume is independent of the pressure. This is only the case for ideal solutions.)

Now, it is possible to substitute Eq. (3.2.12), related to the solvent (w) into Eq. (3.2.4):

$$p^{II} - p^I = \frac{RT}{\bar{V}_w} \ln \frac{x_w^I}{x_w^{II}}. \tag{3.2.13}$$

This equation indicates the hydrostatic pressure difference ($p^{II} - p^I$) induced in our system at thermodynamic equilibrium for different values of the mole fractions in the two phases $x_w^I \neq x_w^{II}$. If phase I contains only pure solvent ($x_w^I = 1$), this pressure difference is called osmotic pressure (π), therefore:

$$\pi = \frac{RT}{\bar{V}_w} \ln \frac{1}{x_w^{II}} = -\frac{RT}{\bar{V}_w} \ln x_w^{II}. \tag{3.2.14}$$

This equation and its derivation allow us to describe osmotic pressure as a parameter reflecting a property of a solution. Under isothermal conditions the osmotic pressure of a solution is equal to the hydrostatic pressure, which is required to alter the chemical potential of the pure solvent in such a way that it will be equal to the chemical potential of the solvent in this solution.

Equation (3.2.14) is a precise expression for calculating the osmotic pressure of a solution, provided the mole fraction activity of the solvent x_w^{II} is used to express the concentration of the solvent (w) in the solution. Some simplifications can be made for dilute solutions. First, the mole fraction *concentration* can be used instead of the mole fraction *activity*. According to Eq. (3.1.35), this means:

$$x_w = \frac{n_w}{n_s + n_w}. \tag{3.2.15}$$

Thus, if the sum of the mole fractions of all components of a solution equals 1 [Eq. (3.1.36)], then

$$x_w = 1 - x_s. \tag{3.2.16}$$

The number of molecules of solvent in diluted solutions is much larger than the number of molecules of the solute $(n_w \gg n_s)$. A 0.1 molar solution of the substance (s) in water (w), for example, contains $n_s = 0.1$ mol of solute, but $n_w = 55.6$ mol of water (n_w means moles water per 1000 g solution, therefore, $1000/18 = 55.6$). In this case, combined with Eqs. (3.2.16) and (3.2.15) becomes simply

$$x_w = 1 - \frac{n_s}{n_s + n_w} \approx 1 - \frac{n_s}{n_w}. \tag{3.2.17}$$

In addition, the following holds for dilute solutions:

$$\bar{V}_w = \frac{\partial V_w}{\partial n_w} \approx \frac{V_w}{n_w}. \tag{3.2.18}$$

Thus, $n_w = V_w/\bar{V}_w$. One can also introduce the molar concentration of the solute (s), using $n_s/V = c_s$, and the total volume of the solution $(V \approx V_w)$. Substituting this in Eq. (3.2.17), we obtain:

$$x_w = 1 - \frac{n_s \bar{V}_w}{V} = 1 - c_s \bar{V}_w. \tag{3.2.19}$$

Let us substitute this expression in Eq. (3.2.14):

$$\pi = -\frac{RT}{\bar{V}_w} \ln\left(1 - c_s^{II} \bar{V}_w\right). \tag{3.2.20}$$

Now, we use the following rule to expand logarithmic expressions in series. For any number $|q| < 1$, it holds:

$$\ln(1 - q) = -q - \frac{q^2}{2} - \frac{q^3}{3} - \frac{q^4}{4} - \cdots. \tag{3.2.21}$$

For q we use the expression $c_s^{II} \bar{V}_w$, which is much smaller than 1. In this case we are justified in retaining only the first term of this series. This gives:

$$\pi = -\frac{RT}{\bar{V}_w}\left(-c_s^{II} \bar{V}_w\right), \tag{3.2.22}$$

and

$$\pi = RTc_s^{II}. \tag{3.2.23}$$

This is the *Van't Hoff equation* for osmotic pressure.

It is interesting to note that, in 1877, Van't Hoff derived this relation not in the way shown here, but by considering the analogy with the state equation for an ideal gas. He assumed that 1 mol of an ideal gas, when confined to a volume of 1 liter, would exert a pressure of 2.27 MPa on the walls of the vessel. He reasoned that the molecules of a solute, when dissolved at a concentration of 1 mol/l should behave in the same way as the particles of the gas. This pressure he called "osmotic".

The equation of state for an ideal gas is:

$$p = \frac{n}{V}RT. \tag{3.2.24}$$

This can be transformed into:

$$\pi = \frac{n}{V}RT = cRT. \tag{3.2.25}$$

Both the thermodynamic derivation as well as the analogy used by Van't Hoff show that Eq. (3.2.23) is only applicable for ideal solutions, or with some approximation for solutions which are very diluted. This restriction can be overcome using a correction factor. This is the so-called *osmotic coefficient* (g):

$$\pi = gcRT. \tag{3.2.26}$$

This factor must not be confused with the activity coefficient (f), which was introduced in Section 3.1.2 [Eq. (3.1.34)]. This will be easily understood if f is directly introduced into the thermodynamic derivation of the Van't Hoff equation [say, in Eq. (3.2.3)]. The relation between f and g is somewhat complicated and will not be discussed here.

The osmotic coefficients for NaCl, KCl and sucrose are given in Fig. 3.10 as functions of concentration. With the increase in concentration, the osmotic coefficient (g) deviates more and more from the value 1. This deviation, however, is smaller as for the activity coefficient (f) (see Fig. 3.1).

In addition to the non-ideal behavior of a solution, it must be noted that the osmotic pressure exerted by dissociable substances results from the sum of the osmotic pressures of the ions formed by this dissociation. Thus, if a mono-monovalent salt (such as NaCl) is assumed to be completely dissociated into Na^+ and Cl^- then the osmotically active concentration is twice as great as the salt concentration. Therefore one must distinguish between the *osmolarity* and the *molarity* of a solution. The osmolarity of a 0.1 M solution of NaCl (at 25 °C) is therefore: $2 \times 0.1 \times g = 2 \times 0.1 \times 0.932 = 0.1864$ osmolar. The osmolarity can be pH-dependent for polyelectrolyte solutions because it will change according to the degree of dissociation.

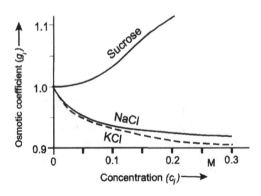

Fig. 3.10. The osmotic coefficient (g) as function of the concentration (c_s) in aqueous solutions

The osmotic pressure of a solution is usually measured in an indirect way. Because the vapor pressure and the freezing point of a solution are basically determined by the same set of laws, instruments which can measure these properties are also convenient for the determination of osmotic pressure. True solutions are most simply and most accurately measured cryoscopically. The use of a vapor pressure osmometer is better for solutions of larger molecules or for polydisperse suspensions.

Osmotic pressure is not only a property of true solutions, but also of colloidal solutions and to some extent also of suspensions. In this context it is referred to as *colloid-osmotic pressure*. In gels, osmotic pressure acts to oppose the tendency of the molecular framework to contract (Parsegian, in Leopold 1986). In this context it should be mentioned that even the water in membrane pores is in osmotic equilibrium with the bulk water. Therefore the diameter of a pore becomes controlled osmotically (Zimmerberg and Parsegian 1986). In view of the fixed charges of all biological surfaces (see Sect. 2.5.4), however, the osmotic behavior of these microphases must be considered as Donnan-osmotic systems, which we will consider later (Sect. 3.2.3).

In cell physiological considerations the hydrostatic pressure, which arises from differences in osmotic pressure, is of special interest. Only in the case of the thermodynamic equilibrium of water in this system, and only if the membrane is really semipermeable in relation to all components of the solution, does the difference in the osmotic pressure equal the difference in the generated hydrostatic pressure.

For solutions with several components and with various permeability properties, one can use the following relation between osmotic ($\Delta\pi$) and hydrostatic (Δp) differences:

$$\Delta p = \sum_{i=1}^{n} \sigma_i \Delta\pi_i. \tag{3.2.27}$$

This equation takes into account that in the system n substances, each with an osmotic difference $\Delta\pi_i = \pi_{i(\text{internal})} - \pi_{i(\text{external})}$, determine the osmotic conditions. Their effectiveness, with regard to the development of hydrostatic pressure, depends on the value of a factor (σ_i) which is known as *Staverman's reflection coefficient*. In contrast to the 'classical' approach, this model takes into account that the membrane is not semipermeable, but permselective. This means that all components of the solution can more or less penetrate the membrane. We will analyze this situation in detail later using the approaches of nonequilibrium thermodynamics (Sect. 3.3.1). We will see that from consideration of the corresponding flux matrix [Eq. (3.3.23)] the following relation can be derived:

$$\sigma_s = \frac{v_w - v_s}{v_w}. \tag{3.2.28}$$

Using the indices of the Van't Hoff's equation, v_w and v_s represent the rate of movement of the solvent (water) and the solute in the membrane. In the case of $v_s \to 0$, the reflection coefficient becomes $\sigma_s \to 1$. This is the 'classical' situation of semipermeability, and therefore $\Delta p \to \Delta \pi$. However, when $v_s \to v_w$, then $\sigma_i \to 0$ and hence $\Delta p \to 0$. This occurs in the case of a membrane which allows the osmotically active substance as well as the solvent to pass through. In this case, no hydrostatic pressure can develop even at initial differences in osmotic pressure. In such a system a thermodynamic equilibrium would finally lead to a homogeneous distribution of the substance i (see Fig. 3.20).

In general, the reflection coefficient for disaccharides, such as sucrose, and for larger molecules equals nearly 1. Smaller molecules, especially those which can directly penetrate the lipid layer of a biological membrane, show lower values (see Table 3.1).

In plant cells, osmotic differences generate the so-called *turgor pressure*, which can be as high as several hundred kPa. This intracellular pressure plays a large role in the mechanical stabilization of plants. It forces the cell membrane of

Table 3.1. Typical values of reflection coefficients of non-electrolytes for human erythrocytes (Giebisch et al. 1979) and for *Nitella flexilis* (Zimmermann and Steudle 1978). Values in parentheses: Levitt and Mlekoday 1983

	Erythrocytes	*Nitella*
Urea	0.79 (0.95)	0.91
Thiourea	0.91	
Ethylene glycol	0.86 (1.0)	0.94
Glycerol	0.88	0.80
Acetamide	0.80	0.91
Propionamide	0.84	
Malonamide	1.00	
Sucrose		0.97
Glucose		0.96
Methanol		0.31
Ethanol		0.34
Isopropanol		0.35
n-Propanol		0.17

the plant cell against the mechanically stabilizing cell wall, which itself is freely permeable for ions and small nonelectrolytes. The turgor pressure can be directly measured by means of special pressure probes that can be introduced into the cell (Zimmermann 1989).

In contrast to plant cells, animal cells (and also plant protoplasts) have no ability to resist an internal hydrostatic pressure. Within certain limits, osmotic changes in the environment can be compensated by alterations in cellular volume, thus maintaining a constant membrane area. Cell swelling, for example, can cause the forming of a sphere or smoothing of the cell surface. As described in Section 2.5.3 (Fig. 2.41), a mechanical expansion of the cell membrane, without introduction of additional molecules, however, is nearly impossible. An osmotic difference of only 1 mosmol can generate a maximum internal pressure of 2.27 kPa. Measurements made on erythrocyte membranes have shown that their tension can, at the most, only withstand an internal pressure of 0.1 kPa. For this reason, these cells have complicated mechanisms for osmoregulation, including a number of coupled receptors and transport systems.

Formally, the swelling of a cell with a constant content of osmotically active substances can be expressed by the following relation:

$$\pi(V - V_i) = \pi'(V' - V_i). \tag{3.2.29}$$

When there is a change in the osmotic pressure of the external medium from π' to π then the cell volume changes from V' to V. The parameter V_i represents an apparent, osmotically inert i.e. constant volume. Solving Eq. (3.2.29) for V, we obtain:

$$V = \frac{1}{\pi}\pi'(V' - V_i) + V_i. \tag{3.2.30}$$

This is a phenomenological expression of the cellular volume (V) as a function of external osmotic pressure (π). If V is plotted against $1/\pi$, a straight line is obtained with a slope equal to $\pi'(V' - V_i)$, intersecting the ordinate at the point V_i. This osmotic dependence of the volume has been confirmed, with reasonable accuracy, for a number of cells. In Fig. 3.11 it can be seen that for human erythrocytes the osmotically inert volume V_i amounts to 48% of the total volume of the cells under isotonic conditions. This value is too high, because these cells contain 71% water. In fact, this apparent volume is the result of the interplay between different processes. Water loss causes an increase in internal concentrations of all components of the cell. This leads automatically to changes in the charge density of the proteins, and thus to changes in ionic conditions and pH, having a retroactive effect on the osmotic pressure. Consequently, it is not possible to explain the volume regulation of a cell without considering the complete ionic metabolism. In addition, it must be taken into account that the osmotic coefficient of the proteins itself is to a great extent concentration dependent (see Fig. 3.12). We will come back to this problem in connection with the Donnan equilibrium (Sect. 3.2.3, Fig. 3.17).

Fig. 3.11. The relative volume (V_{rel}) of human erythrocytes in a NaCl-phosphate solution in relation to the isotonic condition ($\pi_{isoton} = 0.295$ osmol, $V_{isoton} = 1$), plotted as a function of $1/\pi$ (after Cook 1967)

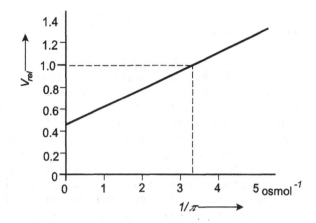

Fig. 3.12. The osmotic coefficient of hemoglobin (g_{Hb}). The *dashed lines* indicate the region of the in vivo concentration. The curve reflects measurements from sheep and cow hemoglobin of Adair as fitted by Freedman and Hoffman (1979)

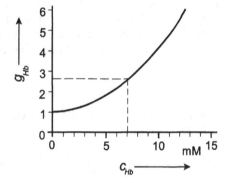

The swelling behavior, as described by Eq. (3.2.29), is also intrinsically demonstrable for mitochondria and chloroplasts. Certainly, the reflection coefficients of their membranes are very different from that of the cell membrane. Additionally, even with minor fluctuations of the transport processes of these extremely small organelles with a relatively large surface area, the internal osmolarity may alter quickly.

Further reading: The philosophy of osmosis: Ogston 1966. Swelling of erythrocytes: Freedman and Hoffman 1979; Glaser and Donath 1984; Solomon et al. 1986. Osmoregulation of animal cells: Hoffmann and Simonsen 1989. Osmotic pressure in plants: Zimmermann 1978; Zimmermann and Steudle 1978

3.2.2
Electrochemical Equilibrium – The Nernst Equation

We will now consider a system consisting of two phases, each of which contains a solution with a salt AB, but with different concentrations in the phases I and II

(Fig. 3.13). Let the salt be completely dissociated into its ions A and B. The membrane is semipermeable in that it allows the ion A, but not the ion B, to pass through.

Before analyzing conditions of thermodynamic equilibrium of this system, we will consider the situation qualitatively. Let us suppose that there is an osmotic equilibration between both phases, compensated by electroneutral components. Thus, alteration of concentrations, induced by volume changes, can be neglected. In this system the electrostatic equilibrium will be disturbed because only the ion A, driven by its concentration gradient, but not its counterpart, the ion B can penetrate the membrane. This leads to an increase in the electrical potential difference across the membrane. Eventually, a strong electric field hinders a further diffusion of ion A. Ion A, consequently, will be subject to two opposing forces: on the one hand, the driving force, induced by the concentration gradient, i.e. the gradient of its chemical potential and, on the other hand, an opposing electrostatic force which only arises as a result of its own diffusion. The following equilibrium will be established: a few ions cross the membrane, inducing an electric field which stops further diffusion.

The basic condition for calculating this equilibrium is the equality of the electrochemical potentials of ion A between phases I and II:

$$\tilde{\mu}_A^I \overset{!}{=} \tilde{\mu}_A^{II}.$$

Substituting the expressions for the electrochemical potentials according to Eq. (3.1.41), one obtains

$$\mu_A^{0I} + RT \ln a_A^I + z_A F\psi^I = \mu_A^{0II} + RT \ln a_A^{II} + z_A F\psi^{II}. \tag{3.2.31}$$

(We will assume isothermal conditions, i.e. $T^I = T^{II} = T$.)

In contrast to the derivation of the equation for the osmotic pressure, in the present case, the standard potentials of these components of phases I and II are equal, because there is no pressure difference ($\mu_A^{0I} = \mu_A^{0II}$).

Taking this into account, and re-arranging Eq. (3.2.31), it gives

$$z_A F(\psi^I - \psi^{II}) = RT(\ln a_A^{II} - \ln a_A^I), \tag{3.2.32}$$

and therefore

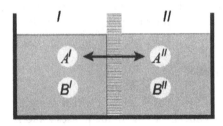

Fig. 3.13. Diagram illustrating the derivation of the Nernst equation. Phases I and II are separated from each other by a membrane which is permeable only for the ion A, but not B, of the salt AB

$$\Delta\psi \equiv \left(\psi^I - \psi^{II}\right) = \frac{RT}{z_A F} \ln \frac{a_A^{II}}{a_A^I}.$$ (3.2.33)

This is the *Nernst-Equation*. It gives the electrical potential difference ($\Delta\psi$) across the membrane as a function of the chemical activities of the permeating ion in both phases at thermodynamic equilibrium.

Equation (3.2.33) can be re-arranged to show the difference of ion activities (a^I and a^{II}) which builds up in two phases, with a given potential difference ($\Delta\psi$) in between:

$$a_A^I = a_A^{II} e^{-\frac{z_A F \Delta\psi}{RT}}.$$ (3.2.34)

Such a relation has already been derived and employed using the Boltzmann equation (Sect. 2.3.15), and applying it to calculate local ion concentrations near charged particles [Eq. (2.4.10)], or in electric double layers [Eq. (2.5.6)]. In these cases, however, the concentrations (c_i) were used instead of the chemical activities (a_i). This is allowed only for ideal, i.e. diluted, solutions, or when the activity coefficients (f_i) are equal in both phases.

The Nernst equation, therefore, permits the calculation on the one hand, of the distribution of ions as a function of the electrical potential [Eq. (3.2.34)] and, on the other hand, the electrical potential, which is induced by an unequal distribution of ions [Eq. (3.2.33)]. For both cases, however, thermodynamic equilibrium is required!

The separation of the two phases of the system by a semipermeable membrane, as discussed here, reflects just one special example. In the case of ion distribution in electric double layers, or ionic clouds, as described in Sections 2.5.4 and 2.5.5, the electric potential gradient is predicted by the fixed charges, and the distributions of the ions are not limited by a membrane. Here, both the anions and the cations in the case of equilibrium, are distributed according to this equation.

All equations derived in the present section are applicable only for thermodynamic equilibria. This means that the Nernst equation cannot be used to calculate the membrane potential of a living cell. Actually, the membrane potential of a living cell is either a diffusion potential (liquid junction potential), or it is generated by electrogenic ion pumps (see Sects. 3.4.2 and 3.4.3). On the other hand, despite the nonequilibrium distribution of the ions in the cell in general, it is quite possible that some types of ions may be distributed passively and then they are actually in equilibrium.

An example of this is the chloride concentration of most animal cells. Because its membrane permeability is rather fast, and because no chloride pumps in the membrane exist, its distribution is mostly passive, and predicted by the existing transmembrane potential. The Nernst equation, therefore, allows the calculation of the internal chloride concentration, if the external chloride concentration and the transmembrane potential are known. On the other hand, knowing, for example, the distribution of chloride inside and outside the cells, one can calculate the transmembrane potential of the cell. In this context, however, it is important to stress that in this case the chloride distribution is just an *indicator* of the membrane potential, but not the *reason* for it!

This consideration allows the establishment of a method for measuring the transmembrane potential of cells without microelectrodes. The chloride distribution can be easily determined using the radioisotope ^{36}Cl. Sometimes one can use other small charged organic molecules, which penetrate the membrane quickly, and are labeled by ^3H or ^{14}C. Knowing the distribution of these ions, the membrane potential ($\Delta\psi$) can be calculated according to Eq. (3.2.33).

Furthermore, the Nernst equation allows the calculation of electrode potentials. If a metal is dipped into an electrolyte solution, then cations are detached from the metal lattice and an electrochemical equilibrium is obtained. In general, this is the same situation as discussed above. The resulting electrical potential difference between the metal and the solution is called *electrode potential*. If the cation from the metal of the electrode forms a poorly soluble salt with an anion of the solution, and if the concentration of this anion is much greater than that of the cation, then the electrode potential, linked through the soluble product of the salt, is directly determined by the chemical activity of this anion. In this way, for example, a silver electrode covered by AgCl can be used to measure the Cl^- activity of a solution (see Fig. 3.14). The voltage of this electrode with respect to a reference electrode, according to the Nernst equation [Eq. (3.2.33)], is proportional to the logarithm of the Cl^- activity in the solution. It is important to emphasize that these kinds of electrochemical methods allow the measurement of ion activities (a_i), in contrast to most chemical methods, indicating its concentrations (c_i).

The electrochemical measurement of ion activity and of many other chemical substances has become an important and universal technique. Using special semipermeable membranes, or water impermeable layers containing special ionophores, electrodes can be made which are highly selective in measuring the activity of special chemical components. In this case, these membranes or microphases separate a reference solution of known composition from the test solution. The potential difference between the reference solution and the test solution is measured as the voltage between two reference electrodes dipped in both phases.

A typical example of this kind of measurement is the usual pH electrode (see Fig. 3.14). In this case, a thin membrane made of special glass allows the protons to equilibrate between both phases. If, as a reference phase, the pH electrode is filled by a buffer holding the pH constant, then the voltage between the pH electrode and the reference electrode indicates the pH of the test solution. Usually, combinations between pH and reference electrodes are used, consisting of one single glass body.

For electrochemical methods of analysis, a pair of electrodes is always necessary. The electromotive force will be measured between a selective measuring electrode and a so-called reference electrode. A *reference electrode* is an electrode that exhibits a constant electrode potential which is independent of the composition of the solution into which it is immersed. If a pair of identical reference electrodes is used, then the electrode potentials will mutually oppose

Fig. 3.14. The construction of an Ag/AgCl/KCl-reference electrode, as well as its application for electrochemical measurements of Cl^- activity and pH, and as a reference electrode in electrophysiology

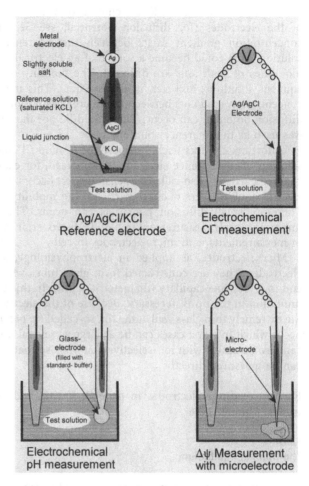

each other and therefore the direct potential difference, i.e. the actual electromotive force between the two phases can be measured. If two different reference electrodes are used, a constant $\Delta\psi$ is superimposed, which can be subtracted from the measured value.

To make a reference electrode, a metal electrode is covered with a slightly soluble salt of this metal. This is immersed into a reference solution of constant composition, which is connected to the test solution by means of a pathway, where little, if any, convection can take place (liquid junction). This liquid junction is formed in different ways. It can be a tube, filled by an agar gel, however, in industrially produced electrodes, it is mostly just an opening in the glass wall, covered by a small filter made from sintered glass. Sometimes a ground glass stopper is used with a liquid film in between.

The most common types of reference electrodes are Ag/AgCl/KCl electrodes (see Fig. 3.14). In this case, a saturated solution containing 4.6 M KCl as a reference solution is used. KCl is chosen to avoid diffusion potentials at the tip of the electrodes (for diffusion potential, see Sect. 3.3.3). In any case, a concentration gradient exists between the reference solution and the test solution. This could become a source of an electrical potential difference if the mobilities of both ions are different. The mobility of the K^+ and Cl^- ions in aqueous solutions, however, is almost equal. This means that even for strong concentration gradients between the reference solution and the test solution, no diffusion potential can occur. This would be not the case if, for example, NaCl was used as the reference solution.

Unfortunately, this useful property of K^+ and Cl^- ions is valid only in pure solutions. If a reference electrode is immersed, for example, in suspensions of charged particles, a so-called *suspension effect* occurs. Under the influence of the electric double layers of these particles, the mobility of the ions may change. Therefore small diffusion potentials can occur. This effect, for example, is important in pH measurement in blood. Moreover, it can be the source of errors in measurements with microelectrodes in cells.

Microelectrodes, as applied in electrophysiology, are usually Ag/AgCl/KCl electrodes. They are constructed from glass tubes which are drawn out at one end to a very fine capillary (diameter < 1 μm). In this case, no further diffusion limitation in the tip is necessary. Because of the electric charge of the glass and the extremely thin glass wall at the tip, so-called *tip potentials* of microelectrodes occur which in worst cases can be as large as several millivolts. Microelectrodes can also be sealed with ion-selective materials so that intracellular ionic activity can be measured directly.

Further reading: Electrodes in general: Varma and Selman 1991; microelectrodes: Amman 1986

3.2.3
The Donnan Equilibrium

The Donnan state represents an equilibrium between two phases, containing not only small anions (A) and cations (C), both of which can penetrate the membrane, but also charged molecules or particles (M) for which the membrane is impermeable (see Fig. 3.15). This type of equilibrium was investigated by F. G. Donnan in 1911.

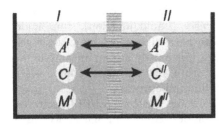

Fig. 3.15. The derivation of the Donnan equilibrium. Phases *I* and *II* are separated from each other by a membrane that is permeable for the anions (A) and the cations (C), but not for the charged molecules M

The Donnan equilibrium, like the Nernst equilibrium which was explained in the previous section, is important for calculating the distribution of ions in the living cell. It requires that these ions are distributed according to the thermodynamic equilibrium. In each case it must be decided which sorts of charged components are exchangeable (like A and C), and which of them are nonexchangeable (like M).

If the cell membrane is opened for ions by an ionophore, by a toxin, or by any other influence, then only organic molecules, like proteins, are in the class of nonexchangeable charged components (M). A living cell will also shift slowly toward a Donnan equilibrium, if, as a result of low temperature or a chemical blocker, the ATP-driven ion pumps are inhibited. An example of this is the distribution of ions in erythrocytes in blood stored over a long period of time.

A Donnan equilibrium can be established also as a sort of quasi-equilibrium state (see Sect. 3.1.4). Figure 3.8 demonstrates the time hierarchy of characteristic rate constants of various biological processes. For illustration, let us consider the following situation. In human erythrocytes, for example, the distribution of Cl^- ions between the internal and external media is passive, i.e. not influenced by an active pump. The same holds for the equilibration of the pH. These processes of equilibration are very fast. The whole erythrocyte itself, however, in vivo is not in equilibrium, because sodium and potassium are pumped by Na-K-ATPase. Nevertheless, the distribution of the quickly exchangeable Cl^- ions and the pH can be calculated according to the Donnan equilibrium, if Na^+ and K^+ are considered as quasi nonpermeable charges (M). This state is called a "quasi" equilibrium state, because actually for a longer time interval, the concentration of potassium and sodium may shift. This type of time hierarchy was discussed already in Section 3.1.4 (Fig. 3.8). The equilibrium of Cl^- and the pH, therefore, follows the shifting Na^+ and K^+ concentration and can be considered as constant only for a limited time of observation.

Let us now consider such a system, consisting of permeable ions (index i), and nonpermeable charged components (index m). Let us denote the concentration, or the activity of the components inside the cell, with c^I or a^I, and in the external solution with c^{II}, or a^{II} respectively. The parameter z_m denotes the number and the sign of charges of the nonpermeable molecule.

It must be decided whether an isobar ($\Delta p = 0$) or whether an isochor ($\Delta V = 0$) system will be considered. The first case is suitable for investigating an animal cell, which is able to compensate for changes in the volume by swelling or shrinking. The second situation, to some extent, can be applied to plant cells, with the possibility of increasing the internal pressure considerably. Further, let us assume Staverman coefficients of the nonpermeable components are $\sigma_i = 1$.

In the case of animal cells, which is easier to handle, the Donnan equilibrium is determined by the following conditions:

– All permeable ions (i) are distributed in equilibrium according to Eq. (3.2.34):

$$a_i^I = a_i^{II} e^{-\frac{z_i F \Delta \psi}{RT}}. \tag{3.2.35}$$

- In both phases the sum of charges is zero (electroneutrality condition):

$$\sum z_i c_i + \sum z_m c_m = 0. \tag{3.2.36}$$

- Water between both phases is distributed according to its equilibrium.

$$\Delta \pi = 0 \quad (\text{for: } \Delta p = 0). \tag{3.2.37}$$

These conditions allow us now to create equations for particular cases and to solve them for the corresponding parameters. The conditions (3.2.35) allow the calculation of the relation of the activities of the permeating ions inside and outside the cell. If the activities of univalent cations are denoted by a_{iC} (for $z_{iC} = +1$), and correspondingly for anions a_{iA} (for $z_{iA} = -1$), it follows:

$$\frac{a_A^I}{a_A^{II}} = \frac{a_C^{II}}{a_C^I} = e^{\frac{F\Delta\psi}{RT}} \equiv r. \tag{3.2.38}$$

The parameter r is known as the *Donnan ratio*. According to Eq. (3.2.38), it is related to the Donnan potential in the following way:

$$\Delta\psi = \frac{RT}{F} \ln r. \tag{3.2.39}$$

The Donnan potential ($\Delta\psi$) is substantially determined by the amount of nonpermeable charged components of the cell, as reflected in the condition (3.2.36). Let us now consider the situation with just a single kind of cation (C, with $z_C = +1$), and a single kind of anion (A, with $z_A = -1$), and a singly charged component (M, with z_M) inside the cell. In this case, the equation of electroneutrality of both phases, according Eq. (3.2.36) can be written easily:

$$\begin{cases} c_C^I - c_A^I + z_M^I c_M^I = 0 \\ c_C^{II} - c_A^{II} = 0 \end{cases}. \tag{3.2.40}$$

This system of equations can be solved by reorganizing it and dividing one by the other:

$$\frac{c_A^I}{c_A^{II}} = \frac{c_C^I + z_M^I c_M^I}{c_C^{II}}. \tag{3.2.41}$$

If the activity coefficients of the ions in both phases are equal, one can substitute the Donnan ratio [Eq. (3.2.38)] in this equation:

$$r = \frac{1}{r} + \frac{z_M^I c_M^I}{c_C^{II}}. \tag{3.2.42}$$

Re-arranging this, a simple quadratic equation for r is obtained:

$$r^2 - \frac{z_M^I c_M^I}{c_C^{II}} r - 1 = 0, \tag{3.2.43}$$

which can be solved in the usual way:

$$r = \frac{z_M^I c_M^I}{2 c_C^{II}} \pm \sqrt{\left(\frac{z_M^I c_M^I}{2 c_C^{II}}\right)^2 + 1}. \tag{3.2.44}$$

The dependence of the Donnan ratio (r) on the expression $z_M^I c_M^I / 2 c_C^{II}$ is shown in Fig. 3.16. Considering the corresponding definition (3.2.38), it is clear that the negative values of the Donnan ratio (r) have no real meaning. Therefore, only the positive roots of Eq. (3.2.44) are of interest. For the case $z_M^I c_M^I = 0$ it becomes $r = 1$ and Eq. (3.2.39) gives $\Delta\psi = 0$. If the impermeable components are negatively charged ($z_M^I < 0$), then $r < 1$. This gives a negative membrane potential ($\psi < 0$). On the other hand, the value $z_M^I > 0$ means $r > 1$ and therefore $\Delta\psi > 0$.

A reduction of the ionic strength in the external solution, i.e. a reduction of c_C^{II}, leads to an increase in the absolute value of the ratio $z_M^I c_M^I / 2 c_C^{II}$. If $z_M^I < 0$, this will lead to a shift of $\Delta\psi$ in a negative direction, and vice versa, if $z_M^I > 0$, a reduction of the ionic strength causes an increase in the Donnan potential. The absolute value of the Donnan potential, therefore, increases when the charge concentration ($z_M^I c_M^I$) is increased, as well as when the ionic strength in the external solution (c_C^{II}) is reduced.

Equation (3.2.44) allows us to understand the behavior of a Donnan system in general, but in fact, it reflects an extremely simplified system. It has not been

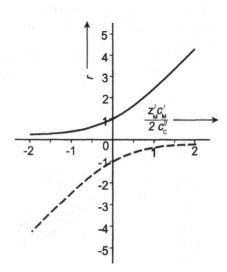

Fig. 3.16. Plot of Eq. (3.2.44). *Solid curve* positive, *dashed curve* negative values of the root expression

taken into account that according to the condition of iso-osmolarity [Eq. (3.2.37)], alterations of the cell volume could occur. This would lead to alterations of all intracellular concentrations.

In addition, as determined by the Donnan potential, a pH equilibrium between the two phases in this system could be obtained. This effect has already been discussed in connection with the local pH values at charged surfaces (Sect. 2.5.4). Retroactively, this pH alteration can influence the Donnan potential, because the charges of organic molecules are usually pH-dependent (see Sect. 2.4.6, Figs. 2.34 and 2.35). In the simplest way, this dependence can be expressed as:

$$z_M = -z_{M0}\left(pH - pH_{iso}\right). \tag{3.2.45}$$

Near the isoelectric point (pH = pH_{iso}) the total molecule is uncharged ($z_M = 0$). Below this point (pH < pH_{iso}), z_M becomes positive, and above it (pH > pH_{iso}), z_M becomes negative.

Considering all of these relations, one obtains a system of non-linear equations that can be solved by iteration. Some basic properties of such a system will be demonstrated for the case of Donnan equilibrium of human erythrocytes. As already mentioned, this state can occur under conditions of blood storage or as a result of other treatment.

The most important non-penetrating charge component in erythrocytes is hemoglobin. In vivo, it has a concentration of 7 mM. Its isoelectric point $pH_{iso} = 6.8$ (at 25 °C) and $z_{M0} = 10.5$ eq/mol. The Donnan potential of erythrocytes in solutions of various NaCl concentrations containing sucrose to correct differences in the osmotic pressure, is shown in Fig. 3.17. The intracellular pH of the erythrocytes depends on the pH value of the external solution which is indicated on the abscissa, and on the Donnan potential itself. At the isoelectric point of hemoglobin, the Donnan potential: $\Delta\psi = 0$. The highest potentials, both negative and positive, are obtained at the greatest distances from the isoelectric point, and in solutions with the lowest NaCl concentrations.

The volume changes in these cells are shown in Fig. 3.18. The volume, as the percentage of the in vivo volume of erythrocytes, is indicated as a function of

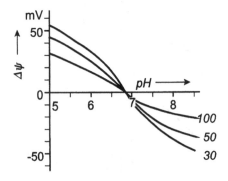

Fig. 3.17. The Donnan potential ($\Delta\psi$) of human erythrocytes in isotonic NaCl-sucrose solutions as a function of the external pH. The curves represent situations for NaCl solutions of 30, 50, and 100 mM (π = 300 mosmol, T = 25 °C). (After Glaser et al. 1980)

the NaCl concentration of the external solution at constant pH = 7.4. It is important to note that independently of the external NaCl concentration, the osmotic pressure is always adjusted to the isotonic osmotic pressure, $\pi = 300$ mosmol. In spite of this, the cells shrink in solutions of low ionic strength! An isotonic condition of an incubation medium, therefore, is no guarantee for the maintenance of a normal cell volume! A normal volume is achieved only in a solution containing about 20 mosmol sucrose and 140 mM NaCl. In this case, the sucrose just compensates the osmotic pressure of the hemoglobin, which has a comparatively high osmotic coefficient (see Fig. 3.12). Furthermore, it is interesting that the volume curve rises steeply at high ionic strengths. Mathematically, no results can be obtained from the equations assuming conditions of pure 152 mM NaCl solutions. In this case, the volume would become infinitely large. The experiments indicate that erythrocytes in solutions of pure electrolytes undergo hemolysis, if the membrane becomes permeable for these ions. This is known as *Donnan osmotic hemolysis*. As indicated in Fig. 3.18, this can occur even in isotonic solutions.

To determine experimentally whether a cell is in a state of Donnan equilibrium, the relation between the internal and external ionic activities must be ascertained. For a Donnan state to be present, the Donnan ratios from Eq. (3.2.38) must correspond. For most cells in vivo, the ratio obtained using the sum of the sodium and potassium ions $[(a_K^I + a_{Na}^I)/(a_K^{II} + a_{Na}^{II})]$ may be close to that for a Donnan equilibrium, but this would not be the case for either of the individual ions. Active transport changes their relative concentrations in opposing directions.

Further reading: Overbeek 1956; Glaser and Donath 1984

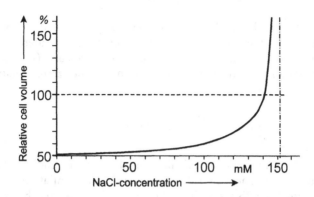

Fig. 3.18. Donnan-osmotic alterations of the relative volume (*V*) of human erythrocytes as a function of external NaCl concentration in isotonic NaCl-sucrose solutions (pH = 7.4, *T* = 25 °C). The osmotic pressure of the solutions with different NaCl-concentrations is always balanced by sucrose according to $\pi = 300$ mosmol. At $c = 152$ mM, the solution contains only NaCl, with no sucrose. The volume is given as the percentage of the in vivo volume of erythrocytes. (After Glaser et al. 1980, redrawn)

3.3
The Thermodynamic Analysis of Fluxes

Before we discuss the particular properties of the transport of molecules and ions through biological membranes, we must consider some general thermo-dynamic equations which underlie all diffusion processes. Later we will see that these equations are also applicable for biological membranes, although simple diffusion processes can be considered as rather exceptional cases in biological membranes.

A flux, as defined in Section 3.1.3, is the amount of a substance which passes in a perpendicular direction through a definite area of a surface in a unit of time. Its basic measure, therefore, is: kg s^{-1} m^{-2}, or better: mol s^{-1} m^{-2}. In the following discussion, the molar flux of a substance will be denoted by the symbol J.

When considering fluxes through cell membranes, difficulties may arise in some cases if the exact surface area of a cell is not known. In these cases modified units are used, such as: 'mol s^{-1} per cell', 'mol s^{-1} per liter of cells', 'mol s^{-1} per liter of cell-water' etc. As unit of time, instead of seconds, minutes, hours, or even days are frequently used. Such units are quite convenient for some physiological investigations. However, it is not possible to substitute such units directly into the thermodynamic equations.

3.3.1
The Flux of Uncharged Substances

Let us first consider the diffusion of an uncharged substance (i), neglecting any coupling with other fluxes. Using Eqs. (3.1.45) and (3.1.49), gives

$$J_i = L_i X_i = -L_i \operatorname{grad} \mu_i = -L_i \operatorname{grad} \left(\mu_i^0 + RT \ln a_i \right). \tag{3.3.1}$$

In order to calculate the gradient of the chemical potential, the dependence of the parameters on their position in space must be considered. Assuming constant temperature (grad $T = 0$), and constant pressure (grad $p = 0$), the gradient of the standard potential (grad μ_i^0) also becomes zero. Equation (3.3.1) therefore can be written as:

$$J_i = -L_i RT \operatorname{grad} \ln a_i = -\frac{L_i RT}{a_i} \operatorname{grad} a_i. \tag{3.3.2}$$

(To explain this rearrangement: the differential operator 'grad' can be handled like a deviation d/dx. Then the rules of sequential differentiation are applied.)

As has already been discussed in Section 3.1.3, a flux can also be expressed by the parameters concentration (c_i) and velocity (v_i) [Eq. (3.1.47)]. From this we came to an equation which includes mobility (ω_i) [Eq. (3.1.53)]. If we consider, furthermore, that X_i is not a simple mechanical force, but results from the gradient of a chemical potential determined by molar concentrations, the Avogadro number (N) must be included.

$$J_i = \frac{c_i w_i}{N} X_i. \tag{3.3.3}$$

By comparing Eq. (3.3.1) and (3.3.3), we obtain:

$$L_i = \frac{c_i w_i}{N}. \tag{3.3.4}$$

Suppose: $c_i \approx a_i$, then the combination of Eq. (3.3.4) with Eq. (3.3.2) gives:

$$J_i = -\frac{w_i}{N} RT \operatorname{grad} c_i = -w_i kT \operatorname{grad} c_i. \tag{3.3.5}$$

Introducing the diffusion coefficient, one gets *Fick's First Law of Diffusion*:

$$J_i = -D_i \operatorname{grad} c_i. \tag{3.3.6}$$

If there is only a 1-dimensional concentration gradient in the x-direction, this equation simplifies to:

$$J_{ix} = -D_i \frac{dc_i}{dx}. \tag{3.3.7}$$

The mobility (ω_i) of an almost spherical molecule can be interpreted mechanically by referring to Stockes' law [Eq. (3.1.55)]. By definition it has the units: m s^{-1} N^{-1}. In Section 2.3.6 we introduced the diffusion coefficient (D_i) in the context of the translational movement of molecules. This led to relations between the diffusion coefficient, mobility (ω_i), and molar mass (M_i) [see: Eqs. (2.3.38) to (2.3.40)].

We introduced the diffusion coefficient as: $D_i = \omega_i kT$. Its unit, therefore, is: m^2 s^{-1}. Using the measure: mol m^{-3} m^{-1} = mol m^{-4} for the concentration gradient in Eq. (3.3.6), or (3.3.7) one gets the units: mol s^{-1} m^{-2} for flux, according to its definition.

In addition to Fick's first law of diffusion, which gives the flux as a function of the concentration gradient, *Fick's second law* allows us to determine the establishment of a concentration gradient of a given substance (c_i) as a function of time (t). It is a partial differential equation of the second order which, for diffusion in one direction, is:

$$\left(\frac{\partial c_i}{\partial t}\right)_x = D \left(\frac{\partial^2 c_i}{\partial x^2}\right)_t. \tag{3.3.8}$$

This equation is used to calculate a concentration gradient which occurs when diffusion takes place in a homogeneous phase.

In contrast to this case where a continuous gradient of concentration occurs, in membrane systems various discontinuities in the concentration are to be expected. This means that the function $c_i(x)$ could become rather complicated. Schematically, this is illustrated in Fig. 3.19. The simplest case is shown by the solid line where phase *I* contains a solution with concentration c^I and correspondingly, c^{II} is the concentration in phase *II*. The concentration in the membranes falls linearly with a slope of $\Delta c/\Delta x$.

The broken line in Fig. 3.19 shows an irregular concentration pattern. In this case effects are considered which can occur at the membrane surface as well as in its interior. The deviation **A** marks an adsorption of the substance at the membrane surface, or a change in concentration of ions in an electrical double layer. The effective thickness of this layer is very small, being less than 10 nm. Inside the membrane (Fig. 3.19B), deviations of the linear concentration profile can occur by differences of the mobility of the substance in the x-direction or, in the case of ion transport, even by dielectric inhomogeneities.

In special cases *diffusion layers*, also called *unstirred* or *boundary layers*, may occur near the surface of membranes (Fig. 3.19C). These are near membrane regions without streaming or convection. In these regions, substances transported through the membrane, or involved in a chemical reaction, can move only by diffusion. This can produce an increase or a decrease in local concentrations. It depends simply on the relationship between the transport or reaction rate through the membrane, introducing substance into this region, and the rate of taking it away by diffusion. In this case a stationary concentration gradient is built up.

In systems with artificial ion exchange membranes, large diffusional layers can be indicated by special interference methods or by microelectrodes. In contrast to some speculations in earlier papers, in vivo such layers are mostly much lower than 1 μm. These layers may substantially affect biochemical reactions or transport processes of biological membranes. They can become important especially in cases where the cell surface is covered by microvilli or special caverns, or where reactions take place in the intercellular space. Even the occurrence of diffusional layers of protons near membranes is discussed.

Let us now consider the simplest case, represented by the solid line in Fig. 3.19 in more detail. Let the concentration gradient (dc_i/dx) inside the membrane be constant and equal $\Delta c_i/\Delta x$, whereas: $\Delta c_i = c_i^{II} - c_i^{I}$. In this case Eq. (3.3.7) becomes

$$J_i = -D_i \frac{\Delta c_i}{\Delta x} \equiv -P_i \Delta c_i. \tag{3.3.9}$$

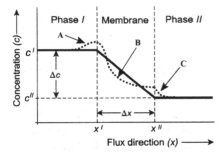

Fig. 3.19. Possible functions $c(x)$ in a membrane system. —— – ideal case, ---- – disturbed case, caused by the following reasons: A adsorption of substance at the membrane surface; B differences in the mobility of the substance inside the membrane; C diffusion layer caused by imperfect stirring at the surface

The parameter $P_i = D_i/\Delta x$ is the permeability coefficient measured in m s^{-1}. The same parameter will be used in Section 3.3.3 to calculate fluxes of ions. It is important to stress that this is the same parameter with an identical definition.

Let us now consider a system with flux interactions. In this case the flux coupling must be taken into consideration as discussed in Section 3.1.3. To demonstrate these approaches, we will consider only the simplest case of a binary system, represented for example by a flux of an uncharged substance (J_s) and its solvent, the water (J_w). The driving forces of both fluxes are the negative gradients of their chemical potentials. To simplify the derivation we will use simply their differences $\Delta\mu_s$ and $\Delta\mu_w$ as driving forces.

In a first step we must write down the equation for the dissipation function according to the rule of Eq. (3.1.64):

$$\Phi = J_w\Delta\mu_w + J_s\Delta\mu_s. \tag{3.3.10}$$

In this case Φ is an integral parameter for the whole membrane thickness Δx.

In the next step we will modify this equation in such a way that instead of the parameters $\Delta\mu_s$ and $\Delta\mu_w$, forces are introduced which are directly measurable. Let us first consider the difference of chemical potential of the water ($\Delta\mu_w$):

$$\Delta\mu_w = \Delta\mu_{wx}^0 + RT \ln\frac{x_w^I}{x_w^{II}}. \tag{3.3.11}$$

In Section 3.2.1 we discussed the chemical potential of water in detail and considered in particular its dependence on the mole fraction (x_w) as well as on pressure (p). Now we will make use of these derivations. Using Eqs. (3.2.12) and (3.2.13), as well as the definition of the osmotic pressure, we come to:

$$\Delta\mu_w^0 = \bar{V}_w\Delta p \quad \text{and} \quad RT \ln\frac{x_w^I}{x_w^{II}} = -\bar{V}_w\Delta\pi, \tag{3.3.12}$$

where $\bar{V}$ is the partial volume of water, Δp the difference of hydrostatic pressure and $\Delta\pi$ the difference of osmotic pressure. Equation (3.3.11) therefore can be rewritten as:

$$\Delta\mu_w = \bar{V}_w(\Delta p - \Delta\pi). \tag{3.3.13}$$

Let us now consider the other chemical potential difference in Eq. (3.3.10), namely $\Delta\mu_s$. This parameter depends on pressure difference in the same way as the difference of the chemical potential of water. According to Eqs. (3.3.11) and (3.3.12) one can write:

$$\Delta\mu_s = \bar{V}_s\Delta p + RT \ln\frac{c_s^I}{c_s^{II}}. \tag{3.3.14}$$

The second term of this equation can be expanded as a series using the common rule for parameters $x > 0$:

$$\ln x = 2\left[\left(\frac{x-1}{x+1}\right)+\frac{1}{3}\left(\frac{x-1}{x+1}\right)^3+\frac{1}{5}\left(\frac{x-1}{x+1}\right)^5+\cdots\right].$$ (3.3.15)

Now we substitute for x the value c^I/c^{II}. With good approximation we are justified in retaining only the first term of this series:

$$\ln\frac{c^I}{c^{II}} = 2\left(\frac{\frac{c^I}{c^{II}}-1}{\frac{c^I}{c^{II}}+1}\right) = \frac{c^I-c^{II}}{\frac{c^I+c^{II}}{2}} = \frac{\Delta c}{\bar{c}}.$$ (3.3.16)

This shows that the logarithm of the ratio of the two concentrations, or even activities, can be replaced by the difference in concentration (Δc) divided by the arithmetic mean ($\bar{c}$) of the concentrations of the two phases. This connection is not only true approximately, as seen by this expansion of the series, but it can be proved mathematically that it is exactly equal.

Relation (3.3.16), applied to Eq. (3.3.14), together with the Van't Hoff equation for the osmotic pressure: $\Delta\pi = RT\Delta c$ [Eq. (3.2.23)], gives:

$$\Delta\mu_s = \bar{V}_s\Delta p + \frac{1}{\bar{c}_s}\Delta\pi.$$ (3.3.17)

Now we have really found reasonable expressions for the differences of the chemical potentials, and using them we can rearrange the equation for the dissipation function [Eq. (3.3.10)]. Introducing Eqs. (3.3.14) and (3.3.17) into Eq. (3.3.10) we get:

$$\Phi = J_w\bar{V}_w(\Delta p - \Delta\pi) + J_s\left(\bar{V}_s\Delta p + \frac{1}{\bar{c}_s}\Delta\pi\right),$$ (3.3.18)

and after rearrangement:

$$\Phi = \Delta p(J_w\bar{V}_w + J_s\bar{V}_s) + \Delta\pi\left(\frac{J_s}{\bar{c}_s} - J_w\bar{V}_w\right).$$ (3.3.19)

The expressions in parentheses can be regarded as new flux variables. Flux was defined as the amount of a substance in moles that traverses a surface in a unit of time. Multiplying this parameter by the partial molar volume ($\bar{V}$), one gets a volume flow. Let us define a total volume flux (J_V) as the sum of individual volume fluxes of all components:

$$J_V = J_w\bar{V}_w + J_s\bar{V}_s.$$ (3.3.20)

To illustrate the meaning of the second term of Eq. (3.3.19), let us remember the relation $J_i = c_i v_i$ [Eq. (3.1.47)]. The term: $J_s/\bar{c}_s$ in Eq. (3.3.19) therefore is an expression for the velocity of the substance (v_s). The volume flow of the solvent ($J_w\bar{V}_w$) can also be considered as the velocity of it (v_w). From this it follows:

$$\frac{J_s}{\bar{c}_s} - J_w \bar{V}_w = \mathbf{v}_s - \mathbf{v}_w \equiv J_D. \tag{3.3.21}$$

The parameter J_D, called *exchange flux* is therefore the difference between the velocity of solute relative to the solvent.

As a result of these transformations we finally get a dissipation function which contains measurable parameters:

$$\Phi = J_V \Delta p + J_D \Delta \pi. \tag{3.3.22}$$

This makes it possible to write down a flux matrix following the pattern of Eq. (3.1.58):

$$J_V = L_V \Delta p + L_{VD} \Delta \pi$$
$$J_D = L_{DV} \Delta p + L_D \Delta \pi. \tag{3.3.23}$$

This equation shows that in case of the transport of a solution with a single uncharged substance, the membrane permeability is determined by four coefficients: L_V, L_{VD}, L_{DV} and L_D. The driving forces are the osmotic ($\Delta \pi$), and the hydrostatic (Δp) pressures.

Some of the coefficients used in these phenomenological equations can easily be illustrated: the parameter L_V shows for example how fast the solution passes through the membrane in response to a hydrostatic pressure difference (Δp). Introducing the conditions $\Delta \pi = 0$ and $\Delta p > 0$ in the upper part of Eq. (3.3.23) it becomes: $L_V = J_V/\Delta p$. This means that L_V is a mechanical *filtration coefficient* or a kind of hydraulic conductivity.

Under the same conditions the substance can also be forced through a filter. In this case the second line in Eq. (3.3.23) gives $L_{DV} = J_D/\Delta p$. Because of this, L_{DV} is called *ultrafiltration coefficient*.

The flux matrix (3.3.23) makes it possible to describe the time dependence of an osmotic system. Figure 3.20 shows how the pressure in a Pfeffer's cell,

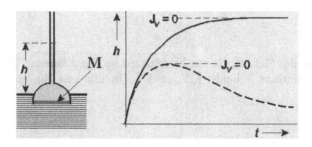

Fig. 3.20. The height of the water column in a Pfeffer osmometer depending on time in two experiments with different conditions: Pfeffer's cell is closed by a semipermeable membrane (M) (———) and solvent as well as solute can penetrate the membrane (- - -). The conditions with stationary pressure in this manometer ($dh/dt = 0$, i.e. $dp/dt = 0$) simultaneously means stationary volume ($dV/dt = J_V = 0$)

measured by the height of the water column in the vertical tube, changes with time. Only in the case of a semipermeable membrane (solid line), will a thermodynamic equilibrium with a constant pressure difference be achieved. If, in addition to the solvent, the membrane allows some of the solute to pass through then there will be a decline in the pressure after the initial rise. The osmotic pressure inside the osmometer will continuously fall. In this case a state without any volume flux ($dV/dt = J_V = 0$) is achieved only for a short time. For this situation it follows from the first line of the flux matrix [Eq. (3.3.23)]:

$$L_V \Delta p + L_{VD} \Delta \pi = 0, \tag{3.3.24}$$

and furthermore:

$$(\Delta p)_{J_V=0} = -\frac{L_{VD}}{L_V} \Delta \pi. \tag{3.3.25}$$

In Section 3.2.1 we introduced Staverman's reflection coefficient (σ). Relating Eq. (3.2.27) with Eq. (3.3.25), the following connection between the reflection coefficient and the coupling coefficients results:

$$\sigma = -\frac{L_{VD}}{L_V}. \tag{3.3.26}$$

For a semipermeable membrane with: $\sigma = 1$, it therefore holds that:

$$L_{VD} = -L_V.$$

A better illustration of this parameter allows the following consideration: a solution will be forced through a membrane by a hydrostatic pressure ($\Delta p > 0$) without any osmotic difference ($\Delta \pi = 0$). Using these conditions and dividing the second equation of the flux matrix (3.3.23) by the first, we obtain:

$$\frac{J_D}{J_V} = \frac{L_{DV}}{L_V} = -\sigma. \tag{3.3.27}$$

According to Eq. (3.3.21), J_D can be replaced by the difference ($v_s - v_w$). For very dilute solutions it holds: $J_w \bar{V}_w \gg J_s \bar{V}_s$. In this case Eq. (3.2.27) can be written as:

$$\sigma = \frac{v_w - v_s}{v_w}. \tag{3.3.28}$$

This equation has already been discussed in Section 3.2.1.

The deviations shown here clearly demonstrate that the consideration of more complex systems, for example solutions with more components, would lead to a huge number of parameters and relations.

This theory of coupled fluxes has been widely used to explain the stationary volume and stationary pressure observed in experiments with cells under non-equilibrium conditions. Such experiments have been carried out mostly on red blood cells and on plant cells. Human red blood cells, placed initially in an isotonic solution, shrink immediately when the osmotic pressure of the external solution is increased by adding a permeable substance i ($\sigma_i < 1$). Then, however, if the substance passes through the membrane, the cell usually reverts to its initial volume, as illustrated in Fig. 3.20 by the dashed line. Such experiments with fast volume changes are undertaken to determine the reflection coefficient using stop-flow techniques.

Further reading: Katchalsky and Curran 1965; Kotyk et al. 1988; Stein 1990; Zimmermann 1989. Recent papers on unstirred diffusion layers: Barry 1998; Evtodienko et al. 1998; Pohl et al. 1998

3.3.2
Fluxes of Electrolytes

The diffusion of ions is governed by the same fundamental laws as fluxes of uncharged substances [see Eq. (3.3.1)].

$$J_i = L_i X_i = \frac{c_i w_i}{N} X_i = -\frac{c_i w_i}{N} \operatorname{grad} \tilde{\mu}_i. \tag{3.3.29}$$

In the case of electrolyte fluxes, the driving force is induced by the negative gradient of the electrochemical potential ($\tilde{\mu}_i$). The coupling coefficient (L_i) can again be considered as a product of concentration (c_i), and mobility (ω_i/N).

Let us first consider the expression 'grad $\tilde{\mu}_i$.' For a concentration gradient only in the x-direction, instead of the differential operator 'grad', the differential quotient can be applied:

$$\frac{d\tilde{\mu}_i}{dx} = \frac{d}{dx}\left(\mu_i^0 + RT \ln a_i + z_i F \psi\right). \tag{3.3.30}$$

In order to simplify the equation, let the solution be close to ideal, i.e. let the activity coefficient be: $f_i \approx 1$, and therefore $a_i \approx c_i$. Furthermore, let the system be under isobaric (grad $p = 0$), and isothermic (grad $T = 0$) conditions. In this case Eq. (3.3.30) becomes

$$\frac{d\tilde{\mu}_i}{dx} = \frac{RT}{c_i}\frac{dc_i}{dx} + z_i F \frac{d\psi}{dx}, \tag{3.3.31}$$

which, when combined with Eq. (3.3.29), results in

$$J_i = -\frac{c_i w_i}{N}\left(\frac{RT}{c_i}\frac{dc_i}{dx} + z_i F \frac{d\psi}{dx}\right), \tag{3.3.32}$$

or, modified using the diffusion coefficient according to Eq. (3.3.6):

$$J_i = -D\left(\frac{dc_i}{dx} + \frac{z_i F c_i}{RT}\frac{d\psi}{dx}\right).$$ (3.3.33)

This is the *Nernst-Planck equation*. It contains the differential quotients of the concentration $[c_i(x)]$, and of the electrical potential $[\psi(x)]$. These differential quotients can be integrated only if the corresponding functions are known. This problem has already been discussed with regard to the concentration $[c_i(x)]$ in Section 3.3.1 (see Fig. 3.19). In contrast to Fick's equation, here the function $\psi(x)$ must be known in addition.

The simplest approach is the consideration of linear gradients. This corresponds, for example, to a membrane with large, water filled, and non charged pores where the ions can move freely as in bulk water. Integrating Eq. (3.3.33) for these conditions, the differential quotients are replaced by simple ratios of differences, and instead of the concentration c_i, the mean concentration of both phases $[\bar{c}_i = (c^I + c^{II})/2]$ appears.

$$J_{ix} = -D\left(\frac{\Delta c_i}{\Delta x} + \frac{z_i F \bar{c}_i}{RT}\frac{\Delta\psi}{\Delta x}\right).$$ (3.3.34)

Or, using the permeability coefficient: $P_i = D/\Delta x$ [see Eq. (3.3.9)]:

$$J_{ix} = -P_i\left(\Delta c_i + \frac{z_i F \bar{c}_i}{RT}\Delta\psi\right).$$ (3.3.35)

In 1943 D. E. Goldman integrated the Nernst-Planck equation, supposing only the so-called *constant field conditions*, i.e., assuming: $\mathbf{E} = -\text{grad }\psi = \text{const}$. The concentration profile results from the bulk concentrations in both phases and from its passive distribution in the electric field. Integrating the Nernst-Planck equation (3.3.33) with these conditions, one gets the following expression:

$$J_i = -P_i\beta\frac{c_i^I - c_i^{II}e^\beta}{1 - e^\beta} \quad \text{with: } \beta = \frac{z_i F}{RT}\Delta\psi.$$ (3.3.36)

The function $J = f(\Delta\psi)$ is illustrated in Fig. 3.21. It considers the flux of a monovalent cation ($z_i = +1$) penetrating the membrane with a permeability $P_i = 10^{-7}$ m s^{-1}. Let the flux (J_i) be positive if it is directed phase $I \Rightarrow$ phase II. A negative J_i, therefore, means that the flux is in the opposite direction. The potential gradient is negative per definition, if ψ decreases from phase I to phase II.

The dotted line in Fig. 3.21 represents a cation flux which is driven only by the electrical potential. ($c_i^I = c_i^{II} = 100$ mM). In this case, the flux vanishes ($J_i = 0$) if there is no potential difference ($\Delta\psi = 0$).

If there is an additional driving force resulting from a concentration gradient, the curve is displaced towards either of the solid lines. In this case an ion flux exists even if $\Delta\psi = 0$. If $c_i^I = 100$ mM, and $c_i^{II} = 10$ mM then, when $\Delta\psi = 0$, there

Fig. 3.21. The flux (J_i) of a univalent cation ($z_i = +1$, $P_i = 10^{-7}$ m s^{-1}) driven by a concentration, as well as an electrical potential gradient according to the Goldman equation (3.3.36). The concentrations in the phases I and II are listed on the curves. Description in the text

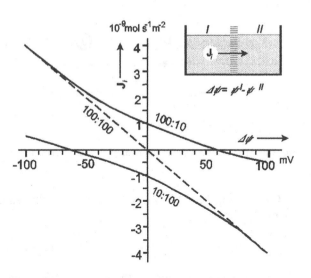

is a flux from I to II (i.e. $J_i > 0$). When the concentrations are reversed, J_i becomes negative. The solid lines cut the abscissa at $+60$ mV and -60 mV respectively. These points mark where the driving force due to the electric field exactly balances the driving force due to the concentration gradient. This corresponds to the equilibrium situation with $J_i = 0$, described by the Nernst equation [Eq. (3.2.34)].

In fact, the Goldman equation is a good approach even in cases where the linearity of the function $\psi(x)$ is not exactly realized. At the end of the following Section we will consider the role of surface charges.

Let us consider at this point an equation that makes it possible to calculate the ratios of ionic fluxes independently of the functions $c(x)$ and $\psi(x)$. In this context a term must be defined which we need again in further considerations: the flux which is measurable directly by changes of chemical concentrations is called the *net flux* (J). Using radioactive isotopes it is possible to indicate that this net flux in fact results from the difference between two opposite *unidirectional fluxes* J_{12} and J_{21}:

$$J = J_{12} - J_{21}. \tag{3.3.37}$$

In contrast to the unidirectional fluxes, the directions of which are indicated by subscripts (for example: J_{12}, J_{21}), the net flux will be defined as positive, as in the example of Fig. 3.22, if it takes place in the direction corresponding to Eq. (3.3.37).

It must be stressed that these are examples of simple diffusion processes based on thermodynamic fluctuations. Later in Section 3.4.1 (see Fig. 3.23) we will consider systems of co-transport through biological membranes, which show the same kinetics but which must be treated using completely different thermodynamic approaches. These two kinds of tranport processes therefore must not be confused.

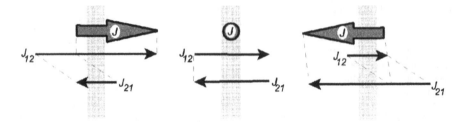

Fig. 3.22. The net flux (J), as the sum of the unidirectional fluxes J_{12} and J_{21}

Considering Eq. (3.3.37), one can write Fick's equation [Eq. (3.3.9)] in the following way:

$$J = J_{12} - J_{21} = -P\Delta c = P(c^I - c^{II}) = Pc^I - Pc^{II}. \tag{3.3.38}$$

From this we can conclude with certain justification:

$$J_{12} = Pc^I; \quad J_{21} = Pc^{II}. \tag{3.3.39}$$

(We will use these formulations again in Section 5.2.1 for compartment analysis.)

To calculate unidirectional fluxes of ions one could use the same approaches as in Eqs. (3.3.39), but just introducing a kind of electrochemical concentration ($\tilde{c}$). This parameter, as we will indicate later, has no further importance. The definition results from the following equation:

$$\tilde{\mu} = \mu^0 + RT \ln c + zF\psi \overset{!}{=} \mu^0 + RT \ln \tilde{c} + zF\psi^0. \tag{3.3.40}$$

Therefore: $\tilde{c} = c$, if $\psi^0 = \psi$. This means that a kind of zero-potential will be established, whatever it is. From this definition follows:

$$\ln \tilde{c} = \ln c + \frac{zF}{RT}(\psi - \psi^0), \tag{3.3.41}$$

and:

$$\tilde{c} = c e^{\frac{zF}{RT}(\psi - \psi^0)}. \tag{3.3.42}$$

This equation is of little use since the reference potential ψ^0 is unknown. However, if this relation is substituted in the Eqs. (3.3.39), and if the ratio of the unidirectional fluxes are calculated, all unknown parameters cancel:

$$\frac{J_{12}}{J_{21}} = \frac{c^I}{c^{II}} e^{\frac{zF}{RT}(\psi^I - \psi^{II})}. \tag{3.3.43}$$

(To avoid the accumulation of subscripts we ignored in these derivations the index for the specific substance i.)

This formula is known as *Ussing's equation* which relates unidirectional fluxes to concentrations and electrical potential differences. All parameters of this equation can be measured and the validity of the equation can therefore be checked experimentally. As preconditions for this equation only concentration gradients and electrical potential gradients are considered. If the relation between two measured unidirectional fluxes does not agree with the results calculated by these gradients, then additional driving forces are involved, such as in processes of active transport.

Further reading: Friedman 1986; Goldman 1943; Kotyk 1988; Stein 1986, 1990; Syganow and vonKlitzing 1999

3.3.3
The Diffusion Potential

The diffusion of an electrolyte, in general, can be considered as a separate diffusion of the dissociated ions along their own electrochemical gradient. Cations and anions, however, may have different mobilities (w_i). The slower diffusion types of ions will lag behind the faster ones. A local potential difference called *diffusion potential* builds up, retarding the faster ions and speeding up the slower ones. Diffusion potentials can occur in homogenous solutions as well as between two phases separated by a membrane which is permeable for the ions.

An equation for the diffusion potential can be derived, postulating the electroneutrality of the sum of all ion fluxes. In this case the cation flux (J_C) induces a flow of charge ($J_C z_C$). In the case of electroneutrality, it must be compensated by the charge transport of anions ($J_A z_A$):

$$J_C z_C + J_A z_A = 0. \tag{3.3.44}$$

Now, we can use the flux equations derived in the previous Section 3.3.2. If linear functions $c(x)$ and $\psi(x)$ are proposed, the simplest flux equations [Eq. (3.3.35)] can be used. Inserting them into Eq. (3.3.44), one gets:

$$z_C P_C \Delta c_C + \frac{z_C^2 F \bar{c}_C P_C}{RT} \Delta \psi + z_A P_A \Delta c_A + \frac{z_A^2 F \bar{c}_A P_A}{RT} \Delta \psi = 0. \tag{3.3.45}$$

Let us take into account that the concentration of the ions (c_C, c_A) depends on the concentration of the salt (c), whereas $c_C = \nu_C c$, and $c_A = \nu_A c$. Introducing this, one can rearrange Eq. (3.3.45) and resolve it for $\Delta\psi$ in the following way:

$$\Delta\psi = -\frac{RT}{F} \left(\frac{z_C P_C \nu_C + z_A P_A \nu_A}{z_C^2 P_C \nu_C + z_A^2 P_A \nu_A} \right) \frac{\Delta c}{\bar{c}}, \tag{3.3.46}$$

or:

$$\Delta\psi = \frac{RT}{F}\left(\frac{z_C P_C v_C + z_A P_A v_A}{z_C^2 P_C v_C + z_A^2 P_A v_A}\right)\ln\frac{c^{II}}{c^{I}}. \tag{3.3.47}$$

The relation: $\Delta c/\bar{c} = \ln(c^{I}/c^{II})$ has already been introduced in Section 3.3.1 in relation to Eq. (3.3.16). It is easy to understand that Eq. (3.3.47) will be transformed into the Nernst Equation [Eq. (3.2.33)], if the membrane becomes semipermeable, i.e. if $P_C = 0$, or if $P_A = 0$.

A better approach for the conditions of membranes will be the flux equation, derived by Goldman for conditions of constant electric field (see Sect. 3.3.2). Introducing this Goldman equation [Eq. (3.3.36)] into the equation for electroneutrality of fluxes [Eq. (3.3.44)], one gets the following expression:

$$P_C\frac{F\Delta\psi}{RT}\left(\frac{c_C^{I} - c_C^{II}e^{\frac{F\Delta\psi}{RT}}}{1 - e^{\frac{F\Delta\psi}{RT}}}\right) + P_A\frac{F\Delta\psi}{RT}\left(\frac{c_A^{I} - c_A^{II}e^{-\frac{F\Delta\psi}{RT}}}{1 - e^{-\frac{F\Delta\psi}{RT}}}\right) = 0. \tag{3.3.48}$$

This equation can also be rearranged and solved for $\Delta\psi$. For this we first transform the denominators of the fractions, using the expression:

$$1 - e^{-x} = -e^{-x}(1 - e^{x}).$$

This leads to:

$$\frac{F\Delta\psi}{RT\left(1 - e^{\frac{F\Delta\psi}{RT}}\right)}\left[P_C\left(c_C^{I} - c_C^{II}e^{\frac{F\Delta\psi}{RT}}\right) - P_A e^{\frac{F\Delta\psi}{RT}}\left(c_A^{I} - c_A^{II}e^{-\frac{F\Delta\psi}{RT}}\right)\right] = 0. \tag{3.3.49}$$

When $\Delta\psi \Rightarrow 0$ the expression in front of the square brackets will not approach zero. Therefore the sum inside the parentheses must be equal to zero:

$$P_C c_C^{I} - P_C c_C^{II}e^{\frac{F\Delta\psi}{RT}} - P_A c_A^{I}e^{\frac{F\Delta\psi}{RT}} + P_A c_A^{II} = 0. \tag{3.3.50}$$

Which gives, after some rearrangements:

$$e^{\frac{F\Delta\psi}{RT}} = \frac{P_A c_A^{II} + P_C c_C^{I}}{P_A c_A^{I} + P_C c_C^{II}}, \tag{3.3.51}$$

and:

$$\Delta\psi = \frac{RT}{F}\ln\frac{P_A c_A^{II} + P_C c_C^{I}}{P_A c_A^{I} + P_C c_C^{II}}. \tag{3.3.52}$$

This is the *Goldman-Hodgkin-Katz equation* which is commonly used in electrophysiology to calculate diffusion potentials in living cells (mostly it is just named the *Goldman equation*). This expression also becomes a Nernst equation [Eq. (3.2.33)], introducing the conditions of a semipermeable membrane ($P_A = 0$, or $P_C = 0$).

It is possible to extend this equation to systems containing more than one salt, i.e. various monovalent ions. To take into account the non-ideal character of solutions, chemical activities (a_i) instead of concentrations (c_i) can be used. In this case the *Goldman-Hodgkin-Katz equation* can be written as follows:

$$\Delta\psi = \frac{RT}{F}\ln\frac{\sum_{\text{Anions}} P_A a_A^i + \sum_{\text{Cations}} P_C a_C^e}{\sum_{\text{Anions}} P_A a_A^e + \sum_{\text{Cations}} P_C a_C^i}. \tag{3.3.53}$$

When considering cells, the superscript i in this formula means "internal", the superscript e "external" concentrations. Correspondingly: $\Delta\psi = \psi^i - \psi^e$.

In a large number of papers experiments are described indicating the applicability of the Goldman equation for various cells and membranes. However, some limitations become obvious when one tries to apply this equation in relation to real cellular conditions. Mostly these are already included in the conditions of the applied flux equation.

The most important limitation comes from the assumption considering a free diffusion of the ions in a homogeneous and constant electric field. Even in the case of large pores in the membrane this assumption is valid only with some approximations. It must be taken into account that the thickness of a biological membrane is just as large as the Debye-Hückel radius of an ion. Additionally, the coefficients of permeability are defined for large systems and can be used for considerations of molecular systems only approximately.

In order to calculate fluxes and transmembrane potentials, there are approaches that take into account the local concentrations of ions in the electrical double layer at the membrane boundary. If the surface potential ψ_0 is known, it is possible to introduce surface concentrations into the Goldman equation instead of the concentration of ions in the bulk phase, using Eqs. (2.4.10) or (3.2.34). In this case, however, the value $\Delta\psi$ of Eq. (3.3.53) no longer means the potential difference between the two bulk phases, as measured in electrophysiology by microelectrodes, but just between the inner and outer surface (see Fig. 2.48). Thus, the diffusion potential of a membrane can be controlled by local surface charges.

Further reading: Goldman 1943; Katchalsky and Curran 1965; Syganow and vonKlitzing 1999

3.4
The Nonequilibrium Distribution of Ions in Cells and Organelles

In recent decades, our knowledge on transport mechanisms in biological membranes has developed rapidly. New insights into structures and molecular mechanisms of transport proteins and ionophores show the occurrence of complicated systems of control and regulation. Furthermore, the large number of different transport processes in a single cell appear to be closely meshed together. By pH and volume regulation, and by control of internal Ca^{++} concentrations etc. transport processes may furthermore control complex cellular functions, such as protein expression, cell proliferation etc. In this respect, the transmembrane potential also becomes a subject of particular interest, regarding its generation as well as its role in cellular processes.

3.4.1
Ion Transport in Biological Membranes

Figure 3.23 illustrates a classification of various types of ion transport mechanisms in the membranes of cells and organelles. It shows that the simple diffusion of ions, as described in Section 3.3 occurs only in special cases. Processes of simple electrodiffusion are postulated for example for Na^+ and K^+ channels in membranes of nerve and muscle cells. Certainly, to some extent they occur also as *leak flux* in other cells, and of course in the extracellular space of tissues. Furthermore, membrane pores produced by antibiotics, toxins, virus penetrations etc. produce diffusion paths which sometimes indicate a high degree of selectivity. We will come back to the functional role of selective changes of ion permeability of biological membranes later in Section 3.4.2.

Ion transport in biological membranes is mainly achieved through proteins or protein complexes which transport simultaneously two or more ions in a constant stoichiometrical relationship. Such stoichiometrically coupled fluxes are called *co-transports*. There are two kinds of co-transport systems: in the case of the *symport*, a strongly coupled flux of anions and cations occurs in the same direction. An example of this could be a complex which transfers simultaneously one Cl^- and one K^+ ion through the membrane in the same direction. An *antiport*, in contrast to this, is a system transporting simultaneously two ions of identical charges in opposite directions, for example one K^+, against one H^+. If, in a co-transport system an equal number of charges is transported, then the process becomes electroneutral, and therefore does not depend directly on electric field conditions. If not, an electrical current will be the result of the transport. We will call this sort of process *rheogenic*, i.e. "current producing". Rheogenic co-transport processes can be recognized by their electrical conductivity, a property which they have in common with simple diffusion processes. They can be controlled by electric fields.

An *active transport* is a sort of pump, transporting ions or even uncharged molecules against their own chemical or electrochemical gradients. It is "uphill transport", which uses metabolic energy (ΔG, in Fig. 3.23). In most cases this involves so-called transport ATPases, using the energy of the hydrolytic reaction: ATP $\Rightarrow$ ADP. This mechanism can also run in the opposite direction. In chloroplasts and mitochondria, for example, ATP is synthesized by a "downhill" proton flux (see Sect. 2.2.3, Fig. 2.7).

	Passive transport				Active transport	
	Diffusion		Cotransport		Transport ATPase	
	PORE	CHANNEL	SYMPORT	ANTIPORT		
	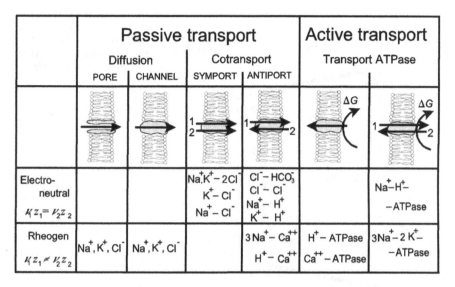					
Electro-neutral $\nu_1 z_1 = \nu_2 z_2$			$Na^+K^+-2Cl^-$ K^+-Cl^- Na^+-Cl^-	$Cl^--HCO_3^-$ Cl^--Cl^- Na^+-H^+ K^+-H^+		Na^+-H^+- $-ATPase$
Rheogen $\nu_1 z_1 \neq \nu_2 z_2$	Na^+,K^+,Cl^-	Na^+,K^+,Cl^-		$3Na^+-Ca^{++}$ H^+-Ca^{++}	$H^+-ATPase$ $Ca^{++}-ATPase$	$3Na^+-2K^+-$ $-ATPase$

Fig. 3.23. Classification of various systems of ion transporters in biological membranes

Active transport can also be rheogenic. In this case the transport directly induces electric currents and fields (see Sect. 3.5.2, Fig. 3.32). Frequently such types of transport are also called *electrogenic* which means: "generating an electrical membrane potential". Looking at these two terms accurately, they are not identical! Even an electro-neutral pump can be "electrogenic" if it produces a concentration gradient of ions which subsequently generates a diffusion potential. On the other hand, a rheogenic pump may influence the transmembrane potential only if there is a sufficiently high membrane resistance.

The number of different transport paths in a single membrane of a cell can be rather high. Let us illustrate this for the case of cells of a renal proximal tubule (Fig. 3.24). It is obvious that the fluxes are coupled with each other by the concentrations of their common ions. Co-transport systems not only couple together fluxes of particular ions, but additionally in some cases they are responsible for permeation of small uncharged molecules. This, for example, is the case in the uptake of monosaccharides in the epithelium of the small intestine. In this case a rheogenic Na^+ glucose symport occurs. In a similar way, fluxes of amino acids are coupled with transport of Na^+, or H^+ ions. In many cases the electrochemical gradient produced by active transporters is the source of energy driving the uphill transport of metabolites.

Modern biotechnological approaches allow new possibilities for studying the molecular mechanisms of these processes. It is possible to isolate the corresponding proteins, to clone them, and to modify specific groups. A large number of kinetic models have been proposed to explain these transport mechanisms. These models treat the processes of transport like enzymatic

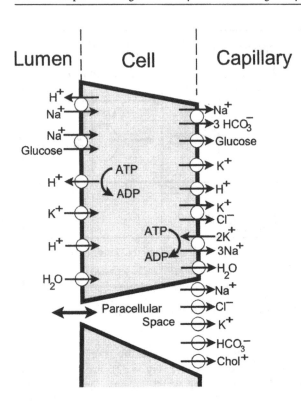

Fig. 3.24. Various types of ion transporters in cells of a renal proximal tubule and in the paracellular space. (After Verkman and Alpern 1987)

reactions. They propose a sequence of binding, of molecular conformation, and of releasing processes. Mechanistic approaches, like a revolving door, or a ferry boat, in fact are not realistic. This would require the rotation or the transmembrane shift of a large molecule, or at least of a part of it in the membrane during the transport process. More likely, periodic oscillations of small side chains of the transporting molecule complexes would occur. Depending on the kind of transport, 1 to 10^6 ions could be translocated per second. The side chains which are responsible for the transport therefore must oscillate in a frequency range up to 1 MHz. We will come back to this number later, discussing effects of external AC fields (Sect. 4.4.5).

The existence of transporters with strongly fixed stoichiometry forces us to rethink the requirement of flux electroneutrality which we postulated in Section 3.3.3 [Eq. (3.3.44)]. Considering rheogenic symports, it is not the electroneutrality of a single flux which is required, but rather the electroneutrality of all fluxes in a membrane of a single cell together. The calculation of the balance of charges and ions in a cell therefore is possible only considering all fluxes. Additionally, changes of fixed charges inside the cell induced by internal pH alterations should be taken into account.

The existence of co-transporters in cells rather than simple diffusion is a sign of optimization. If ionic transport was based on simple processes of electrodiffusion, then an occasional alteration of the membrane potential would cause an immediate change of all electrolyte fluxes, and subsequently a shift of the internal concentration of all ions in the cell. In contrast, the system of electroneutral co-transporters is independent of the transmembrane potential and will protect the cell against such disturbances.

Simple balance calculations can be applied to describe active transport thermodynamically. Such approaches, of course, do not help to clarify the molecular mechanisms, but they give some insights into the processes of energy transformation. Let us illustrate this, asking the question: how much power is required for an ion pump to maintain a steady state of ions in the cell?

A system like that of Fig. 3.4B is determined by the active transport (J_A) as well as by the passive flux (J_i). In Section 3.1.4 we introduced the dissipation function $\Phi = \sigma T$ [Eq. (3.1.64)], which must be larger than 0. According to Eq. (3.1.64) for our system it amounts to:

$$\Phi = J_A X_A + J_i X_i \quad \text{for:} \quad \Phi > 0. \tag{3.4.1}$$

Let us look at a real example. The glucose uptake (J_G) of a cell in the intestine is driven by the passive influx of sodium (J_{Na}). Using Eq. (3.4.1) we obtain:

$$J_G X_G + J_{Na} X_{Na} > 0. \tag{3.4.2}$$

If v equivalents of sodium ions are transported for each mole of glucose then:

$$J_G = v J_{Na}.$$

Introducing this into Eq. (3.4.2) and assuming that both fluxes are not equal to zero, it follows that:

$$v X_G + X_{Na} > 0, \tag{3.4.3}$$

and thus:

$$X_{Na} > -v X_G. \tag{3.4.4}$$

Let us now replace the forces (X) by the differences of the corresponding chemical or electrochemical potential (see Sect. 3.3.1), we obtain:

$$-v \Delta \mu_G < \Delta \tilde{\mu}_{Na}. \tag{3.4.5}$$

Using Eqs. (3.1.33) and (3.1.41), and the conditions: $\Delta T = 0$ and $\Delta p = 0$, we get:

$$v RT \ln \frac{a_G^i}{a_G^e} < -\left(RT \ln \frac{a_{Na}^i}{a_{Na}^e} + F \Delta \psi \right), \tag{3.4.6}$$

where $\Delta \psi = \psi_i - \psi_e$, and after rearrangement:

$$\left(\frac{a_G^i}{a_G^e}\right)^v < \frac{a_{Na}^e}{a_{Na}^i} e^{-\frac{F\Delta\psi}{RT}}. \tag{3.4.7}$$

This equation allows us to calculate the maximal rate of enrichment of glucose in the cell which can be achieved for a given electrochemical gradient of sodium ions. Assuming that the membrane potential of the cell is: $\Delta\psi = -50$ mV, and the relation of sodium ions: $a_{Na}^e/a_{Na}^i = 10$ ($T = 300$ K), it follows:

$$\left(\frac{a_G^i}{a_G^e}\right)^v < 69. \tag{3.4.8}$$

If the fluxes are coupled 1:1 ($v = 1$), this process can give a maximum enrichment of glucose by a factor of 69, when the pump is performing optimally.

Similar calculations can also be applied to types of active transport which are driven by chemical reactions, as for example transport ATPases. In this case in the equation of the dissipation function [Eq. (3.4.1)], the reaction rate (as a sort of scalar flux) and the chemical affinity of the energy supplying reaction [Eq. (3.1.75)] must be included.

This calculation shows that the intensity of the pump which is necessary to build up a certain concentration gradient depends both on the coupling stoichiometry of the fluxes and on the passive back flow. This means that it is not only the power of the pump which is responsible for the achieved level of steady state, but also the conductivity or the permeability of the substance being considered, allowing it to flow backwards. This is illustrated in the scheme of Fig. 3.4B: the power of the pump must be higher if a greater difference in the levels of the vessels is required, or if the outflow is to be faster.

Let us briefly consider a problem which has been discussed for a long time, but which is not yet resolvable because of lack of experimental approaches. Can the condition: grad $T = 0$ really be valid for biological membranes? It is interesting to remember that very small temperature differences between the two sides of a cell membrane may cause a considerable temperature gradient. Let us suppose that a fast exothermic enzymatic reaction on one side of the membrane continuously produces heat, and in this way increases the local temperature by only 0.01 °C in relation to the opposite membrane surface. In this case a temperature gradient of $0.01/10^{-8} = 10^6$ K m^{-1} is built up. This could be sufficient to explain active transport as a kind of thermodiffusion. A pressure difference of 130 kPa could be induced in this way (Spanner 1954). These are, of course, just interesting speculations because despite some model experiments, we cannot yet check this experimentally (see Harth and Vaupel 1971; Tasaka 1975).

Further reading: Bernhardt 1994; Fleischer et al. 1989; Hille 1992; Hoffmann and Simonsen 1989; Kotyk et al. 1988; Läuger 1991; Stein 1986, 1990; Yeagle 1991

3.4.2
The Network of Cellular Transporters. The Cell as an Accumulator
of Electrochemical Energy

If a cell were only a poly-electrolyte system without metabolically driven ion pumps it would remain in a state of Donnan equilibrium. This means that there would be a Donnan distribution of the mobile ions, and as a result, a Donnan osmotic pressure (see Sect. 3.2.3). In the living cell, however, active transport systems driven by metabolic energy (Fig. 3.23), modify this ionic composition, as shown schematically in the model of Fig. 3.4B. The living cell therefore reaches a steady state, i.e. a stationary state of nonequilibrium (see Fig. 3.6).

The function of this particular ionic state in the process of life can be considered from several aspects. One of these is that the ion pumps raise the electrochemical energy of the cell to a higher level. Thus the cell can be regarded as a kind of electrochemical energy store which can be easily tapped. Furthermore, the shift of a system from the equilibrium into a nonequilibrium state is the precondition for its homoeostatic regulation. This, by the way, is also the reason for the increased temperature in homeothermic animals. The setting up of a concentration gradient of ions across the membrane makes the cells able to control and regulate an intracellular environment which is the precondition of various cellular processes.

What are the immediate effects of ionic pumps on the cell?

- They control and regulate the internal ionic milieu. In this way, steep gradients of the electrochemical potentials of particular ions are built up, essentially without changing the total internal ionic concentration. The internal potassium concentration of animal cells, for example, is usually much higher than the external one. Simultaneously, however, the sodium concentration is lower to the same degree. The sum of both of these ions in the cytoplasm, taken together, is nearly the same as in the external medium.
- In the case of rheogenic pumps, they directly induce transmembrane potentials. In this case the pumps are called electrogenic.
- They can produce a direct osmotic effect, changing the concentration of osmotically active substances.
- They can establish particular internal ionic conditions, controlling, for example, the extremely low intracellular calcium concentration.

Some direct consequences of the active transport processes can be construed from the effects of stopping the pump by the use of specific inhibitors. In this case effects can be observed like Donnan-osmotic swelling, internal pH shifts, an increase in the internal calcium concentration, a change of transmembrane potential etc. Mostly, using such inhibitors, the overall internal ionic conditions are altered.

In Fig. 3.24 for example, the multitude of transporters is illustrated for kidney tubule cells. There are 13 different transport systems shown which determine the cellular milieu and another five other fluxes between the luminal and serosal surfaces of the epithelium across the paracellular gap. This picture in fact is incomplete as, for example, Ca^{++} fluxes are not shown, and the diagram does not include the intracellular organelles with their own transporters.

Using this example, we will illustrate the interconnections of these transport properties qualitatively, following for example one particular path: transport ATPases pump protons out of the cell, others decrease the internal sodium content, and in the same way enrich the cytoplasm with potassium. Extruding positive charges, both processes induce an inside negative transmembrane potential. Simultaneously, an electrochemical sodium gradient is generated which drives a "passive" sodium flux outside-in. This influx, however, is made by a glucose-sodium co-transporter. It therefore leads automatically to a glucose accumulation in the cell. The glucose finally diffuses in its own concentration gradient on the opposite side of the cell layer from the cytoplasm into the capillary.

All these manifold transporters occurring in a single cell respond to different stimulants. Some of them become active only if a particular pH exists, others if the internal calcium concentration is raised. There are voltage sensible transporters responding to particular transmembrane potentials, or others which respond to mechanical stress. The electroneutral $Na^+ H^+$ antiporter which is present in most animal cells, merits particular attention. Under physiological conditions, at neutral pH_i it is inactive. However, if the internal pH increases, it becomes activated. This property qualifies it to be a volume regulating system. This mechanism was demonstrated in the case of lymphocytes. It has also been shown that this $Na^+ H^+$ antiporter can be activated by a multitude of substances including hormones, growth factors, lectins etc. These substances alter the above mentioned pH threshold. This seems to be an important control mechanism for the regulation of complex biological phenomena.

After this overview of some more or less cell-based physiological considerations we will now return to biophysical aspects and will look into the origin of the transmembrane potential of living cells. We started in Section 2.4.1 with the general definition of the electrical potential. According to this, the electrical potential $[\psi(x, y, z)]$ is a scalar state parameter in 3-dimensional space, similar to temperature (T), or pressure (p). The function $\psi(x)$ is mostly used to characterize the function of the potential along a line, perpendicular through the membrane (Figs. 2.21, 2.48). As *transmembrane potential* $(\Delta\psi)$ the potential difference is defined between two points, one on the inside, the other on the outside of the membrane, each at a suitable distance from it (Fig. 2.48). The sign of this difference results from its definition:

$$\Delta\psi = \psi_i - \psi_e. \tag{3.4.9}$$

Names such as Donnan potential, diffusion potential, Nernst potential, are just expressions describing the mechanisms which can give rise to the electrical transmembrane potential and do not refer in any way to different kinds of potentials which might exist simultaneously. There is only one electrical potential $\psi(x, y, z, t)$ at a given point in space (x, y, z), and at a given time (t). In Fig. 2.46, the function $\psi(x)$ illustrates this function in a very simplified way. It includes the transmembrane potential and the two surface potentials at both boundaries.

We have already learned that processes of active transport can be rheogenic (Fig. 3.23). If the transported charges can be rapidly neutralized by other fluxes, as for example by the Cl^- exchange in the membrane of human erythrocytes, then a rheogenic pump has no direct electrical consequences for the cell. If, however, no such short-circuit flux exists, the transported net charges build up a transmembrane potential, and the rheogenic pump becomes electrogenic.

Classical electrophysiology assumed that the transmembrane potential was essentially a K^+ diffusion potential. This is likely to be true for some cells, however, it has been established that for many cells electrogenic pumps are exclusively responsible for the membrane potential. Inhibition of the pumps in this case immediately leads to changes of $\Delta\psi$. Additionally, in such cells transmembrane potentials appear to be independent of external potassium concentrations (see: Bashford and Pasternak 1986).

In any case, the Na^+ K^+ ATPase, occurring in nearly all cell membranes, generates an electrochemical gradient of sodium and potassium. For most animal cells a relation near 1:10 occurs for $a_K^i > a_K^e$, and $a_{Na}^i < a_{Na}^e$. Chloride ions are distributed mostly passively, according to the Nernst equation. This nonequilibrium distribution of the cations can lead to a diffusion potential which can be calculated by the Goldman equation [Eq. (3.3.53)] as follows:

$$\Delta\psi = \frac{RT}{F}\ln\frac{P_{Cl}a_{Cl}^i + P_K a_K^e + P_{Na}a_{Na}^e}{P_{Cl}a_{Cl}^e + P_K a_K^i + P_{Na}a_{Na}^i}. \tag{3.4.10}$$

Even if the internal ion activities a_K^i and a_{Na}^i remain constant, the diffusion potential ($\Delta\psi$) can vary widely because of changing permeabilities (P_i). The limits of such variations can be easily obtained from Eq. (3.4.10):

For $P_K \gg P_{Na}, P_{Cl}$ Eq. (3.4.10) reduces to:

$$\Delta\psi_K = \frac{RT}{F}\ln\frac{a_K^e}{a_K^i}, \tag{3.4.11}$$

and for $P_{Na} \gg P_K, P_{Cl}$ it follows that

$$\Delta\psi_{Na} = \frac{RT}{F}\ln\frac{a_{Na}^e}{a_{Na}^i}. \tag{3.4.12}$$

The Goldman equation [Eq. (3.4.10)], therefore, reduces to a Nernst equation [Eq. (3.2.33)] which was derived for such kinds of semi-permeable membranes. If the typical relations of activities for sodium and potassium, as mentioned before, are inserted into Eqs. (3.4.11) and (3.4.12), then it is easy to understand that $\Delta\psi_K < 0$, and $\Delta\psi_{Na} > 0$.

This situation is illustrated in Fig. 3.25. The electrochemical gradients of potassium and sodium which are generated using metabolic energy can be

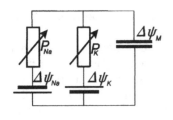

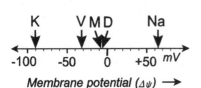

Fig. 3.25. An electrical circuit as a model illustrating the Na^+-K^+ diffusion potential of a cell as the result of a sodium ($\Delta\psi_{Na}$), and a potassium ($\Delta\psi_K$) battery. In the lower part of the figure, possible potential alterations are illustrated for the case of human erythrocytes in a solution containing 145 mM NaCl and 5 mM KCl. K $\Delta\psi_K$, V valinomycin induced diffusion potential, M potential of untreated erythrocytes corresponding to $\Delta\psi_M$, D position of the Donnan potential, Na $\Delta\psi_{Na}$

considered as storage batteries, or electrical accumulators having opposite polarities. The permeability characteristics of the ions are expressed in this model as conductivities of the variable resistors, or potentiometers through which these accumulators are discharged. If the resistance is low, then a large discharge current would flow, and if the accumulator is not recharged continuously, it would soon be empty. In fact, the permeabilities P_{Na} and P_K are usually so low that the electrochemical gradient of the living cell will persist for hours or even days. The effective membrane potential in this model is represented by the voltage difference across the capacitor $\Delta\psi_M$. This capacitor represents the capacity of the membrane (see Sect. 2.5.5). If P_{Na} and P_K have about the same value, then $\Delta\psi_M$ will be very small. If they differ, a membrane potential ($\Delta\psi_M$) will be established according to Eqs. (3.4.11) and (3.4.12).

Figure 3.25 demonstrates membrane potentials which can be induced in human erythrocytes. In this case the Nernst potentials for potassium and sodium give the limits of these possible shifts. They range approximately between −95 and +65 mV. The actual membrane potential of these cells in vivo is found to be −9 mV (M), and is only a little greater than the Donnan potential (D) which would result if the cell achieved a thermodynamic equilibrium (see Fig. 3.17). If the cells are treated with valinomycin, the membrane potential falls to about −35 mV (V). Valinomycin is an ionophore that is rapidly incorporated into the membrane causing a highly selective increase of potassium permeability. It will not reach the limiting value of the Nernst potential of potassium, because the values of P_{Cl} and P_{Na} are not negligible, as was assumed for Eq. (3.4.11). However, it is shifted in this direction.

Even if these sorts of potential alterations are possible without a significant change of concentration profiles, they must in fact be accompanied by a certain transmembrane shift of charges. It is easy to show that this charge flux is extremely small. For this we calculate the charge transfer across the membrane capacitor, which is required to adjust these potential differences ($\Delta\psi_M$ in Fig. 3.25). Let us ask the question: how many charges must be displaced in the cell membrane with a specific capacity of 10^{-2} F m^{-2} (see Sect. 2.5.5) in order to generate a transmembrane potential $\Delta\psi_M = 0.1$ V?

Equation (2.5.19) gives the corresponding relation for a capacitor. This enables us to calculate this surface charge density (σ) as a function of the transmembrane potential ($\Delta\psi$) and specific capacity (C_{sp}):

$$\sigma = C_{sp}\Delta\psi = 10^{-3} \text{ C m}^{-2}. \tag{3.4.13}$$

This value can be converted into charge equivalents of ions, using the Faraday constant (F):

$$\frac{\sigma}{F} = \frac{10^{-3}}{9.65 \cdot 10^4} \approx 10^{-8} \text{ charge equivalents} \cdot \text{m}^{-2}.$$

The resulting charge density, so far, is very small. Considering a certain geometry of the cell, for example a sphere, or in the case of a neuron, a cylinder, one can easily transform this number into a concentration shift. This type of calculation shows that the amount of ions which really must be exchanged to induce an action potential is negligibly small in relation to the total amount of ions in the cell.

This example demonstrates a most important element in the functional arrangement of the living cell: an ion pump driven by metabolic energy accumulates electrochemical energy by generating a concentration gradient of sodium and potassium. This electrochemical energy can be converted into electrical energy, alerting the membrane permeabilities (for example: P_K and P_{Na}). In this way a wide-ranging control of the electric field in the cell membrane is possible. It must be emphasized that this control is possible without any sizeable input of energy and can be realized in milliseconds.

Such permeability changes can be induced by the cell itself as well as by external influences.

As mentioned before, there are many ion-selective transporters in the cell which are controlled by internal calcium concentrations, by internal pH, by mechanical tension of the membrane, or by modifications of other parameters. Diffusion potentials may also result from an interaction between the cell and specific drugs, or may be triggered locally through mechanical contacts with surfaces or particles, such as viruses. These alterations of membrane potentials caused by local permeability changes can induce electric fields in the plane of the membrane itself.

In the next section we will consider the action potential of nerve cells as a classical example of the feedback loop between an electric field and ionic permeability in more detail. Recently, the interest in the transmembrane potential of the cell as a regulator of cellular events has greatly increased. This concerns the size of the membrane potential in various cells, as well as its time dependence. Although action potentials have a special significance in signal transfer in neurons, they occur also in many other cells.

In Fig. 3.26 quite interesting correlations of membrane potential and the state of various animal cells are illustrated. In contrast to cells with active proliferation rates like cancer cells or cells of embryos, indicating a transmembrane potential between −10 and −30 mV, non-dividing cells, like neurons or skeletal muscle cells show membrane potentials between −70 to −90 mV. The transmembrane potential of cells which pass through a state of proliferation falls before mitosis takes place. It is not yet clear whether this reflects a regulatory mechanism of the cell, or whether it is only a phenomenon that accompanies such a mechanism.

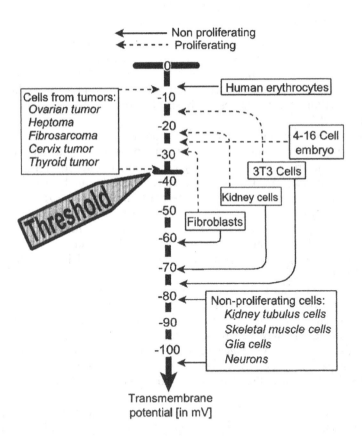

Fig. 3.26. The transmembrane potential of normal animal cells (*right*) and transformed tumor cells (*left*). It can be seen that proliferating cells indicate a membrane potential which is above the threshold value of −37 mV. Cells transiently arriving at a proliferating state lower their absolute amount of the potential. The human erythrocyte, as a non-nucleated cell with special physiological functions appears to be an exception. (Drawn according to values from Bingeli and Weinstein 1986)

In fact, in many cases alterations in the electrical field of a membrane seem to be of functional importance. The following mechanisms may cause this:

- The transverse component of an electrical field in the membrane may affect the functional state of intrinsic molecules. Dipole orientations, for example, may modify the function of transport or other functional proteins, phase transitions in the lipid components of the membrane can be influenced by the field, or a transverse shift of small charged molecules can occur.
- The lateral component of the field can cause a displacement in its mosaic structure. This could lead to a local change in the mechanical properties of the membrane causing vesiculation, spike formation etc.
- The electric field can influence local ionic concentrations, as well as local pH values in close proximity to the membrane which, in turn, could affect transport processes, biochemical reactions at the membrane surface and receptor properties.

Further reading: Role of membrane potential: Glaser 1996; voltage dependent transports: Keynes 1994; Sigworth 1994

3.4.3
The Action Potential

In the previous section we described the possibility of cells using the electrochemical gradient of potassium and sodium ions which is built up by active transport, to trigger various amounts of membrane potential simply by changing their permeabilities. This mechanism is expressed most clearly in nerve and muscle cells. This is why excitation phenomena were detected first in these cells.

Particular progress was achieved following the rediscovery of the giant axon of the squid in 1937 by J. Z. Young, and its subsequent introduction for biophysical measurements by K. S. Cole. The use of these giant axons have made it possible to apply the method of the voltage clamp. In this technique, the membrane resistance is determined at various fixed transmembrane potentials, generated by microelectrodes. Recently, using patch-clamp techniques it has been possible to investigate the kinetics of these permeability alterations in extremely small membrane areas.

Because of the large number of existing textbooks on nerve physiology we will concentrate here just on some biophysical principles of the process of nerve excitation.

The action potentials of various nerve and muscle cells as illustrated in Fig. 3.27, can be qualitatively explained using the electrical scheme of Fig. 3.25 which was discussed in the previous section. The non-exited nerve shows a very low sodium permeability (P_{Na}), its resting potential, therefore, was determined chiefly by the diffusion potential of potassium which is negative. After excitation the membrane permeability for ions increases abruptly, whereas the sodium permeability rises more quickly than that of potassium. For a short time,

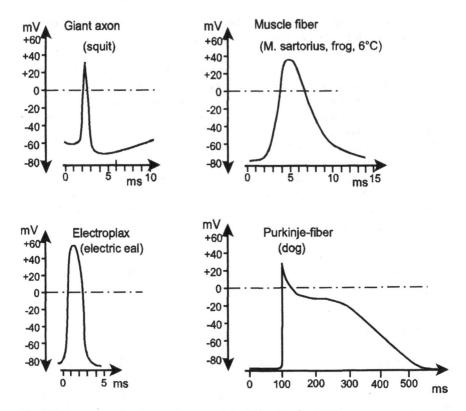

Fig. 3.27. Examples of various action potentials. (After Penzlin 1991)

therefore, the diffusion potential of sodium becomes dominant. This has the opposite polarity to the potassium potential which explains the spike of the action potentials.

As we demonstrated in the previous section, the amount of charge which is needed for this kind of depolarization is extremely low. This was checked by flux measurements in excited nerves. During the generation of an action potential, therefore, no significant alterations of the internal ion concentration occur. A nerve can generate action potentials for a long time after the ion pumps have been blocked. Only after hours does the electrochemical battery of the cell become empty.

In 1952 Hodgkin and Huxley proposed a theoretical model of this process, based on intensive experimental investigations. Its form is of a purely kinetic nature and does not contain information about concrete molecular mechanisms taking place in the membrane.

The basic equation describes the kinetics of the current in an electrical circuit, similar to the scheme in Fig. 3.25. The current density (j) in such a system can be described by the following equation:

$$j = C' \frac{\mathrm{d}(\Delta\psi_M)}{\mathrm{d}t} + (\Delta\psi_M - \Delta\psi_K)G'_K + (\Delta\psi_M - \Delta\psi_{Na})G'_{Na}. \tag{3.4.14}$$

$\Delta\psi_M$ is the electrical membrane potential, whereas the symbols $\Delta\psi_K$ and $\Delta\psi_{Na}$ indicate the Nernst potentials of potassium and sodium according to Eqs. (3.4.11) and (3.4.12). C' is the capacity of the membrane, and G'_K and G'_{Na} the potassium and sodium conductivities, always corresponding to a unit of area in the membrane. The conductivity of the membrane for individual ions cannot be measured electrically but can be obtained from experiments in which the kinetics of radioactive tracer ions are measured.

The first term of Eq. (3.4.14) gives the current density which leads to the charge of the membrane capacitor (Fig. 3.25). The following terms represent the current densities associated with potassium and sodium fluxes.

The conductivities G'_K and G'_{Na} are not constant, but are functions of the electric field in the membrane, that is of the membrane potential. The potentiometers in Fig. 3.25, therefore, are controlled directly by $\Delta\psi_M$. From the molecular point of view this means that these conductivities are the result of voltage dependent channels. It is therefore necessary to make statements about field dependents of these conductivities, i.e. the functions $G'_K(\Delta\psi_M)$ and $G'_{Na}(\Delta\psi_M)$.

To describe the behavior of these channels, a statistical approach is used. It is assumed that the channels can obtain two discrete states: "open", or "closed". The phenomenological conductivities (G'_K, G'_{Na}) then represent the average of the functional states of a large number of such channels. If all of the channels are open then the maximal conductivities $G'_{K\,max}$ and $G'_{Na\,max}$ are established.

Furthermore, it is assumed that the potassium channel will be open when exactly four events take place simultaneously, all having the same probability of occurrence (n). The real nature of these events is not explained. It could be, for example, the presence of four potassium ions near the entrance of the channel. This assumption leads to the following equation:

$$G'_K = G_{K\,max}n^4. \tag{3.4.15}$$

The probability n is a function of time and can be characterized by rate constants α_n and β_n as follows:

$$\frac{\mathrm{d}n}{\mathrm{d}t} = \alpha_n(1 - n) - \beta_n n. \tag{3.4.16}$$

Concerning the sodium permeability, it is assumed that the channel will be open when three events, each having the probability m occur simultaneously, and if another inhibitory event having the probability h has not taken place. This leads to the expression:

$$G'_{Na} = G_{Na\,max}m^3 h. \tag{3.4.17}$$

For the parameters m and h the following kinetic equations can also be written:

$$\frac{dm}{dt} = \alpha_m(1 - m) - \beta_m m, \qquad (3.4.18)$$

$$\frac{dh}{dt} = \alpha_h(1 - h) - \beta_h h. \qquad (3.4.19)$$

The voltage dependence of the channels is proposed to be the result of influences on the rate constants α and β:

$$\alpha_n = \frac{0.01(\Delta\psi + 10)}{e^{\frac{\Delta\psi+10}{10}} - 1}, \quad \beta_n = 0.125 e^{\frac{\Delta\psi}{80}};$$

$$\alpha_m = \frac{0.1(\Delta\psi + 25)}{e^{\frac{\Delta\psi+25}{10}} - 1}, \quad \beta_m = 4 e^{\frac{\Delta\psi}{18}}; \qquad (3.4.20)$$

$$\alpha_h = 0.7 e^{\frac{\Delta\psi}{20}}, \quad \beta_h = \frac{1}{e^{\frac{\Delta\psi+30}{10}} + 1}.$$

(In these equations, the potentials are in mV!)

These equations were obtained from a purely empirical approach.

It is easy to see that if the relations given in Eq. (3.4.20) are substituted into Eqs. (3.4.16), (3.4.18), and (3.4.19), a system of non-linear differential equations will be obtained. The solution of these equations can be substituted into Eqs. (3.6.15) and (3.6.17), and eventually, into the basic Eq. (3.4.14). An analytical solution of this system of differential equations is not possible. Computer simulations of these equations, however, indicate a good accordance with experimental results.

Figure 3.28 shows the calculated time courses for the changes in sodium and potassium conductivities at different membrane potentials. This also corresponds well with experimental findings. These curves illustrate the mechanism described above for the generation of an action potential. These conductivities from Fig. 3.28 illustrate the time dependent changes of the potentiometers shown in Fig. 3.25, whereas the conductivities are directly proportional to the permeabilities. Within the first millisecond following the stimulus, the sodium potential is dominant because of the rapid increase in G'_{Na} (and thus P_{Na}). This will then be counteracted by the increasing potassium potential.

Beside the direct mechanism of membrane excitation in nerve cells, the translation of this point of excitation along the axon of a nerve cell is of particular interest. In unmyelinated axons of nerves the process of pulse transmission is based on a lateral spreading of excitability by the electric field of excitation itself. A local action potential triggers the neighbor proteins in the membrane. In fact the impulse can proceed only in one direction, because of the refractory period of several milliseconds which the proteins need after an excitation to become excitable again.

Figure 3.29 illustrates the advantage of this kind of impulse propagation in relation to the transmission of a voltage pulse in an electric cable. In contrast to

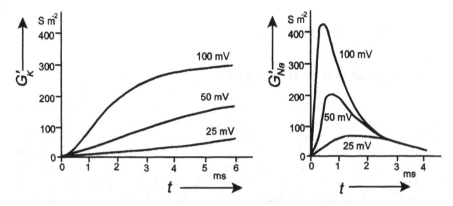

Fig. 3.28. The time dependence of the conductivities G'_k and G'_{Na} for various membrane potentials, corresponding to the theory of Hodgkin and Huxley. (After Waterman and Morowitz 1965)

Fig. 3.29. The time course of a voltage pulse which is put at time $t = 0$ at the point $x = 0$, by transmission in an isolated cable (*dotted line*) and in an unmyelinated nerve (*solid line*)

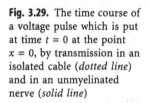

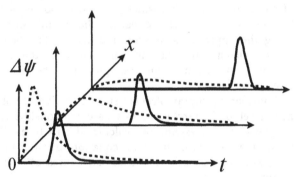

the cable, the time characteristics of the nerve pulse remain constant, even after a certain distance of transmission. On the other hand of course, the absolute velocity of pulse transmission in a cable is much faster than in an axon of a nerve. This advantage of simple electrical conductivity is used in many vertebrate, and in a few invertebrate axons which are surrounded by a fatty sheath, composed of myelin. Such nerves are called *myelinated*. This sheath is interrupted at intervals of some millimeters by so-called *nodes of Ranvier*. In the myelinated regions simple electric conductivity of the pulse occurs, as in a cable. The nodes of Ranvier represent membrane areas which are excitable in a normal way. If a certain node of Ranvier is excited, then the pulse propagates by simple electric conduction along the myelinated length and excites the subsequent node. This so-called *saltatory conduction* is a way of pulse amplification leading to a faster transport of information.

Further reading: Ferreira and Marshall 1985; Hille 1992; Markin et al. 1987; Sigworth 1994

3.5
Electric Fields in Cells and Organism

3.5.1
The Electric Structure of the Living Organism

In Section 2.3.3 we pointed out that the term "structure" must be used in a very broad sense, not limiting it to visible details of a biological system. There exists not only a morphological structure of the biological system, but also structures which are based on other physical properties. Using the definition of the electrical potential as given in Section 2.4.1, and considering its distribution in a protein, a cell, an organ etc., what we will call an "electric structure" of a biological system becomes clear. In the following we will consider chiefly the electric structure on the supracellular and organismic level of organization. Firstly, however, we will take a look at this problem in general.

As shown in Fig. 3.30, from the physical point of view a hierarchic order of structures is possible which fully corresponds to the usual classification in biology. In Section 2.1, we learned that atomic interactions can be explained using the approaches of wave mechanics. The Schrödinger equation allows us to calculate the electric parameters in the dimensions of atoms and therefore determines the energies of chemical bonds and molecular interactions.

Considering supramolecular structures like membranes, we arrived at the next level of hierarchy. In this case we considered charges and dipoles as elements of the system, and calculated structures like electric double layers, using the Poisson-Boltzmann equation of statistical thermodynamics [Eq. (2.4.13)]. We discussed electrical structures at this level of organization in Section 2.5.5.

In a next step we considered the cell as a thermodynamic system, consisting of various phases with particular properties. We found that differences exist in the electrical potential between these phases, and directed our attention to the transmembrane potential in particular. To calculate this, we used the approaches of phenomenological thermodynamics. The Nernst-Planck equation [Eqs. (3.3.32), (3.3.33)], for example, allows us to calculate flows of ions and the resulting diffusion potentials. The role of the transmembrane potential was discussed mainly in Sections 3.4.2 and 3.4.3.

In the following text we enter a further region in the hierarchic structure. We will consider electric fields in the extracellular space, in tissues and organs. This is the region of classical electrodynamics. In this case the Maxwell equations allow us to calculate electric fields in inhomogeneous dielectrics. The question arises: how does the field spread inside the body, which consists of organs with different conductivities, like bones, soft tissue, air filled cavities etc? We will come back to this question again in Section 4.4., where we will discuss the influence of electromagnetic fields on biological systems.

Fig. 3.30. The hierarchic system of the electric structure of the living organism and the corresponding physical approaches.

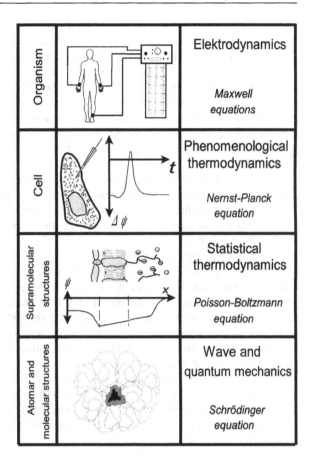

3.5.2
Electric Fields in the Extracellular Space

It has long been known that electric fields exist in the living body. Such fields are measured as electrocardiograms (ECG), electromyograms (EMG), and electroencephalograms (EEG) in medical diagnosis using simple metal electrodes. In the case of ECG potential differences between the limbs are in the order of several millivolts. In the case of EEG, because of the isolating role of the skull, they are much lower.

Furthermore, it is possible to find DC potentials between various parts of the animal body and even in plants. One has, however, to be very careful to avoid measuring electrode artefacts since in these cases absolute values in the order of a few millivolts are measured, in contrast to relative alterations of the potential as in case of EEG, ECG and EMG. Many measurements of these potentials have suffered due to use of inaccurate electrodes.

In the last 20 years it has been possible to measure electrostatic fields in the vicinity of cells using the vibrating probe techniques (see Fig. 3.32). In this case a glass microelectrode is used, the top of which consists of a metallic sphere with a diameter of only a few micrometers. This electrode vibrates at a frequency of about 500 Hz causing this sphere to be displaced periodically at an amplitude of 10–30 μm. Using a very sensitive low noise amplifier it is possible to measure voltage differences near $1 \text{ nV} = 10^{-9} \text{ V}$ between these two points. In this way, DC electric fields around a cell or an organ can be detected. A large number of various biological objects have already been investigated with this method, such as single cells (Fig. 3.32), various growing and wounded parts of plants, follicles of insect eggs, growing embryos, muscle fibers and various other tissues and organs.

The origin of extracellular electric fields and currents in biological organisms can be different. One must differentiate between fields which are generated directly by living cells, and others which are induced by mechanical deformations of tissue and motions of fluids inside the body and, in swimming fishes, also on their outer surface.

The transmembrane potential, i.e. the electrical membrane potential difference between the internal and the external phase of a cell usually does not induce an external field. This does happen however, if the transmembrane potential of the same cell differs at different microscopic locations of the membrane. Another possibility of influencing the extracellular space is by the rheogenic pumps which induce a certain transmembrane current. In both cases, the extracellular induced electric currents and fields are the result of electrochemical nonequilibria.

Of course, lateral electric fields also may be the result of a lateral mosaic of surface charges (see. Sect. 2.5.5). These fields, however, like the strong electric fields in a double layer, are part of an electrochemical equilibrium system. Consequently, they do not generate electrical currents and do not influence the system outside the electrical double layer.

The axon of a nerve cell is the best example for the case where local differences in the membrane potential will induce external fields. If an action potential is propagated along the axon, small areas of this membrane will subsequently become depolarized. In contrast to the parts of the axon with resting potential, where the external membrane is positively charged in relation to the inner one, the polarity of membrane areas with an action potential is opposite. The result is an electrical current from one point of the membrane, laterally, to another one (Fig. 3.31). Such local depolarization occurs not only in nerves, but also in muscle and other cells.

The second possibility of inducing an extracellular electric field is depicted as an example in Fig. 3.32. In this case, a rheogenic calcium pump is localized at a particular point in the membrane of an egg of a fucoid seaweed *Pelvetia*. This pump induces a flow with a current density up to 0.3 A m^{-2}. This current flows in the direction in which the first rhizoid of the cell will grow. Experiments in which such eggs were exposed to artificial fields indicate that, in fact, the field determines the direction of growth.

Examples of the above mentioned extracellular electric fields not generated by living cells are those which are induced by piezoelectricity (Sect. 2.2.2), and by

Fig. 3.31. Schematic illustration of a snapshot of an excited nerve. The *dark areas* represent the actual region of depolarized membrane (see also Fig. 3.27)

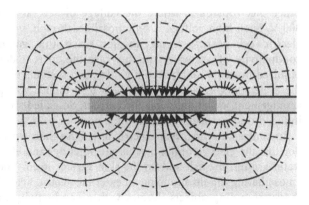

Fig. 3.32. Current (——), and equipotential lines (- - - -) near a growing egg of the brown algae *Pelvecia*, caused by a local rheogenic calcium transport, which marks the location where the rhizoid is formed. The external field was measured using the vibrating probe method. The corresponding electrode in two extreme positions during vibration is shown in relation to the size of the cell in approximately correct dimensions. (According to data from Jaffe 1979)

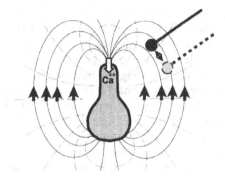

electrokinetic effects (Sect. 2.5.4). Both of them occur in bones and cartilage during in vivo deformations. Piezoelectric potentials result from deformations of charged materials, in this case chiefly by deformation of collagens. Furthermore, such deformations cause a flow in the narrow channels of the bone, carrying negative surface charges. This additionally causes streaming potentials. In contrast to the electric fields in the whole body induced by nerves and muscles, these fields are rather low at larger distances. On the other hand, they seem to be important locally in the process of bone growth and bone remodeling in vivo. This circumstance is used in therapies to stimulate bone repair by applying electric field pulses or ultrasound.

A special source of electric currents and fields are the so-called *wound potentials*. The reason for them is a permanent electrostatic potential difference between different cavities in the body, caused for example by pumps, differences in electrolyte composition and others. If the isolating properties of an epithelium are damaged by a wound, then a current between the phases of two cavities occurs and a lateral field. In most cases the wound represents the positive polarity in this battery. In Section 4.4.2 we will indicate how cells may use this field by way of galvanotaxis to tend the wound.

In a special adaptation, electric fishes use extracellular fields induced by excited cells. In the electric organs specialized muscle cells are organized in so-called electroplaques, where the voltage of several cells adds up to values of 800 V. Such high voltage is only possible, though, in fresh water fishes like the electric eel. Marine fishes do not reach this value because of the high conductivity of sea water. In this case the current density, and therefore the required power would be too high. In such systems it is not the arrangement of these elements in series which leads to a higher voltage, but rather a parallel arrangement which increases the current. So-called strong electric fishes use the induced electric fields to catch their prey, or to defend themselves.

In contrast to this, weak electric fishes use their electric fields for orientation. For this the electric organ discharges potential and electric field patterns in the water around the fish and on the skin surface itself. The fish therefore becomes an oscillating dipole, inducing an external electric field. This field is disturbed by the surrounding objects. Bodies with a conductivity higher than the environment, such as animals or submersed plants, focus the field, others with lower, or even no conductivity, like for example stones, induce regions with divergent field lines. Over each cycle of the electric organ discharge, the fish measured the electric field amplitude at any point near its surface. As a result, the relatively simple electric organ discharge leads to a fairly uniform pattern of electrosensory stimuli along the body surface, which may facilitate central processing of electrosensory images. (For electroreception see Sect. 4.4.2.)

These considerations indicate that the role of extracellular electric fields in the body is rather heterogeneous. This results from the different electric field strengths and current densities which have been measured in various kinds of growing tissues, embryos etc., which amounts to values of up to 1 A m^{-2}. Wound potential can induce current densities which are even larger. The electric fields near cells, indicated by use of the vibrating probe techniques are in the order of 0.1 V m^{-1}. This method, however, has no high resolution. It only allows the measuring of fields at a distance of more than 30 μm from the source. One can assume that the field close to the membrane of an excited nerve, for example, may be much higher. The field strengths and current densities of usual electrocardiograms (ECG), or electroencephalograms (EEG) at a certain distance from the heart, or in the brain, are obviously much lower.

Recently, the role of extracellular electric fields has been discussed in the context of processes of differentiation and growth. So, for example, wound potentials may lead to galvanotaxis of embryonal cells and therefore influence the process of healing (for galvanotaxis, see Sect. 4.4.2). Electrophoresis of growth factors in DC fields in the body probably also occur, controlling processes of morphogenesis (for theoretical models of morphogenesis see Sect. 5.3.2).

Very little is known about the electric fields in the direct vicinity of cells and in intercellular clefts. The mosaic of surface charges creates a static field, and additionally rheogenic transporters induce various nonequilibrium situations which may lead to electric currents. Furthermore, streaming potentials and streaming currents can be induced which could become important locally. In this respect, the control of lateral translocations of membrane proteins is discussed (see also Sect. 2.5.5).

The use of electric fields in the body in medical diagnosis raises the question: how do electric fields, generated for example by the beating heart, spread in the body and, vice versa, what can an electrocardiogram tell a doctor about the function or dysfunction of the heart? Modern techniques of ECG measurement allow a reasonable reconstruction of the electric field in the body using parallel recordings of the ECG simultaneously in different parts of the body. These fields were not induced by single cells but rather by cellular arrangements excited simultaneously. In the case of an ECG, definite parts of the cardiac muscle are excited in the rhythm of the beating heart. Consequently, the heart becomes an oscillating dipole, the orientation of which changes according to the orientation of the excited parts of the muscle.

The first person to propose a method to evaluate ECGs was Willem Einthoven who was awarded the Nobel Prize for this in 1924. He proposed that it should be possible to localize the excited parts of the heart leading the potentials from three points of the body, which are more or less equally distant from the heart. This so-called *Einthoven triangle* uses leads from both arms and the left leg. This kind of recording is used even today for routine measurements of the ECG, though an exact correlation between the resulting dipole vectors of the field with the anatomical position of various heart muscles is impossible because of the electrical heterogeneity of the body.

In Fig. 3.33 this method is illustrated schematically. The three cardiograms, led by the corresponding points, indicate in a periodically repeated sequence P, Q, R, S, and T waves. They represent the sequence of excitation of different parts of the heart muscle. It starts with the excitation of the atria (P wave). If all the atrial fibers are in the plateau phase, the PQ segment is reached. Subsequently, the excitation spreads over the ventricles, beginning on the left side of the ventricular septum, spreading toward the apex (QRS waves), finally reaching the ventricular recovery phase (T wave). As a result of the projection of these curves corresponding to the geometry of an equilateral triangle, a rotating vector appears, the origin of which lies in the crossing point of the three axis. The arrowhead moves periodically along the dashed line.

Further reading: The role of extracellular fields: Hotary and Robinson 1990; Nuccitelli 1995; the electric field of ECG: Reilly 1998; the electric field in bones: MacGinitie 1995

3.5.3
Passive Electrical Properties of Tissue and Cell-Suspensions

To describe the electric structure of cells and organisms in view of electrostatics, or of ionic currents induced by rheogenic pumps, parameters such as membrane resistance, conductivity of cytoplasm etc. were introduced. Furthermore, the membrane capacity was used for explanation of membrane potentials in Section 3.4.2. In contrast to membrane potentials that will require active biological processes, the latter parameters can be summarized as *passive electric properties*.

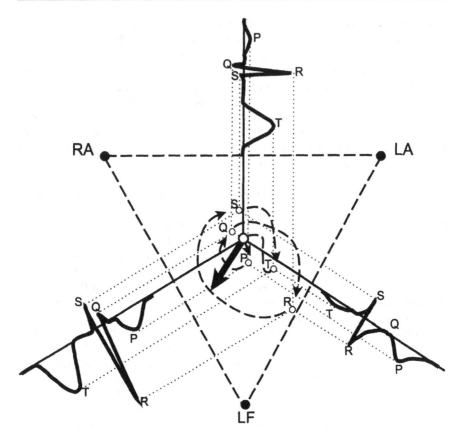

Fig. 3.33. Construction of a vector cardiogram according to Einthoven's triangle. Using the time course of potentials curves, orientated according to the sides of an equilateral triangle, a rotating dipole can be constructed in the center of it. P, Q, R, S, T – means the corresponding waves of the ECG. RA – the lead from the right arm, LA – from the left arm, and LF – from the left foot

In this Section we will consider consequences of passive electric properties for the frequency dependence of effects in AC fields.

To derive the basic parameters and equations, let us first consider the electrical properties of a plate capacitor filled with material of certain conductivity and dielectric constant (Fig. 3.34). This system can be described by an equivalent circuit consisting of a conventional capacitor and a resistor in parallel. This system, therefore, has an effective AC resistance which is called *impedance*, whereas its reciprocal is the AC conductance or *admittance* (Y^*).

Applying an AC voltage, the current will flow through the resistor and periodically recharge the capacitor. The admittance of this system, therefore, is determined on the one hand by the static conductance (G) of the resistor, and on the other hand by the displacement current passing through the capacitor, which depends on the frequency (ω). The admittance (Y^*) of this RC circuit is therefore:

Fig. 3.34. A capacitor consisting of two parallel plates of area A and a mutual distance d filled with an inhomogeneous medium described by conductivity and a dielectric constant as an example for the impedance of a biological tissue

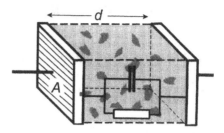

$$Y^* = G + j\omega C. \tag{3.5.1}$$

This equation can be modified using the geometrical parameters of the electrode: area (A) and mutual distance (d), together with the material constants specific conductivity (g) and the permittivity ($\varepsilon\varepsilon_0$):

$$G = \frac{A}{d}g \quad \text{and} \quad C = \frac{A}{d}\varepsilon\varepsilon_0. \tag{3.5.2}$$

Introducing these relations into Eq. (3.5.1), one gets:

$$g^* = g + j\varepsilon\varepsilon_0\omega, \quad \text{whereas:} \quad Y^* = \frac{A}{d}g^*. \tag{3.5.3}$$

In this equation g^* is the complex specific admittance of the system.

Why is the imaginary number $j = \sqrt{-1}$ used in this equation, transforming some of these parameters into complex numbers marked by the superscript *? The reason for this is a particular time dependence of these parameters which must be taken into account.

In describing the behavior of an RC circuit in an AC field, the amplitude of the sine current as well as the phase shift are important. To understand this, let us consider an AC voltage having the following time function:

$$U = U_{max} \sin \omega t. \tag{3.5.4}$$

An applied voltage (U) therefore oscillates with an angular frequency ($\omega = 2\pi v$) at an amplitude of U_{max}. According to Ohm's law [Eq. (3.1.51)], in a circuit containing only a resistance, the current (I) possesses the same time behavior:

$$I = I_{max} \sin \omega t. \tag{3.5.5}$$

As already mentioned, in an RC circuit, the displacement current of the capacitor must also be taken into account. The charge (q) of a capacitor is determined by the following equation:

$$q = UC. \tag{3.5.6}$$

Since the current is defined as the time derivative of charge, the displacement current (I_C) is:

$$I_C = \frac{dq}{dt} = \frac{d(UC)}{dt} = C\frac{dU}{dt}. \tag{3.5.7}$$

Introducing Eq. (3.5.4) into Eq. (3.5.7) and leaving C = constant, one gets:

$$I_C = C\frac{d(U_{max}\sin\omega t)}{dt} = CU_{max}\omega\cos\omega t = CU_{max}\omega\sin\left(\frac{\pi}{2}+\omega t\right). \tag{3.5.8}$$

Further, considering that:

$$I_{C\,max} = CU_{max}\omega, \tag{3.5.9}$$

it follows:

$$I_C = I_{C\,max}\sin\left(\frac{\pi}{2}+\omega t\right). \tag{3.5.10}$$

Comparison of Eqs. (3.5.5) and (3.5.10) clearly indicates a phase shift of $\pi/2$ in between current and voltage.

Consequently, in Eq. (3.5.1) in the case of an AC current, the time course of the capacitive component of the sum must be considered. This circumstance is taken into account using a Gauss plane containing imaginary numbers.

The term ωC in Eq. (3.5.1) is multiplied by the imaginary number $j = \sqrt{-1}$ and therefore plotted on the ordinate in a Gauss plane (Fig. 3.35).

In the same way as the complex specific admittance (g^*) in Eq. (3.5.3), a complex dielectric constant (or relative permittivity; ε^*) of the system can be introduced. Starting with the equation for the effective AC resistance of a capacitor ($R = 1/\omega C$), one can introduce a complex capacitance (C^*) as follows:

$$C^* = \frac{Y^*}{j\omega}. \tag{3.5.11}$$

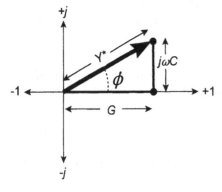

Fig. 3.35. Representation of the conductance (G) of the resistor and the admittance of the capacitor ($j\omega C$) of the analog circuit from Fig. 3.34 in the Gauss plane of complex numbers. The admittance of the system (Y^*) corresponds to the length of the resulting vector. ϕ is the resulting phase shift

Inserting the parameters of [Eqs. (3.5.1) and (3.5.2)] into this equation, and considering $C^* = \varepsilon^* \varepsilon_0 \, A/d$, and $1/j = -j$, one gets:

$$\varepsilon^* = \varepsilon - j \frac{g}{\omega \varepsilon_0}. \qquad (3.5.12)$$

This derivation of the basic equations [Eqs. (3.5.3) and (3.5.12)] allows us to understand the properties of complex dielectrics in AC fields.

These parameters are used to calculate, for example, the distribution of AC fields in the body as illustrated in Fig. 3.30, which is important in medical and environmental biophysics. We shall come back to this problem in Section 4.4.3.

The electrical properties of biological tissues containing a multitude of various cells with various sizes and intercellular clefts are, of course, too complicated to be described by a simple RC circuit as we did before. An enormous network of elementary RC elements with different parameters would be required to build an appropriate model. This, of course, is impossible for practical use. For this reason one uses the same simple RC circuit which is depicted in Fig. 3.34 except that the properties of the resistor as well as of those of the capacitor are now frequency dependent. This simply results from the dielectric heterogeneity of the system, even though the individual dielectric parameters of any particular constituent of the systems, like water, lipid membranes etc. can be considered as frequency independent.

Figure 3.36 indicates mean values of the complex dielectric constants (ε^*) and the specific admittance (g^*) of various tissues in a broad frequency range. Such data can be measured by electrodes as illustrated schematically in Fig. 3.34. It can be seen that the specific admittance increases with increasing frequencies. The main reason for this is the increase of the membrane admittance. On the other hand, the dielectric constants of the tissue at low frequencies have enormous values up to the MHz range. These

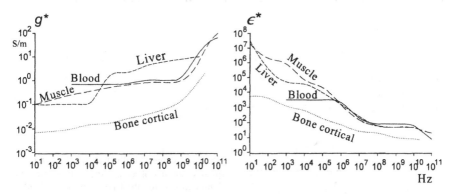

Fig. 3.36. Complex specific admittances and complex dielectric constants of various tissues as functions of frequency. (According to averaged data from Gabriel et al. 1996)

constants drop below the standard value of the dielectric constant of water ($\varepsilon \approx 80$) only at microwave frequencies.

These curves do not continuously increase or decrease with frequencies, but change in a characteristic step-wise fashion. The reasons for this, formally, are the properties of various kinds of RC-circuits, which are included in the system. These frequency regions, in the first instance have formally been designated as α-, β-, and γ-dispersions with increasing frequency (see Fig. 3.37).

Various phenomena are summarized as α-*dispersion*, which for cell-sized objects occur in the frequency range below 10 kHz. Mostly, reactions in the electric double layer of cell membranes, such as various electrokinetic phenomena (see Sect. 2.5.4) e.g. the deformation of the ionic clouds, are responsible for this. Also, artefacts related to polarization phenomena of the measuring electrodes must be expected in the α-range. It is difficult to determine the individual contributions of these different phenomena.

The β-*dispersion* is mostly based on processes connected with the membrane as a dielectric barrier. In general, these are related to *Maxwell-Wagner dispersions* which are caused by structural inhomogeneities of the material, such as tissue inhomogeneities, cells and organelles. Some authors subdivide the β-region into the β_1-, and the β_2-range which refer to the dispersion of the cell membrane (β_1) and cytoplasm (β_2) polarization respectively.

The γ-*dispersion* at higher frequencies is caused by so-called *Debye relaxations* of various molecular dipoles. At frequencies of the γ-dispersion region, even the resistivity of the internal and external milieu of the cell cannot simply be described by ohmic resistances (see small RC-circuits in Fig. 3.38).

Assuming that the dispersion of water dipoles occurs at 18.7 GHz, one might define for this a special δ-region. This, in fact, is true only for free water. Recent measurements indicate that the dispersion of bound water may occur just in the γ-dispersion region. This is important because of possible effects of high frequency electromagnetic fields on biological tissue (Sect. 4.4.4).

Further reading: Gabriel et al. 1996; Gimsa and Wachner 1998; Pethig 1979; Pethig and Kell 1987; Polk and Postow 1996; Riu 1999; Schwan 1957

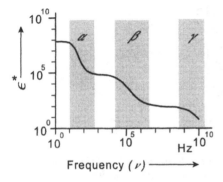

Fig. 3.37. Schematic illustration of the frequency regions of α-, β-, and γ-dispersion as demonstrated for the frequency dependence of the complex dielectric constant of a biological tissue

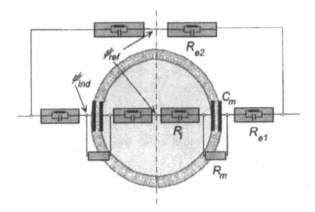

Fig. 3.38. A simplified analog circuit demonstrating current ways through, and around a spherical cell. R_{e1} and R_{e2} – resistances of the external medium, R_i – resistance of the cytoplasm, R_m – membrane resistance, C_m – membrane capacity, ψ_{ref} – reference potential at the symmetry plan of the system, ψ_{ind} – induced potential at the membrane surface. At higher frequencies RC properties must be attributed to the resistances R_{e1}, R_{e2}, and R_i too (internal and external resistors were split to demonstrate the line of the reference potential ψ_{ref})

3.5.4
Single Cells in External Electric Fields

Figure 3.38 indicates a simplified electronic circuit, describing the passive electric properties of a spherical cell in an electrolyte solution. As in Fig. 3.25, the membrane is represented by a capacitor (C_m) in parallel with a resistor (R_m) simulating the membrane conductivity. In DC fields and low frequency AC fields the conductivities of external and internal media can be described by simple resistors(R_{e1}, R_{e2}, R_i).

The high specific conductivity of the cytoplasm and the external medium on the one hand, and the high resistance of the membrane on the other results in R_m being more than seven orders of magnitude higher than R_e or R_i. Applying Kirchhoff's law, a low frequency current therefore does not flow through the cell, but around it.

This is demonstrated schematically in Fig. 3.39 where a spherical cell is shown with the electrical properties as described above. Since in the membrane the field cannot be resolved at this scale, it will be discussed in the next figure. Because of the high membrane resistance, at DC fields and extremely low frequency AC fields, the induced field inside the cell is negligible, and the external field becomes deformed. As already discussed in Section 3.5.3 with increasing frequencies of the external field, the membrane capacitor will become short circuited. This leads to a change of the external field distribution (Fig. 3.39B and C). If there is no large difference in the permittivities between the medium and cell plasma, the amount of field penetrating the cell depends only on the relation of the inside-outside conductivities. In any case, the field penetration increases

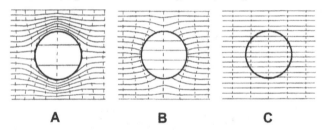

Fig 3.39. Current lines (——), and equipotential lines (- - -) for a spherical cell in a homogeneous electric AC field. In contrast to Fig. 3.40 the polarization charges at the membrane are not depicted. **A** the cell in a low frequency, **B** and **C** – in a high frequency AC field. For **B** the conductivity of the external medium is lower than that of the internal one, in **C** the cell is surrounded by a physiological medium where the permittivities of the internal and external milieus are the same

at high frequencies. Therefore, membrane polarization decreases and polarization of the cytoplasm increases.

To discuss the influence of external fields on the membrane potential, let us now consider a spherical cell in a DC field (Fig. 3.40). Neglecting the conductivity of the membrane, which in fact is very low, it can be understood that the external field leads to an accumulation of charges, especially in membrane areas that are orientated perpendicular to the current lines. This yields a deformation of the external electric field (Fig. 3.39) and also produces charging of the membrane capacitor. The polarization of the membrane induces an opposite intracellular field which significantly mitigates the field influence from outside (short arrows inside the cell). Because of the low field strength inside the cell, polarization of the membranes of cell organelles can be neglected at least in DC, or low frequency AC fields.

The transmembrane potential difference $\Delta\psi_{ind}$ which is induced by an external field, as depicted in Fig. 3.40, corresponds to the difference $\psi_{ref} - \psi_{ind}$ of Fig. 3.38. Therefore, there is no potential difference across the resistor R_i, and no current flows inside the cell.

Because of the deformation of the external field by the cell itself (Fig. 3.39A), the potential ψ_{ind} is somewhat different from the potential at an identical x-coordinate away from the cell. This was taken into account by introducing the resistor R_{el} (Fig. 3.38) and by bending the lines of the extracellular potential function near the membranes in the lower part of Fig. 3.40. The extent of this deviation depends on the shape of the cell. In case of a spherical cell, 1.5r has to be considered.

The polarity of the induced potential difference ($\Delta\psi_{ind}$) on both sides of the cell is the same as that of the external field, whereas the undisturbed in-vivo transmembrane potential $\Delta\psi_M$ is always polarized inside-out. Therefore, on one side the cell is polarized in the same direction as the induced potential difference ($\Delta\psi_{ind}$), and on the other side in an opposite direction. This means that the resulting potential differences ($\Delta\psi_1$, $\Delta\psi_2$) differ from one another on both sides of the cell. The transmembrane potential at locations in the cell perpendicular to

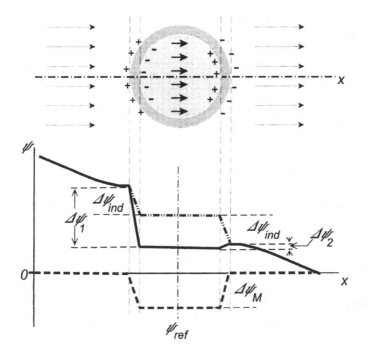

Fig. 3.40. The distribution of charges and potentials at a spherical cell in an external DC field. In the *upper* part the *arrows* indicate direction and strengths of the field vectors. The charges near the membrane are the result of polarization by the external field. In the *lower* part of the figure the potential profile over the x-axis through the center of the cell is depicted: - - - membrane potential without an external field influence including the undisturbed in vivo transmembrane potential ($\Delta\psi_M$); ·······- potential induced by the external field, not considering the biological membrane potential, with the induced potential difference ($\Delta\psi_{ind}$); —— actual potential function, as the sum of both functions, and the resulting membrane potential differences on both sides of the cell ($\Delta\psi_1$, $\Delta\psi_2$). ψ_{ref} is the reference potential according to Fig. 3.38. For simplicity, in contrast to Fig. 2.48, the electric double layer is not taken into account

the current lines will not be influenced by the external field at all. For this reason, in Fig. 3.38 the reference level for the induced potential (ψ_{ref}) is identical inside and outside the cell.

To calculate this induced membrane potential (ψ_{ind}), a small column oriented in the field direction which processes the cross-sectional area A, and cut out along a line through the cell's center can be considered. The characteristic length up to which cell polarization enhances the external medium is $r_e = 1.5r$. Furthermore, we shall consider that the Ohmic current flow through the membrane resistor R_m (Fig. 3.38) can be neglected. Considering an external field of the strength E, the potential difference outside the cell, from point ψ_{ref} to $\psi_{1.5r}$ amounts to 1.5rE. The same potential must drop over the column. Using the proportionality between the impedance, as effective AC resistance, and the potential drop, and considering the impedance of the membrane capacitor: $Z^* = 1/Y^* = -j/\omega C$, one can write:

$$\frac{\Delta\psi'_{ind}}{1.5Er} = \frac{-j/\omega C}{-j/\omega C + R_1 + R_{el}},$$ (3.5.13)

where ψ'_{ind}, in contrast to ψ_{ind} means the induced potential just in the membrane section of this column. Introducing:

$$R_i = \frac{r}{g_i A}; \quad R_{el} = \frac{0.5r}{g_e A} = \frac{r}{2g_e A}; \quad C_m = \frac{C}{A},$$ (3.5.14)

and noticing that:

$$|j + a| = \sqrt{1 + a^2},$$ (3.5.15)

one gets:

$$\Delta\psi'_{ind} = \frac{1.5rE}{\sqrt{1 + \left[r\omega C_m\left(\frac{1}{g_i} + \frac{1}{2g_e}\right)\right]^2}}.$$ (3.5.16)

This is the maximum of the membrane potential induced by the external field at the point of the spherical cell, where the membrane is perpendicular to the field direction. At an angle perpendicular to the field (Fig. 3.40) the induced transmembrane potential vanishes. To consider the induced potential at all points of the sphere, the radial coordinate α can be introduced, and the [Eq. (3.5.16)] must be multiplied by $\cos \alpha$:

Let us consider a further aspect: what is the time constant (τ) of the charge of the membrane capacitor? Again, applying the scheme of Fig. 3.41 we can write:

$$\tau = (R_i + R_{el})C = r\,C_m\left(\frac{1}{g_i} + \frac{1}{2g_2}\right).$$ (3.5.17)

Introducing this into [Eq. (3.5.15)] and considering the angular coordinates, one gets the equation describing the induced potential (ψ_{ind}) for every location:

$$\Delta\psi_{ind} = \frac{1.5Er\cos\alpha}{\sqrt{1 + (\omega\tau)^2}} = \frac{1.5Er\cos\alpha}{\sqrt{1 + (2\pi\nu\tau)^2}}.$$ (3.5.18)

Using common cell parameters, like for example: $r = 10^{-5}$ m, $C_{sp} = 10^{-2}$ F m^{-2}, $g_i = 0.5$ S m^{-1}, and $g_e = 1$ S m^{-1} in Eq. (3.5.17) a relaxation time of $\tau = 2.5 \cdot 10^{-7}$ s can be obtained. Introducing this parameter into Eq. (3.1.18) it is easy to demonstrate that for low frequency AC fields ($\nu < 10^5$ Hz), the

Fig. 3.41. A column with the area A, cut out from the cell (Fig. 3.38), and the electric scheme for a corresponding circuit. The resistor R_m is so high that it is disregarded in this scheme. (Corresponding to the approach of Gimsa and Wachner 1999)

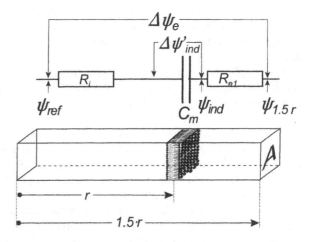

denominator of this equation will approach 1. For DC fields and low frequency AC fields, therefore, this equation reduces to:

$$\Delta\psi_{\text{ind}} = 1.5Er\cos\alpha. \tag{3.5.19}$$

As already mentioned, this, as well as [Eq. (3.5.18)], is correct only for very high membrane resistance which is justified in most cases. Using this equation one can for example calculate that in a spherical cell with a diameter $r = 10$ μm a transmembrane potential of $\Delta\psi_{\text{rel}} = 1.5$ V will be induced at position $\alpha = 0°$ by an external low frequency, or DC field of $E = 100$ kV m^{-1}.

Further reading: Gimsa and Wachner 1999; Grosse and Schwan 1992

3.5.5
Manipulation of Cells by Electric Fields

Using various types of electric fields, it is possible to move, separate, fuse, perforate, or deform cells. Nowadays, these methods are broadly applied in biotechnology. In contrast to the effects of weak electric and electromagnetic fields, which we will discuss later (Sect. 4.4), in these techniques rather strong fields are used, the energy input of which is significantly larger than the energy of thermic noise.

Let us first consider the influence of strong electric field pulses on cells, leading to considerable hyper-, or hypopolarization of the membrane. We discussed the electrostatic properties of the cell membrane in Sections 2.4.1, 2.5.4, 3.4.2, and 3.5.1. There we focused on the high electric field strength in the cell membrane in vivo, and its role in regulation of structure and distribution of membrane components. An externally applied DC field increases the membrane potential ($\Delta\psi_1$ in Fig. 3.40) on one side of the cell, and consequently, raises the internal electric field strength in the membrane at this point. If the transmembrane potential reaches an amount of approximately 1 V, the internal field strength will destabilize the membrane, resulting in a so-called *electric break down*.

The application of such strong fields in solutions with considerable conductivities, as in a physiological milieu will, of course, induce significant Ohm's heating. The electrical power absorbed in a medium per volume equals E^2g [see Eq. (4.4.7)]. This parameter, by the way, is the so-called specific absorption rate (SAR) which we will use in Section 4.4.3. Considering a specific conductivity of $g = 1$ S m^{-1}, and the above-mentioned field strength of $E = 10^5$ V m^{-1}, a power density of 10^{10} W m^{-3} or 10^7 W kg^{-1} will be absorbed. In fact, electric fields with such a high field strength can be applied only for a short period of time (1 to 500 µs).

In case of a so-called electric break down the membrane loses its property as a diffusional barrier; its electrical resistance breaks down, which may even lead to a full destruction of the cell by subsequent lysis. However, by selecting correct treatment parameters, i.e. using a moderate shape, amplitude and duration of the pulse at optimal temperature and electrolyte conditions, a reversible electric break down can be induced. For this, or also for processes of electroinjection it is very important to avoid a Donnan-osmotic cytolysis of the cells (see Sect. 3.2.3).

This kind of treatment of cells is used for so-called *electroinjection* or *electropermeabilization*. When cells suspended in a solution or in a suspension of macromolecules or small particles are pulsed in this way, the short moment of the induced membrane destabilization is enough to allow for membrane penetration of particles with sufficient probability. Additionally, such disturbances of the membrane structure can also increase the ability of the cell to absorb particles by phagocytosis. Electroinjection is used, for example, to introduce drugs, genetic materials, or even small organelles into cells.

If two cells are attached to each other during electric pulse application, or during a short time after the pulse, the points of induced membrane instability may interact, leading to a fusion of the lipid phases of both cells. Subsequently, osmotically governed process of fusion of the cytoplasm may follow.

This process of so-called *electrofusion* of cells has recently been explained by several mechanisms which cannot be discussed here in detail. It is just necessary to point out that membrane fusion and subsequent cell fusion may be triggered by electric pulses. However, these processes need a longer time than the pulse duration itself.

Electrofusion of cells is broadly applied in biotechnology. In contrast to the induction of cell fusion by some chemical agents, electrofusion has the advantage of being much more selective. Using sufficiently small electrodes, it is possible to fuse even two particular cells. Meanwhile, methods have been developed to bring specific cells into contact with each other. Usually, cells are arranged for fusion by dielectrophoretic aggregation.

This leads us to cell translocations, caused by polarization of cells in AC fields. In the following we will consider in particular the effects of dielectrophoresis and electrorotation.

Dielectrophoresis is the translation of objects in inhomogeneous AC fields (Fig. 3.42). *Electrorotation* is the rotation of particles in a rotating electric field (Figs. 3.43, 3.44).

Fig. 3.42. Dielectrophoresis of cells in an inhomogeneous AC field between a flat and a pointed electrode

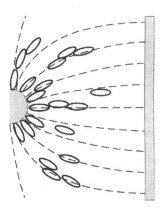

Fig. 3.43. In the same way as depicted in Fig. 3.40, an external field leads to charge separations near the membrane (**A**). In accordance with this polarization, an effective dipole can be constructed for the low frequency region, where the membrane has a low conductivity (**B**). In contrast to this, in high frequency fields, where the membrane capacitor is fully short circuited, the dipole is formed only by the polarization of the cytoplasm (**C**). In both cases the induced dipole follows the rotating external field with some delay (*large arrows*). This delay in one case (**B**) leads to a repelling, in the other case (**C**) to an attractive force resulting in anti-field rotation (**B**), or in cofield rotation (**C**)

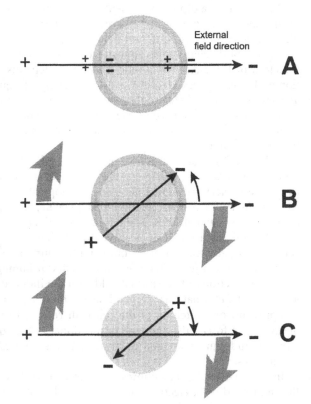

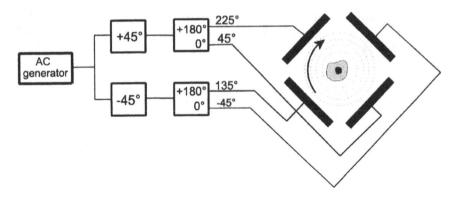

Fig. 3.44. Principle scheme of a four-electrode system for electrorotation. A generator produces an AC signal with adjustable frequency. This signal is split into four signals of particular phase relations. Using a four-electrode chamber, a field rotating at the frequency of the generator is obtained. The rotation of cells or other objects in the chamber is much smaller than that of the field. Therefore, the rotation can be observed under a microscope

Let us first consider the process of dielectrophoresis. A homogeneous electric field may induce a dipole and orientate it, but it will not lead to a translocation. In an inhomogeneous field, however, both charges of the dipole are located at different positions with different field strengths. Therefore, the forces which are acting on both charges in opposite directions are no longer balanced. The direction of the dielectrophoretic translation depends on the kind of polarization of the cell. As depicted in Fig. 3.45, there exist frequency regions of negative as well as positive dielectrophoresis. The driving force of dielectrophoresis is proportional to the square of the applied field strength. This results from the double role of the field: on the one hand it induces the dipole, on the other hand, it drives the object.

In Fig. 3.42 the dielectrophoresis of yeast cells is shown schematically in an inhomogeneous field between the peak of a needle, shown as a small hemispheric electrode, and a flat electrode opposite. The strength of the applied AC field increases in the direction of the needle, as indicated by the increasing density of the current lines. The figure shows the accumulation of the cells near this electrode as a result of positive dielectrophoresis. The polarization of the cells additionally leads to their mutual attraction and subsequently to a sort of *pearl chain formation*. This effect also occurs in homogeneous fields.

The interaction of the external field with the induced cellular dipole may also lead to cell deformations. Depending on the frequency of the applied fields, compression or stretching of the cell with respect to the field direction might occur. These forces also increase proportional to the square of the applied field strength and can be applied to investigate mechanical properties of cells and membranes.

In contrast to dielectrophoresis, which can be described by the real part of the induced dipole, electrorotation is related to its imaginary component (see

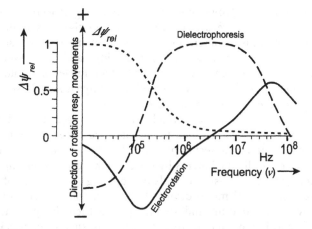

Fig. 3.45. An example for electrorotation (——) and dielectrophoresis (- - -) of a cell in a solution of low conductivity dependent on the applied frequency. Positive values of the electrorotation spectrum mean spinning in the direction of field rotation (cofield rotation), or in the case of dielectrophoresis a movement into the direction of higher field intensity, and vice versa. Additionally, the parameter $\Delta\psi_{rel} = [1+(2\pi\nu)^2]^{-1/2}$ is depicted as function of frequency (····) according to Eq. (3.5.18) standing for the influence of the induced membrane potential. (With the help of Wachner and Gimsa)

Fig. 3.35). Figure 3.43 shows that a permanent torque acts due to the permanent phase shift between the dipole field and the rotating field which is directed opposite to the direction of the external field (*antifield rotation*), or in the same direction (*cofield rotation*). In Fig. 3.45 the frequency dependence of the electrorotation of a cell in a low conductivity medium is depicted. It shows that the antifield rotation occurs at frequencies of about 100 kHz, followed by a point of zero rotation and finally, in the MHz region a maximum of cofield rotation occurs at about 50 MHz.

To calculate these two types of movement, model considerations are applied. For the simplest approach a single shell model is used which considers a dielectrically homogeneous sphere, covered by a membrane. It represents a spherical cell without electrically significant organelles (Fig. 3.38). Non-nucleated swollen erythrocytes can be modelled by this approach. In this case the radius of the cell and the membrane thickness as geometrical parameters are included in the corresponding equations, as well as the conductivities and permittivities of the membrane, the cytoplasm, and the external medium. These model considerations indicate that the peak of the antifield rotation in its frequency position, and its amplitude, reflects the membrane properties of the cell. This first characteristic frequency is determined by the time constant given by Eq. (3.5.17). If the property of the membrane as a diffusion barrier is changed, this peak vanishes. The maximum of the cofield rotation, i.e. the second characteristic frequency, indicates the conductivity and permittivity of the cytoplasm in relation to that of the external milieu.

Electrorotation is an approved method for measuring dielectric properties of individual living cells. There are a number of investigations indicating that electrorotation can measure alterations of cell properties which are induced by drugs, toxic agents, virus attacks, cell activations and other events. Under special conditions even properties of organelles like nuclei or vacuoles can be measured. Automatic video systems and methods of dynamic light scattering are applied to register the induced movement.

In contrast to electrorotation, which is mostly used for cell analyses, dielectrophoresis can also be applied for preparative cell separation and other biotechnologically interesting techniques. As already mentioned, the formation of cell chains, or cell-cell attachments by dielectrophoresis, for example, is used to perform selective electrofusion of cells by applying short pulses of high field intensity. Recently, special micro devices have been constructed to investigate single cells by electrorotation and to manipulate them by dielectrophoretic translations. According to the large relative surface of such chambers in relation to their volume, heating of the samples was minimized. This allows one to apply dielectrophoresis and electrorotation in physiological solutions of relatively high conductivities. As demonstrated in Fig. 3.46, it is possible to produce microscopic field traps, in which cells, lifted by dielectrophoretic force can be held in a stable position without any surface contact. Using electrode arrangements inducing traveling waves (Fig. 3.47), it is also possible to translocate cells on microchips. This new technique opens enormous possibilities for biotechnological applications.

Further reading: For electrorotation and dielectrophoresis: Georgiewa et al. 1998; Gimsa and Wachner 1998; Fuhr et al. 1996; Fuhr and Hagedorn 1996; electromanipulation and electrofusion: Cevc 1990; Lynch and Davey 1996;

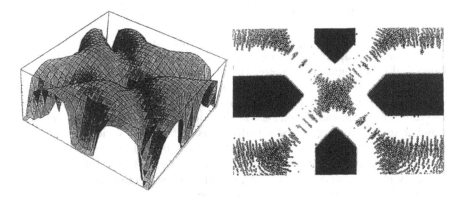

Fig. 3.46. On the *left*: surface of constant force potential in a high frequency field trap. Four electrodes (see also *left*) produce a time average electric field funnel. On the *right*: Trapping, levitation and pearl chaining of latex particles at 1 MHz in the chamber according to the calculated force distribution. (After Fuhr et al. 1996)

Fig. 3.47. Cells moving in a system of inter-digitated travelling-wave electrodes. Note that the direction of field propagation is opposite to the direction of cell motion according to negative dielectrophoresis. (After Fuhr et al. 1996)

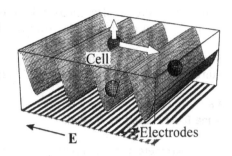

Zimmermann 1996; dielectric cell deformation: Sukhorukov et al. 1998; Ziemann et al. 1994

3.6
Mechanical Properties of Biological Materials

Biomechanics is a branch of biophysics which explains the mechanical properties of anatomical stability, the movements of limbs as well as the mechanics of walking, flying, swimming, the mechanics of blood flow, of mechanoreceptors and many others. In fact it is one of the oldest branches of biophysics. It has developed in strong connection with the development of the physical mechanics itself. In the period of renaissance the pioneers of mechanics, like Gallileo Galliley, René Descartes, Isaac Newton and many others were always also interested in the mechanics of animals. The first classical book on biomechanics, Alfonso Borelli's *De Motu Animalium* was printed in Rome in 1680. It already contained basic physical considerations on swimming, flying, and movement of animals as well as various calculations of moments of human limbs and the spine under conditions of loads. As we already mentioned in Section 1, this book marked the beginning of medical physics, which was called at that time iatro-physics. Furthermore, D'Arcy Thompson's book *On Growth and Form*, published first in 1917 must be mentioned. This book analyzed for the first time basic processes of cell mechanics, shape formations, and many other biomechanical fields.

In recent decades biomechanics has become increasingly important in diagnostics and therapeutics, especially the biophysics of blood circulation (hemorheology). The biomechanics of the skeleton, as well as of limbs and joints are also of central interest. These investigations of complex mechanical processes require a profound knowledge about the viscoelastic properties of blood and tissues. The investigation of these parameters in vitro as well as in vivo has become an important task in biomechanics.

A further interest in special questions of biomechanics comes from sport. Here various kinds of complex body motions are of interest to optimize its outcome. There is also interest in ecological aspects of biomechanics, for example life in moving fluids, wind resistance of plants etc.

Further reading: On biomechanics in general: Alexander 1983; Biewener 1992; Fung 1993; Mazumdar 1992; Niklas 1992; Skalak and Chien 1987; Vogel 1983

3.6.1
Some Basic Properties of Fluids

Many parameters characterizing the physical property of a fluid are defined for the case of laminar flow. A flow is called *laminar* if particles of the fluid move in parallel layers to each other (more on properties of laminar flow in Sect. 3.7.1). Figure 3.48 illustrates examples of three kinds of laminar flows. In contrast to a turbulent flow, processes in laminar flows can be calculated by linear thermodynamics approaches (see Sect. 3.1.3, Fig. 3.3).

To understand the approaches of fluid mechanics, and to characterize laminar fluid profiles, we firstly need to introduce the *velocity gradient* (γ), also known as *shear rate*. It is the derivation of the streaming velocity ($\mathbf{v}$) with respect to a coordinate (z), perpendicular to the direction of the flow.

$$\gamma = \frac{d\mathbf{v}}{dz}. \tag{3.6.1}$$

The measure of this velocity gradient, therefore, is s^{-1}.

The shear rate of a moving fluid is at a maximum near a solid surface, or in the case of a tube, near the wall. Far away from the surface this gradient becomes zero.

If two parallel plates slowly move, one in relation to the other, they generate a laminar flow in between, the velocity gradient of which is always constant, and is equal to the relative velocity of the plates, divided by their mutual distance ($\gamma = \Delta\mathbf{v}/\Delta z$).

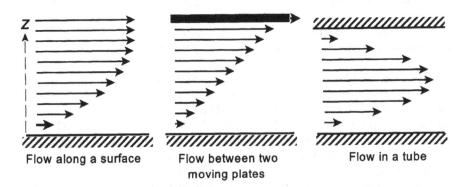

| Flow along a surface | Flow between two moving plates | Flow in a tube |

Fig. 3.48. Various examples of laminar velocity profiles

The force (F) driving a plate with a surface A in the case of laminar flow is proportional to the velocity gradient (γ) between the two plates:

$$F = \eta \gamma A. \tag{3.6.2}$$

In this equation a friction coefficient (η) is introduced which is called *viscosity*. The viscosity therefore determines the force which is required to move a plate with an area of 1 m^2, at a 1 m distance from a parallel surface at a velocity of 1 m s^{-1}, if a laminar flow of a liquid between these surfaces is induced. Thus, the measuring unit for the viscosity is: N $\cdot$ s $\cdot$ m^{-2}, or: Pa $\cdot$ s. Sometimes an older unit P (Poise) is used, where: 1 P = 0.1 N s m^{-2}.

Parallel to the viscosity (η), another parameter, the *fluidity* ($\phi = 1/\eta$) is used, as well as the *kinematic viscosity* ($v = \eta/\rho$), a parameter containing the density (ρ) of the fluid. The force deforming a body in a streaming fluid with a velocity gradient (γ) is given by the *sheer stress* (τ):

$$\tau = \eta \gamma. \tag{3.6.3}$$

The viscosity depends to a high degree on the temperature. Especially for aqueous solutions, this is caused by the cluster structure of water (see Sect. 2.4.2). In contrast to the viscosity of pure water at $T = 0$ °C, which is 1.79 mPa s, at 25 °C it amounts to only 0.89 mPa s, and at 100 °C, finally 0.28 mPa s (see Fig. 2.22 in Sect. 2.42).

In order to characterize the influence of dissolved or suspended substances on the viscosity of a fluid, the following derived parameters are used:

Relative viscosity: $\eta_{rel} = \dfrac{\eta}{\eta_W}$,

Specific viscosity: $\eta_{sp} = \eta_{rel} - 1$,

Reduced viscosity: $\eta_{red} = \dfrac{\eta_{sp}}{c}$ (in: l mol^{-1}),

Intrinsic viscosity: $[\eta] = \lim\limits_{c \to 0} \eta_{red}$ (in: l mol^{-1}),

where η is the viscosity of the solution of suspension, η_W is the viscosity of the pure solvent, and c is the molar concentration of the solute.

The viscosity increases with increasing concentration of the dissolved or suspended substances. As already pointed out in the definition of the intrinsic viscosity $[\eta]$, even the reduced viscosity of a solution is a function of the concentration. The intrinsic viscosity contains information on the structure and the molecular mass of a substance.

For diluted suspensions of small rigid spherical particles the Einstein relation can be applied:

$$\eta_{sp} = 2.5 \, V_{rel} \quad (\text{for:} \quad V_{rel} < 0.1). \tag{3.6.4}$$

The relative volume (V_{rel}) is the volume of all particles in the suspension together in relation to the volume of the suspension. For suspension of cells (sperms, erythrocytes etc.) the term *cytocrit* (or specifically, spermatocrit or hematocrit) is used. It should be emphasised that neither the absolute size of an individual particle, nor the homogeneity of the diameters of all particles in the suspension are of importance for this relation. The Einstein equation, however, is correct only for very diluted suspensions.

Newtonian fluids are those in which viscosity is independent of the velocity gradient (γ). In contrast, *non-Newtonian fluids* alter their viscosity depending on this parameter.

In Fig. 3.49 the behavior of various kinds of non-Newtonian fluids are demonstrated. *Dilatant* fluids are mostly suspensions of solids, like quartz particles. These particles produce entropy by mutual friction as much as the shear rate of fluid increases. The *Bingham-plastic* behavior occurs, for example, in suspensions of non-spherical particles. In this case velocity gradients lead to their orientation, which decreases the viscosity of the suspension. At certain points, if the particles are orientated at a maximum these suspensions behave like Newtonian fluids. The same behavior is to be expected if the particles tend to aggregate in the resting fluid, but disaggregate at low shear stress.

The most common property of biological fluids is the *pseudoplastic* behavior. It occurs for example in blood (Fig. 3.51) and many other biological fluids with heterogeneous composition. Different components of these fluids, for example blood cells, proteins and other macromolecules, aggregate, orientate, and deform at various shear gradients. The resulting function, therefore, does not come to a saturation level at reasonable values of γ.

These shear-induced processes of course need some time to become established. In the case of spontaneous induction of a shear gradient, a time function of the viscosity can be observed. The diminution of the viscosity of a fluid after a sudden decrease of the velocity gradient is called *thixotropy*. The inverse indicates *rheopectic* behavior.

These properties of fluids must be taken into account when choosing instruments to measure the viscosity. Some of the most common methods are depicted in Fig. 3.50. In a capillary viscosimeter (Fig. 3.50A) the time is

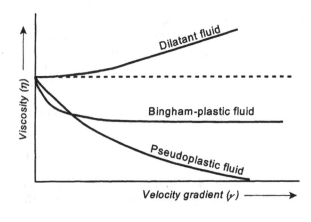

Fig. 3.49. The dependence of the viscosity (η) on the velocity gradient (γ) for a Newtonian fluid (- - - -), and for various types of non-Newtonian fluids (——). (After Glaser 1986)

Fig. 3.50. Various setups to measure the viscosity of fluids. A capillary viscosimeter after Ostwald, B viscosimeter with falling sphere, C coaxial type rotational viscosimeter, D cone type rotational viscosimeter

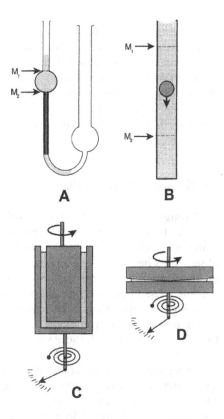

measured which a given fluid needs to pass through a capillary under a known pressure. For this, a certain volume of the fluid is placed in the left part of the U-tube and the time is measured, the fluid must pass the two marks M_1 and M_2. The time measured is proportional to the viscosity. In this way, the viscosity can be measured after calibration of the setup by a fluid with known viscosity. This so-called Ostwald viscosimeter was later modified by Ubbelohde in such a way that a further vertical tube arranged at the end of the capillary produces an interruption of the fluid. Another way to measure the viscosity of a fluid is the use of falling bodies (Fig. 3.50B). Both methods have the advantage of being simple and, in relation to other equipment, cheap to purchase. On the other hand, however, they are only suited for measuring the viscosity of Newtonian fluids because the velocity gradient of the streaming fluid in the capillary, as well as between the falling sphere and the wall of the tube, are not at all constant (see Fig. 3.48).

To investigate the viscosity of non-Newtonian fluids, measuring instruments are required which allow one to establish well-defined shear gradients. As demonstrated in Fig. 3.48, this is possible between plates moving parallel to each other. This principle is used in so-called rotational viscosimeters. In this case the fluid is placed in the gap between two coaxial cylinders (Fig. 3.50C), or between a flat plate and a truncated cone (Fig. 3.50D). By moving one part of this equipment, the viscosity of the fluid transmits a torque which can be measured. Usually, one part is rotated with adjustable speed, and the torque of the opposite part is measured by a sensitive instrument. By changing the speed and the thickness of the gap, one can produce various velocity gradients. In the case of the rotating cone the tangential velocity is compensated for by an increasing width of the gap. Therefore, in this case a constant velocity gradient is also established. Curves like that of Figs. 3.49 and 3.51 are produced using such instruments.

3.6.2
The Viscosity of Biological Fluids

Through the development of sensible rotational viscosimeters it has been possible to investigate viscous properties of a large number of biological materials, such as blood, lymph, various secretions, synovial fluids, and many others. This has led on the one hand to a better understanding of the processes of blood flow, of the mechanics of joints etc., and on the other hand it has become a useful indicator in diagnostics.

In Fig. 3.51 the viscosity of a suspension of red blood cells is depicted as a function of the velocity gradient. In contrast to the blood plasma which appears to be a Newtonian fluid, these suspensions, as well as the whole blood, indicate pseudoplastic thixotropic behavior. The reason for this is complex: native erythrocytes, suspended in blood plasma, aggregate at low shear rates. These aggregates are not very stable and disaggregate at somewhat higher shear rates. This disaggregation lowers the viscosity of the suspension. A further increase of the shear rate leads to a deformation of the erythrocytes, furthermore decreasing

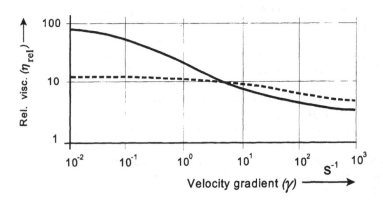

Fig. 3.51. The relative viscosity (η_{rel}) depending on the shear rate (γ) of human blood (———), and heat-hardened erythrocytes, which were resuspended in plasma (- - - -). The differences between the curves in the region of low shear rates are achieved by the aggregation of the native red blood cells, the differences at higher shear rates, by differences in cell deformability. (After Lerche and Bäumler 1984)

the viscosity. Looking at the behavior of erythrocytes which were hardened by fixation, neither the aggregation, nor deformation by shear stress occurs. Living erythrocytes in the shear gradient become elongated ellipsoids which are orientated along the streaming vectors. With increasing shear stress they become more and more elongated. Whereas slow elongations are reversible, up to a certain degree of elongation, irreversible alterations in the membrane occur. Eventually hemolysis occurs as a result of maximal shear stress.

The energy of the streaming fluid which leads to deformation of the cells corresponds to the shear stress (τ) according to Eq. (3.6.3). This means that it depends not only on the shear rate (γ), but additionally on the viscosity (η) of the fluid. To investigate shear induced shape deformations, therefore, high viscosity solutions are usually used, for example, solutions of high molecular weight dextrans of various concentrations.

For biomechanical problems in orthopaedics the properties of the synovial fluid which is located in the joint cavity and surrounded by the joint capsule are of particular interest. In the language of tribology, as the study of the effects of friction on moving machine parts, the mechanism of joints are a kind of depot lubrication with a porous surface. The expression "depot lubrication" points to the synovial fluid which is accumulated in the bursa of the joints. "Porous surface" refers to the articular cartilage which covers the cortical bone in the joint capsule in a 0.3–0.5 mm thick layer. The joints are not only burdened by movement, i.e. by a shear stress of the synovial fluid, but additionally in some cases by a considerable static pressure. It must be guaranteed that these loads do not press the synovial fluid out of the gap. In fact synovial fluid has thixotropic pseudoplastic properties. A large viscosity of this fluid in the joints (η between 1 and 40 Pa s) prevents its exclusion from the gap by hydrostatic pressure. If the joint is moving, however, shear rates up to 10^5 s^{-1} appear. In this case, the viscosity of the synovial fluid falls to 10^{-2} Pa s, leading to a highly effective lubrication of the joint. This particular property of the synovial fluid is caused by a special structure of proteoglycans, which are high molecular weight glycoproteins with an intriguing structure.

This leads us to considerations of the viscosity of microscopic, or even supramolecular structures. It must be pointed out that the definition of viscosity, as given in the previous Section 3.6.1, comes from continuum physics. It does not take into account the behavior of molecules and supramolecular composition of the liquids. It just considers a homogeneous, continuous fluid without any structures. Already, the non-Newtonian behavior discussed above indicates some limitations of this assumption. Moreover, problems arise if we consider the viscoelastic properties of cells and their constituents. At the beginning of the twentieth century, at a time when the physical properties of cells were first discussed, many investigators measured the viscosity of the cytoplasm in relation to various biological functions. This question from the modern point of view is outdated because of our knowledge of the highly organized structure of this region. Even in cells with pronounced cytoplasmic streaming, as in amoebas or plant cells, the question of the origin of these movements and of the molecular mechanisms driving these flows, is of more central interest than phenomenological models using plasma viscosity in general.

This problem of viscosity in molecular and supramolecular dimensions is important to a high degree in relation to the "viscosity" of the cell membrane (see also Sect. 2.5.3). Using molecular probes, various physical parameters of the membrane can be measured which allow calculation of viscosity parameters. So, for example, a lateral diffusion constant of fluorescent labels in the membrane can be measured and, using the Einstein equation [Eq. (2.3.39)], the viscosity of the surrounding medium can be calculated. In the same way, with specific spin-probes which are bound to particular membrane molecules, the electron-spin resonance techniques (ESR) allow determination of their rotation diffusion constant, which is also a function of the viscosity of the surroundings.

In all cases the viscosity which is determined by these methods is a typical effective parameter. In the same way as, for example, the hydration radius of an ion (see Sect. 2.4.2), this parameter depends on the techniques and the physical phenomenon used for its determination. One should not consider these "quasi-viscosities" in the same sense as the phenomenological viscosity of a fluid which we discussed before. Moreover, the mechanical anisotropy of the membrane must be taken into account, which means that the mobility of particles is not equal in all directions of the space.

Further reading: Leyton 1974; Skalak 1987

3.6.3
Viscoelastic Properties of Biomaterials

The simplest kind of deformation of a body is its stretching. In Fig. 3.52 this is illustrated by a stress-strain diagram. The parameters are defined as follows:

$$\text{Stress:} \quad \sigma = \frac{F}{A} \quad (\text{in: } N\,m^{-2}), \tag{3.6.5}$$

$$\text{Strain:} \quad \varepsilon = \frac{\Delta l}{l}, \tag{3.6.6}$$

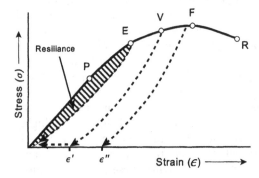

Fig. 3.52. Stress-strain diagram. P – limit of proportionality, E – limit of elasticity, V – limit of reversible viscoelastic deformation, F – point of floating deformation, R – rupture point, ε', ε'' – residual strains, *dashed area* – resilience

where F is the force, A is the cross-sectional area of the body, l is its length and Δl is the difference between the resting and the extended material. The strain (ε) is just a relation and therefore does not have a measuring unit.

In the region of minimal strain up to the limit of proportionality (P) Hooke's law is valid stating that the strain (ε) is directly proportional to the stress (σ). The ratio of these two properties is called the *modulus of elasticity*, or *Young's modulus* (Y).

$$Y = \frac{\sigma}{\varepsilon}, \tag{3.6.7}$$

where Y has the units N m^{-2} or Pa. Mostly the unit 1 MPa = 10^6 Pa is used.

The linear relationship between stress and strain does not hold for large stress behind the point P. The deformation however, is reversible up to the elastic limit (E). This means that the body will spontaneously and quickly return to its original length when the deforming force is removed. In this case, however, the relaxation curve (dotted line) did not follow the extension curve (full line). The area between these two lines is called *resilience*. In general the area in this plot, i.e. the product of stress and strain has the unit of energy. The resilience therefore represents the thermal energy which is dissipated during the process of extension and relaxation. In terms of irreversible thermodynamics it is the dissipation function (Φ) of this process [see: Eq. (3.1.64)]. It results from the viscous friction within the body and is an expression of the non-ideal behavior of the system.

If the strain is taken beyond the elastic limit (E), the body begins to deform irreversibly. Up to the limit of viscoelastic deformation (point V), the body quickly relaxes to some residual strain (ε') which eventually may slowly vanish. Overcoming this point, an irreversible deformation (ε'') persists after the stress has been removed. A further extension leads to the point where the body begins to show spontaneous flowing elongation, in spite of further increase in stress until it rips up at point R.

This stress-strain diagram indicates that the deformation of a body outside the limit of proportionality depends not only on elastic, but also on viscose properties of the material. For this the term *viscoelasticity* is used. Biological tissue shows viscoelasticity to a great extent. In this case the stationary stress-strain function, as demonstrated in Fig. 3.52, as well as the kinetics of deformation become important.

It is possible to simulate the strain behavior of elastic and viscoelastic bodies by mechanical systems made up of elasticity elements (springs, Fig. 3.53A) and viscosity or damping elements (pistons in cylinders, Fig. 3.53B). In contrast to the spring, which, if it is massless, elongates immediately if a force is applied, the viscous damping element elongates with a constant velocity and remains in position if the force vanishes. A damping element combined in series with a spring is called a *Maxwell element* (Fig. 3.53C). In the case of an applied rectangular function of force, the spring elongates immediately, as in the case of Fig. 3.53A, then a further continuous elongation of the system proceeds. If the

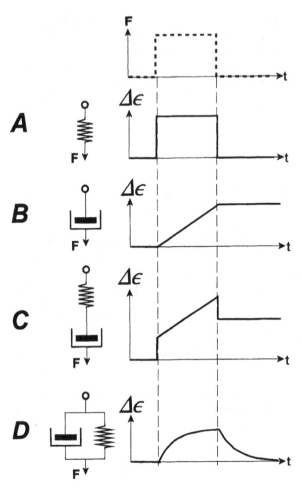

Fig. 3.53. Mechanical models to demonstrate kinetic properties of viscoelastic elements: **A** – elastic element, **B** – damping element, **C** – Maxwell element, **D** – Voigt element. The graphs at the *right* side indicate the strain [$\Delta\varepsilon(t)$] of these systems, if the force [$F(t)$] is applied at their ends corresponding to the rectangular function (·····) in the graph depicted at the top

force vanishes, the spring contracts but the viscous part of the Maxwell element remains elongated. If a spring and a damping component are connected in a parallel arrangement we obtain a *Voigt element* (Fig. 3.53D). In this way a sudden increase of the force (**F**), leads to an exponential elongation of the system which will contract in the same way if the force vanishes.

Real systems must regarded as being made up of both Maxwell and Voigt elements in various combinations. Correspondingly, complicated strain-relaxation graphs and complex kinetic behavior is to be expected. Even complicated viscoelastic systems respond elastically to short-term mechanical stress, provided that the time constant of the viscous elements are large in comparison to the time factor of the mechanical impulse. The schematic graphs of Fig. 3.53 with their different combined mechanical elements are somehow similar to the electrical RC circuits in Figs. 3.25 and 3.38. In fact there are some similarities in their kinetic treatment. In analogy to the electrical impedance which we have discussed in Section 3.5.3, a kind of mechanical impedance can be formulated as the response of a viscoelastic system on time-varying mechanical forces. Measurements of this kind give information on basic mechanical properties of biological tissue.

Comparison of the elastic properties of several materials (Table 3.2) shows that they differ in many degrees of magnitude. In general, two groups of materials with typical elasticity properties can be found – rubber and elastin with very low parameters of Y on the one hand, and steel, bone and flax on the other hand, with an elasticity modulus which is ten-thousand or even a hundred-thousand times larger. Corresponding differences occur in the limits of elasticity.

The reason for this difference are two completely different molecular processes leading to these deformations. In the case of the strain of a steel wire, the atomic spacings in the crystal lattice are found to change. The atoms are moved from their equilibrium position of minimal internal energy and are raised to a higher energy level. The free energy that is stored up in the strained state is released during relaxation. This is the reason for the high elasticity modules of this type of behavior which is called *steel elasticity*.

In contrast to this, materials with a low elasticity modulus ($Y = 0.1\text{–}10$ MPa) show so-called *rubber elasticity*. A rubber thread, of course, can be extended much more easily and to a much larger extent than a steel wire. We discussed this type of elasticity in Section 2.3.6 (Fig. 2.17). This is the characteristic property of macromolecular materials. The strain leads to a partial deconvolution of the molecules, and as a result, to a reduction of their entropy content. The subsequent relaxation is the result of the increase in entropy back to its maximum according to the second law of thermodynamics. This mechanism is the reason why this type of stretching mechanism is also called *entropy elasticity*. In addition, it explains the strong temperature dependence of the elasticity modulus of these kinds of materials.

Most biological materials indicate rubber elastic properties. Steel elasticity occurs in bones, and to some extent in materials with low water content such as hair, wood or dry plant materials, as for example flax (Table 3.2). In these cases a devolution of the macromolecules is not possible.

Table 3.2. Examples of elasticity properties of some materials

Material	Modulus of elasticity Y, in MPa	Limit of elasticity ε_E	Tear resistance σ_R in MPa
Steel	$2 \cdot 10^5$	$1.2 \cdot 10^{-3}$	$5 \cdot 10^2$
Rubber	1	3	
New Zealand Flax (*Phormium tenax*)	$1.5 \cdot 10^4$	$1.4 \cdot 10^{-2}$	$2 \cdot 10^2$
Bone	10^4	10^{-2}	10^5
Elastin	0.6		

A number of structural proteins like, for example collagen, elastin or resilin are responsible for the rubber elastic properties of tendons and ligaments. These substances also play an important role in the storage of mechanical energy in some periodic or suddenly occurring movements. Although these processes also occur in mammals, they are investigated best in insect jumping and flying . In jumping the muscles apply tension to a system of tendons which take up the energy and subsequently release it, by means of a trigger mechanism, to achieve an increase in power. In this way power, i.e. transformed energy per unit time can be generated which would be impossible to obtain directly from the muscle. In fact, this is the same mechanism that we use for archery. To jump, a grasshopper needs a power of about 5 kW per kg muscle. This would exceed the maximal output of a muscle by ten times. Similarly, during flight a grasshopper stores about 20–30% of the energy of the oscillating movements of the wings using passive elastic elements.

The following special characteristics of the viscoelastic behavior of cells and tissues should be noted:

Occurrence of different regions of elasticity in the stress-strain diagram. Frequently cells and elastic fibers in a tissue are interconnected with each other, forming a network. In this case the viscoelastic properties of the system result not only from the intrinsic properties of these fibers, but first of all from the construction of this network, and in the viscosity of the fluid within it. Such a network can be stretched easily as long as it is deformable as a whole. This is the first region in the stress–strain diagram with low Young's modulus. If the components of the network are eventually fully orientated by the applied stress, a further strain is possible only by stretching the fibers themselves. This means that the Young's modulus increases suddenly. The resulting stress-strain diagram, therefore, does not indicate a flattening of the curve as in Fig. 3.52, but on the contrary it becomes steeper at a certain degree of strain. This sort of behavior occurs, for example, in the walls of blood vessels, where the situation becomes even more complicated due to the activity of the smooth muscles.

Regulation of elasticity behavior. The viscoelastic properties of the network as described above can be controlled by biological processes. This, for example, is possible by loosening and fastening the points of connections between the components. The result would be a sort of viscous prolongation of the body. Another way to control the viscoelastic behavior is the alteration of the water content of the tissue. Changing the cell volume or the intercellular space in the network would change its elasticity modulus. This sort of control is best investigated in the case of the viscoelastic properties of the uterine cervix at various periods of pregnancy.

Mechanical anisotropy. In many biological systems the elasticity modulus depends on orientation. This property is best investigated in bones. Depending on the angle of measurement, Young's modulus in bones can vary by a factor of two. The reason for this is the particular structure of the bone. The struts of the cancellous bone, the so-called trabeculae, are orientated according to its loading in vivo. They are orientated according to the trajectories of pressure and tension of the bone in the skeleton. This is the result of an adaptation, which was first investigated in 1892 by J. Wolff, who formulated a law of bone remodeling.

The special mechanical properties of the cell membrane has been discussed in detail in Section 2.5.3 (Fig. 2.41). A lipid membrane can be considered as a two-dimensional crystal. The head groups of the phospholipids show similar behavior to the atoms in the 3-dimensional crystal of steel. Therefore, they show a kind of steel elastic behavior in the plane. The elasticity modulus of these membranes is high and their limit of rupture low. The membrane proteins with their rubber elastic properties have no significant influence on this behavior. In contrast to technical materials like rubber sheets, the biological membrane can be easily deformed in an isoplanar way, but cannot resist expansion. This property is important for the dynamics of cell shape and for cell swelling.

Further reading: On viscoelastic properties of biomaterials: Bereiter-Hahn et al. 1987; Fung 1993; Schmid-Schönbein et al. 1986; Skalak and Chien 1987

3.6.4
The Biomechanics of the Human Body

Investigations of biomechanical conditions of the human body, its movement, its carriage under conditions of loading etc. are important tasks in orthopaedics, surgery, sports and occupational safety. What kinds of loads of the muscles and joints result from a particular loading and a particular movement? How can pathologic deviations be cured by surgical operations? How can diseased bones and joints be replaced by artificial materials? In this case immunological tolerance must be realized as well as the adaption of the applied material to the viscoelastic properties of the living bone the artificial joints will be connected to.

Recently, greater effort has been directed toward the construction of mathematical models of the locomotor system including human walking movements. With computer simulation programs, hopefully surgical corrections can be performed with optimal success. Of course, in this context sports efforts should also be mentioned which optimize various techniques of jumping, running etc.

In contrast to the structure of plants which can be modeled statically by systems of flectionally and torsionally loaded bonded structures, for animal systems dynamic body models are required. They are composed of combinations of elements that are stable in relation to pressure and bending with tendons and muscles as elements of tensile stability and contraction. This is illustrated in Fig. 3.54. The head of the horse, for example, is held by a system of elastin tendons, the ligamentum nuchae, as the tensile element and the cervical part of the spinal column as the component in compression. In the abdominal region of quadrupeds the compression element is located dorsally and the tension element is located ventrally. The carriage of the body is stabilized by tendons and it is maintained in an upright position without an additional supply of energy.

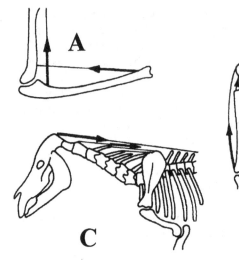

Fig. 3.54. Stabilization of human forearm (**A**) and thigh (**B**), as well as the head of a horse (**C**) by combination of bending and compression stable bones, and tension stable muscles and tendons. (After Glaser 1989)

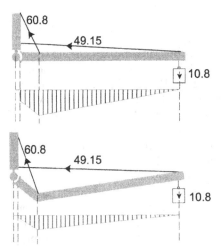

Fig. 3.55. Two steps on the way to optimization of the bending load of the forearm. The two muscles which hold the forearm bone must generate forces of 60.8 N and 49.15 N to compensate the loading force of 10.8 N at its end. These muscle forces are identical in both cases. The bending forces of the bone, expressed in the graphs below, however, are quite different. Compare this scheme with the real situation in Fig. 3.54. (Redrawn after Pauwels 1980)

During evolution, the systems of combined elements which are stable against compression, together with those for tension, have been developed toward optimal use of muscle force, as well as toward maximal stability of the supporting bones. The attachment point of the muscles and tendons determines the vectors of tension and compression in the bone. We already mentioned in the previous section that this induces the orientated growth of the struts in the cancellous bone.

Figure 3.55 shows an example of optimization of the shape of the human forearm. The bending of the bone and the shift of the muscle attachment have led to a significant reduction of the bending force. Similar principles of optimization can also be found in other parts of the muscle-skeleton apparatus.

The bending or the torque of a body can be attributed to stretching and compression of the material. As a measure of the bending, the *radius of bending* (R), or its reciprocal, the *curvature of bending* ($K = 1/R$) is used. If a homogeneous beam or bar bends, a compression of the concave side and a stretching of the convex side occurs. Between these two regions there must be a neutral plane that is neither compressed nor stretched.

Let us consider a section of a bar which is bent between two points that subtend an angle α at the center of curvature (Fig. 3.56). Let R be the radius of curvature measured to the neutral plane. If the angle is expressed in radians then the length (l) of the section at the distance from the neutral plane is:

$$l(x) = \alpha R_i = \alpha(R + x). \tag{3.6.8}$$

If the length of the section along the neutral plane is l_n (at $x_n = 0$; $R_n = R$) then the strain (ε) of any plane parallel to the neutral plane can be calculated by Eq. (3.6.6):

$$\varepsilon(x) = \frac{\Delta l}{l} = \frac{l(x) - l_n}{l_n} = \frac{x}{R}. \tag{3.6.9}$$

As the distance x is measured from the neutral plane, ε can have both positive and negative values. Negative strain in this context means compression of the material and occurs when $x < 0$.

The combination of Hooke's law [Eq. (3.6.7)] and the above relation enables the stress to be calculated:

$$\sigma(x) = Y\frac{x}{R}. \tag{3.6.10}$$

The differential of moment of force (dM) is used to find the force which is necessary to bend the bar. It is calculated from the product of the force (F) and the leverage distance from the neutral plane (x), and it is also related to the differential of an area (dA):

$$dM = x\,dF = x\sigma\,dA = \frac{x^2 Y}{R}\,dA. \tag{3.6.11}$$

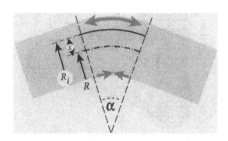

Fig. 3.56. The bending of a bar

The bending moment of the bar is obtained by integration of this equation. If the bar is homogeneous, the modulus of elasticity (Y) is not a function of x.

$$M = \frac{Y}{R} \int x^2 \, dA = \frac{Y}{R} I_A. \tag{3.6.12}$$

In this equation the integral expression has been replaced by the second moment of area (I_A). This is a measure of the bending resistance of a bar made out of a material having a modulus of elasticity (Y). It can be seen that, for a given bending moment, the smaller the value of I_A, the more the bar bends. This means that if the bending moment does not change and the modulus of elasticity remains constant, the second moment of area is directly proportional to the radius of curvature.

It is easily understood from everyday experience that the second moment of area of a bar depends on the plane in which it is bent. A bar having a flat section bends more easily across the flat then it does on edge. This is determined by the position of the neutral plane. It passes through the center of gravity of a cross-section of the bar, and is always perpendicular to the radius of curvature.

The second moment of area for a given cross section, for example a cross section of a tubular bone, can be determined graphically. First it is necessary to find the position of the center of gravity of this kind of profile. This can be done by suspending a cardboard model of the cross-section by a thread. The center of gravity is the crossover point of all lines which can be drawn in a vertical direction from the points of suspension. After the position of the neutral plane has been fixed, the cross section can be subdivided into rectangular areas (Fig. 3.57). The following approximation can be obtained from the definition of the second moment of area:

$$I_A \approx \sum_{1=1}^{n} x_i^2 \Delta A_i. \tag{3.6.13}$$

The individual rectangular areas (ΔA_i) are calculated, the distance of their centers from the neutral plane (x_i) are measured, and the products of these two measurements are summed as required by [Eq. (3.6.13)]. The accuracy of this method will be increased if the areas are made as small as possible. The units of the second moment of area are m^4.

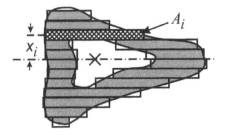

Fig. 3.57. Determination of the second moment of area of a tubular bone by graphical method. The profile of the bone is approximated by many rectangles with individual areas A_i and distances from the neutral plane (-------) x_i. * – center of gravity

For bars with particular geometrical profiles, analytical expressions for I_A are derived. For a compact cylindrical rod with radius r, for example, we have:

$$I_{A(\text{cylinder})} = \frac{\pi}{4} r^4.$$

(3.6.14)

The second moment of area of a tube with an inner radius r_1, and an outer radius r_2 is:

$$I_{A(\text{Tube})} = \frac{\pi}{4} (r_1^4 - r_2^4).$$

(3.6.15)

This equation can be used to show the advantage of a structure made up of hollow tubes, an optimizing principle that has occurred in the construction of bones and some plant stems.

In a similar way to bending, torsional deformation can also be estimated. In contrast to bending, in the case of torsion, all regions are stretched as much as they are away from a central axis of gravity. A *polar moment of area* (I_P) can be formulated if the area elements (dA) are multiplied by the square of the radial distance (r) from the center of gravity, and subsequently, the products are summed. In an integral formulation this means:

$$I_P = \int r^2 \, dA.$$

(3.6.16)

The polar moment of area of a tube is:

$$I_{P(\text{Tube})} = \frac{\pi}{2} (r_1^4 - r_2^4).$$

(3.6.17)

These statements form the basis for calculating the stability of plants, animals and humans. However, it can be seen from the above equations that some simplifying assumptions have to be made. In particular it must be noted that the modulus of elasticity (Y) may vary with the position and also, as illustrated by the properties of bone, with the direction of the applied force. If the viscoelastic properties are also taken into account, the calculations become even more complicated.

Further reading: Biostatics of humans and animals: Fung 1993; Pauwels 1980; Schmid-Schönbein et al. 1986; Skalak and Chien 1987; biomechanics of plants: Niklas 1992; special problems of paleo-biomechanics: Rayner and Wooton 1991

3.7
Biomechanics of Fluid Behavior

Streaming of viscous fluids occurs at all levels of biological organization. It starts with water flow through pores of membranes, streaming of cytoplasm in plants, streaming of blood in vessels, and includes the flow of water and air around animals, i.e. the problems of flying and swimming. In this section we will concentrate on some medically important problems of hemodynamics, or hemorheology. We will further include some aspects of flying and swimming because this is one of the classical problems of biomechanics.

3.7.1
Laminar and Turbulent Flows

When a liquid flows along the surface of a thin plate, a flow profile is formed over this boundary which changes its character as the distance (l) from the leading edge of the plate (at: $l = 0$) increases. This effect is illustrated in Fig. 3.58.

Directly at the surface of the plate there is a trapped layer, i.e. a non-moving film of air or liquid. At increasing distance from the plate (z), the velocity of the flow increases. Layers ("lamina") of the fluid slide over one another. This boundary layer is characterized by the velocity gradient ($dv/dz \neq 0$). There is no sharp transition between this layer and the region of undisturbed flow ($dv/dz = 0$, $v = v_\infty$). Usually the limit of this laminar boundary layer is denoted by the distance $z = \delta_L$, when $v = 0.99 v_\infty$ (see Fig.

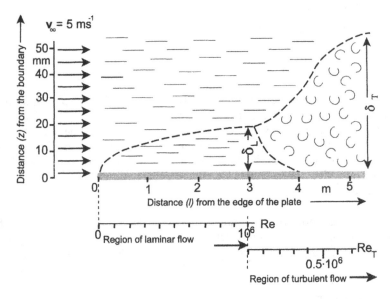

Fig. 3.58. Formation of laminar and turbulent flow profiles near a planar plate. As an example, the numbers correspond to a flow of air with a velocity $v_\infty = 5$ m s^{-1}. The thickness of the laminar (δ_L); and the turbulent boundary layer (δ_T) are calculated according to the equations in Table 3.3

3.59). As can be seen in Fig. 3.58, δ_L gets larger as the distance from the leading edge of the plate increases.

As the boundary layer gets thicker it becomes increasingly unstable. At a critical point, this leads to the spontaneous appearance of turbulence. Turbulent flow is the consequence of non-linear processes (see Sect. 3.1.3, Fig. 3.3). The resulting whirls can be considered as a kind of a dissipative structure. Phenomenologically, the transition from laminar to turbulent flow is accompanied by an increase in friction and a substantial increase in the thickness of the boundary layer ($\delta_L < \delta_T$).

In Table 3.3 some basic equations are listed for calculating parameters of laminar and turbulent flow. These equations are based partly on empirical observations, particularly those for turbulent flow.

It is not possible to determine exactly the critical point at which the transition from laminar to turbulent flow takes place. In fact, this is a stochastic process of state transition of the system which occurs as a result of increasing destabilization. The position of the critical point where the laminar flow abruptly transforms into a turbulent one can only be calculated as a probability. It depends on the flow velocity (v), the viscosity (η), the density of the medium (ρ) and a characteristic streaming distance (l). These parameters are connected in the so-called *Reynolds number* (Re) which plays a crucial role in rheology:

$$Re = \frac{lv\rho}{\eta} = \frac{lv}{v}. \tag{3.7.1}$$

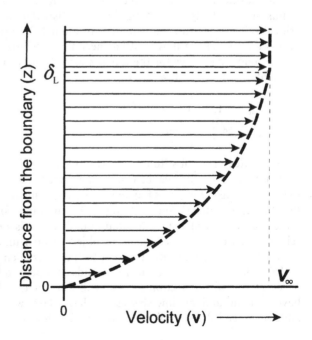

Fig. 3.59. Laminar velocity profile near a surface. The velocity vectors are depicted at various distances (z) from the surface; at $z = 0$, $v = 0$; δ_L is the thickness of the laminar boundary layer

Table 3.3. Equations for parameters of laminar and turbulent flow near boundaries. Symbols (see also Fig. 3.59): $\mathbf{v}$ – velocity, z – distance from the boundary, ρ – density of the medium, η – viscosity, Re – Reynolds number; subscripts: 0 – at the boundary, ∞ – in the bulk phase

	Laminar flow (subscript L)	Turbulent flow (subscript T)
Sheer stress $\tau(z)$	$\tau_0 (1 - z/\delta_L)$	
τ_0	$0{,}332\rho v_\infty^2 \, (\mathrm{Re}_L)^{-1/2}$	$0{,}023\rho v_\infty^2 \, (\mathrm{Re}_T)^{-1/5}$
Velocity $\mathbf{v}(z)$	$2\mathbf{v}_\infty/\delta_L \, (z - z^2/2\delta_L)$	$\mathbf{v}_\infty \, (z/\delta_T)^{1/7}$
Thickness δ	$5l_L \, (\mathrm{Re}_L)^{-1/2}$	$0{,}376l_T \, (\mathrm{Re}_T)^{-1/5}$
Force of surface friction $F_0(l)$	$0{,}664\rho v_\infty^2 \, l_L \, (\mathrm{Re}_L)^{-1/2}$	$0{,}0366\rho v_\infty^2 l_T \, (\mathrm{Re}_T)^{-1/5}$

In this equation the kinematic viscosity ($v = \eta/\rho$) is used which has already been introduced in Section 3.6.1.

The Reynolds number is a typical parameter of the theory of similarity. Bodies of identical shape show identical flow behavior, for flow conditions with the same Reynolds number. This is independent of, whether the body is large or small, or whether it is moving in water or in air.

The critical Reynolds number characterizing the transition from laminar to turbulent streaming of a flow parallel to a flat surface, as illustrated in Fig. 3.58, is about $\mathrm{Re} = 10^6$. For a sphere with a flow around it, this transition occurs at $\mathrm{Re} \approx 10^3$ (see Fig. 3.3). The critical Reynolds numbers of streamlined bodies with so-called laminar profiles are somewhere between these limits, depending on their exact shapes.

Flow inside a cylindrical tube can be characterized in a similar way. If the radius of the tube is taken for the characteristic length l in Eq. (3.7.1) then the critical value of the Reynolds number is about 10^3. Turbulent flow inside a tube means that the entire flow in it has become destabilized.

The following values for the kinematic viscosity can be used to calculate the Reynolds number for T = 291 K:

$$v_{\mathrm{Water}} = 1.06 \cdot 10^{-6} \, \mathrm{m^2 \, s^{-1}},$$
$$v_{\mathrm{Air}} = 14.9 \cdot 10^{-6} \, \mathrm{m^2 \, s^{-1}}.$$

If these values are substituted into [Eq. (3.7.1)] it is seen that, in the example shown in Fig. 3.58, turbulence would occur not at $l = 3$ m, but at $l = 0.2$ m from the leading edge, if water is flowing across the plate instead of air, as depicted in this figure. A comparison of flying and swimming objects can only be made under conditions of equal Reynolds numbers. It is quite senseless simply to relate velocities to the length of an object (which unhappily is often done in popular scientific publications much to the amazement of the readers!).

Table 3.4 shows some typical Reynolds numbers for the movement of quite different animals. In Section 3.7.3 we will show that, although they may possibly have optimal laminar flow shapes, the large, fast swimming fishes and aquatic animals exceed the critical Reynolds number for laminar flow.

Table 3.4. Characteristic Reynolds numbers of various moving animals

	Characteristic length (l) m	Characteristic velocity (v) $m \cdot s^{-1}$	Reynolds number (Re)
Paramecium caudatum	$2.1 \cdot 10^{-4}$	$1.1 \cdot 10^{-3}$	$1.8 \cdot 10^{-1}$
Mosquito (*Ceratopogonidea*)	$0.9 \cdot 10^{-2}$	$2.5 \cdot 10^{-1}$	$1.5 \cdot 10^{2}$
Chaffinch	$3.6 \cdot 10^{-2}$	$2.1 \cdot 10^{1}$	$5.4 \cdot 10^{4}$
Crane	$2.6 \cdot 10^{-1}$	$2.8 \cdot 10^{1}$	$5.0 \cdot 10^{5}$
Water bug (*Dytiscus*)	$3.0 \cdot 10^{-2}$	$3.0 \cdot 10^{-1}$	$8.4 \cdot 10^{3}$
Stickleback (marine)	$1.0 \cdot 10^{-1}$	$7.2 \cdot 10^{-1}$	$5.5 \cdot 10^{4}$
Shark	$1.5 \cdot 10^{0}$	$5.2 \cdot 10^{0}$	$6.1 \cdot 10^{6}$
Dolphin (*Stenella* spec.)	$2.1 \cdot 10^{0}$	$9.3 \cdot 10^{0}$	$1.5 \cdot 10^{7}$
Blue whale	$3.3 \cdot 10^{1}$	$1.0 \cdot 10^{1}$	$2.6 \cdot 10^{8}$

3.7.2
Biomechanics of Blood Circulation

Biomechanical calculations of blood flow are based on the one hand on physical equations describing the flow through a tube, and on the other hand, processes must be taken into account which lead to the expansion of the vessels by internal pressure.

Laminar flow through a tube may be thought of as the mutual movement of concentric hollow cylinders. During this movement, each of these cylinders with a radius r and a thickness dr experience a frictional force (F_F) which is proportional to the velocity gradient (dv/dr), their surface area ($2\pi rl$), and the viscosity of the fluid (η):

$$F_F = 2\pi rl\eta \frac{dv}{dr}. \tag{3.7.2}$$

The driving force (F_D) behind such a flow can be obtained from the pressure difference (Δp) and the cross-sectional area of the cylinder (πr^2):

$$F_D = \pi r^2 \Delta p. \tag{3.7.3}$$

In the case of stationary movement, both forces are balanced: $F_F = F_D$. Let the radius of the tube be r', and assume that there is a trapped boundary layer of liquid [$v(r') = 0$]. Connecting Eqs. (3.7.2) and (3.7.3), and integrating them according to dv one gets an equation for the velocity profile of this flow:

$$v(r) = \frac{\Delta p}{4l\eta}(r'^2 - r^2). \tag{3.7.4}$$

Thus, $v(r)$ is a parabolic function, whereas v_{max} is the velocity at the center of the tube (see Figs. 3.48, 3.60). In order to calculate the volume flux (J_V) through the tube, the function $v(r)$ must be integrated over the entire profile. This finally leads to the *Hagen-Poiseuille equation*:

$$J_V = \frac{\pi \Delta p r'^4}{8 l \eta}. \tag{3.7.5}$$

The units of J_V are $m^3\ s^{-1}$. Thus, the flow through a tube is proportional to the fourth power of its radius. This means, for example, that slight widening or narrowing of the blood vessels can cause large changes in the blood flow.

These fundamental laws of physical rheology can only be used as a first approximation for what really happens in blood flow through the vessels. The following particularities must be taken into account, leading to an extension of these approaches:

Blood is a non-Newtonian fluid. I.e. its viscosity (η) depends on the shear rate (γ) of the flow (see Sect. 3.6.2, Fig. 3.51). However, integrating Eq. (3.7.2) in order to derive Eq. (3.7.4) we considered the viscosity as constant. The parabolic velocity profile according to Eq. (3.7.4) therefore cannot be applied for blood. Using a function $\eta(\gamma)$ another velocity profile would result, not identical with a simple parabolic distribution.

Blood is not a homogeneous liquid, but a suspension of cells. In capillaries, the diameters of which are of the same order of magnitude as the diameter of erythrocytes, the velocity profile of the plasma is determined by the moving cells which become strongly deformed in these narrow and branched vessels. This is a problem of *microrheology* of circulation. In large vessels, on the other hand, the so-called *Fahraeus-Lindqvis effect* occurs. This leads the erythrocytes to concentrate in regions of minimal shear stress, namely in the center of the vessel. This means that the viscosity of the blood which we found to depend on the hematocrit (Sect. 3.6.1) increases in this region, but decreases near the wall of the vessel. This leads to a lowering of the streaming resistance of the total blood flow. On the other hand, of course, the streaming profile is changed dramatically. The parabola becomes flattened at the center of the vessel and

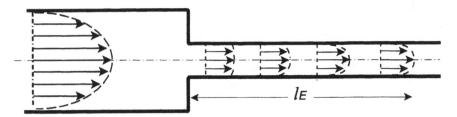

Fig. 3.60. Laminar flow in a tube: a parabolic velocity profile is changing during a sudden narrowing of its radius (entrance effect). Only at a distance l_E is a new parabolic profile established again

steeper near the walls. Furthermore, this effect also leads to a distribution of different sorts of blood cells. In fact, the intensity of the force, shifting the cells by the Fahraeus-Lindqvis effect into regions of lower shear stress, depends on their size. As a result, smaller cells like blood platelets are not as influenced by this effect as erythrocytes with a larger diameter. This leads to a concentration of the platelets near the walls of the vessel. The Fahraeus-Lindqvis effect can be understood as the result of the Prigogine principle of minimal entropy production, as described in Section 3.1.4. It is valid for linear approaches, thus also for laminar flow (see: Sect. 3.6.1). Minimal entropy production for the case of blood flow means that the cells should concentrate at locations of minimal frictional energy dissipation, namely at locations of minimal shear rate.

The diameter of the blood vessels differs along the system of circulation. If a tube suddenly becomes narrow, a so-called *entrance effect* occurs (Fig. 3.60). This means that first the velocity profile of the narrow part of the tube corresponds to that of the central part of the broad tube. Only after a certain distance from the place of narrowing (l_E in Fig. 3.60), will a new profile be established. Usually this occurs at: $l_E = 0.06 r \mathrm{Re}$, where r is the radius of the narrow tube. This effect becomes important at the entrance of blood into the aorta. Furthermore, it occurs in the case of air flow in the lungs.

Blood flow is not stationary, but pulsed. Therefore the condition $\mathbf{F}_F = \mathbf{F}_D$ is no longer valid. This fact is only important for arterial flow. The pulse waves of the heart partly become damped by the elasticity of the walls of the vessels, but they nevertheless proceed as oscillations of pressure and velocity of the blood flow up to the arterioles. One must differentiate between the true velocity of the blood streaming ($\mathbf{v}$) on the one hand, and the velocity of pulse propagation ($\mathbf{v}_P$) on the other hand. This can be calculated using the Moens-Korteweg equation, which is derived for cylindric tubes with thin walls:

$$\mathbf{v}_p = \sqrt{\frac{Yd}{2r\rho}}. \tag{3.7.6}$$

In contrast to the flow rate ($\mathbf{v}$) [Eq. (3.7.4)], the pulse propagation rate ($\mathbf{v}_P$) depends on the density (ρ) of the medium, but not on the viscosity (η). On the other hand, the thickness (d), and the elasticity modulus (Y) of the wall of the vessels are included. This equation does not say anything about the damping of the pulse waves.

To relate various pulsing flows, similar to the Reynolds number, another parameter of similarity (α) is introduced:

$$\alpha = r\sqrt{\frac{\omega\rho}{\eta}}. \tag{3.7.7}$$

In this equation, besides the density (ρ) and the viscosity (η), the pulse frequency (ω) is used. For values $\alpha \geq 3$, the pulse propagation velocity becomes equal to the streaming velocity ($v_p = v$). In the aorta this parameter is higher than that limit ($\alpha = 15$). In the arteria femoralis this limit is reached ($\alpha = 3$). In these vessels, therefore, the pulses are propagated with the same velocity as the total flow rate of the blood. Additionally, at points of branching of the vessels, reflections of the waves can occur.

These considerations already raise the question of elasticity of the blood vessels. We will start to answer these questions again with simple physical approaches. How is the radius of a tube changed by the strain of the tube wall? For the simplest case, let us use Hooke's law [Eq. (3.6.7)], considering an elastic tube with a radius r, and a circumference $2\pi r$. In this case the stress (σ) of the wall of the tube can be written as follows:

$$\sigma = Y \frac{\Delta r}{r}. \tag{3.7.8}$$

We defined the stress (σ) as the stretching force (**F**) per area (*A*) of the stretched material [Eq. (3.6.5)]. Let the thickness of the wall be d, and it length l, so that the cross-sectional area of the stretched wall becomes $A = d \cdot l$. Therefore:

$$\sigma = \frac{F}{ld}. \tag{3.7.9}$$

Defining σ' as stretching force per length, it gives:

$$\sigma' = \frac{F}{l} = \sigma d. \tag{3.7.10}$$

Now, how large would the stress (σ) be in the wall of a tube with an internal pressure p? This can be calculated using the Laplace equation for a tube with radius r:

$$\sigma' = pr. \tag{3.7.11}$$

Equations (3.7.8) to (3.7.11) enable us to derive an equation to calculate the increase of the radius of a tube (Δr) as a function of the internal pressure (p):

$$\Delta r = \frac{pr^2}{Yd}. \tag{3.7.12}$$

Let us remember that we started with the assumption that the vessel wall shows a linear elastic behavior of the material according to Hooke's law. The relation between stress and strain therefore would correspond to Young's modulus [Eqs. (3.6.7) and (3.7.8)]. In fact the wall of the vessels exhibit a particular viscoelastic behavior. The strain is additionally controlled by smooth muscles. This interaction of passive and active properties of the vessel wall is of great importance for the regulation of blood flow.

All these calculations consider the blood flow in the system of circulation as laminar. Is this correct or is there turbulence in human blood flow? The numbers in Table 3.5 indicate that in fact, in large arteries critical Reynolds numbers can occur. We mentioned in Section 3.7.1 that the laminar flow in tubes becomes unstable at Reynolds numbers near 1000. This limit, however, is only correct for smooth, rigid and absolutely cylindrical tubes. In blood vessels, various factors help to stabilize the laminarity even beyond this limit, others induce turbulent flow at lower Reynolds numbers. Factors inducing turbulence are for example the branching of the vessels and inhomogeneities of their walls. These include arteriosclerotic alterations, or alterations following surgical operations. In general the system of blood circulation can be considered as biomechanically optimized.

Ultrasound techniques are mainly used to investigate flow properties of blood vessels in vivo. Furthermore, investigations are undertaken in tubes made of transparent materials which copy particular regions of human vessels (Fig. 3.61).

The model system of Fig. 3.61 shows that near the branching of vessels special properties of flow behavior occur. The critical Reynolds numbers in these regions are lower than in an unbranched vessel. This depends on the angle of bifurcation (α in Fig. 3.61). For $\alpha = 180°$ the laminar flow becomes critical at Re = 350. If the angle is only 165°, the critical point is at Re = 1500. Additionally, the critical Reynolds number depends on the relation of the radius of the branches.

These few aspects of biophysics of blood rheology show us how complicated this situation is in vivo. This branch of biomechanics is developing quickly. This tendency is promoted by the fast progress in surgery of blood vessels on the one hand, and on the construction of artificial cardiac valves, of artificial hearts as well as of various systems of extra-corporal circulation.

Further reading: Fung 1984, 1993; Gabelnick and Litt 1973; Mazumdar 1992; Schneck 1980; Skalak and Chien 1987

Table 3.5. Some rheological parameters of human blood circulation. (Data from Talbot and Berger 1974)

Vessel	Average velocity (m s^{-1})	Diameter (m)	Average wall shear rate (s^{-1})	Reynolds number (Re)
Aorta	$4.8 \cdot 10^{-1}$	$2.5 \cdot 10^{-2}$	155	$3.4 \cdot 10^{3}$
Artery	$4.5 \cdot 10^{-1}$	$4 \cdot 10^{-3}$	900	$5 \cdot 10^{2}$
Arteriole	$5 \cdot 10^{-2}$	$5 \cdot 10^{-5}$	8000	$7 \cdot 10^{-1}$
Capillary	$1 \cdot 10^{-3}$	$8 \cdot 10^{-6}$	1000	$2 \cdot 10^{-3}$
Venule	$2 \cdot 10^{-3}$	$2 \cdot 10^{-5}$	800	$1 \cdot 10^{-2}$
Vein	$1 \cdot 10^{-1}$	$5 \cdot 10^{-3}$	160	$1.4 \cdot 10^{3}$
Vena cava	$3.8 \cdot 10^{-1}$	$3 \cdot 10^{-2}$	100	$3.3 \cdot 10^{3}$

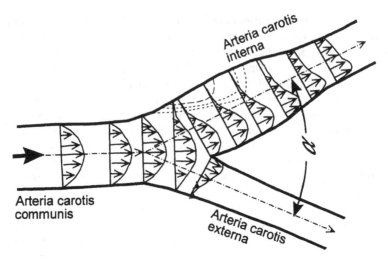

Fig. 3.61. Velocity profile in a model of the human carotis with 70% flow through the arteria carotis interna. α – angle of bifurcation. (After Schneck 1980 modified)

3.7.3
Swimming and Flying

We learned that the Reynolds number enables us to relate flow properties of water with those of air. This gives us the possibility of relating mechanisms of swimming to that of flying.

When a body moves relative to its surrounding medium then a *surface, or skin friction* occurs, and furthermore a drag which is caused by the shape of the body, the so-called *profile, or form drag*. The surface friction arises from phenomena which occur in the boundary layer and which have been discussed in Section 3.7.1. The form drag, on the other hand, is related to the volume of the surrounding medium that is displaced by the moving body. Regions around the body can become influenced which are much larger than those occupied by the boundary layer in a fluid flowing parallel to a flat plate (Fig. 3.58). The critical Reynolds number, indicating that the laminar boundary layer becomes destabilized, is determined, to a great extent, by the shape of the body. This explains why these two components of drag generation, namely the surface drag on one hand, and the form drag on the other hand, cannot be treated separately from one another.

The boundary layer around a moving body is to a great extent influenced by local pressures caused by differences in the local velocities (**v**) at different locations. Because of the law of conservation of energy, the kinetic energy of the moving medium ($1/2\rho\mathbf{v}^2 V$), and the static energy of compression (pV) must be equal at all points of the space:

$$p + \tfrac{1}{2}\rho\mathbf{v}^2 = \text{const.} \tag{3.7.13}$$

This constant is determined by the following conditions: at $\mathbf{v} = 0$ the hydrostatic pressure must be: $p = p_0$, where p_0 is the static pressure of the environment. This leads to the *Bernoulli equation*:

$$p = p_0 - \tfrac{1}{2}\rho\mathbf{v}^2. \tag{3.7.14}$$

The local pressure (p) acting on the surface of a body which moves in relation to a fluid, therefore results from the hydrostatic pressure (p_0) of the environment, lowered by the parameter $1/2\rho\mathbf{v}^2$. If the local velocity $\mathbf{v}$ is great enough, then p can even become negative.

Figure 3.62 indicates the flow profile and the pressure distribution around a streamlined body, i.e. a body with minimal form drag. Additional to the pressure which is given by the Bernoulli equation [Eq. (3.7.14)] an impact pressure at the point occurs where the velocity vector is perpendicular to the body surface. At points where the velocity of flow is greatest, which in the diagram is indicated by maximal density of the lines of flow near the middle of the body, there will be a local reduction in pressure. These local pressures lead to forces which are always directed perpendicular to the surface of the body. At points where this dynamic pressure becomes negative, i.e. the forces are directed away from the surface, a disruption of the flow from the surface of the body can occur. This results in a wake of large eddies that is a region of turbulent flow which occupies a space much larger than that corresponding to the thickness of the turbulent boundary layers given in Fig. 3.58 and Table 3.3 (page 222).

The following flow patterns can occur by increasing the velocity:

- laminar flow around the body,
- turbulent flow in a boundary layer,
- turbulent, disrupted flow forming a wake.

From step to step the drag resistance is increasing.

In Table 3.4 (page 223) Reynolds numbers are indicated for some swimming animals. The small animals, in spite of having unfavorable shapes, are in the region where the flow is always laminar, but for animals where the Reynolds number is in the order of 100 and higher, a hydrodynamic optimization of the shape of the body is required. In contrast to the sphere which allows laminar

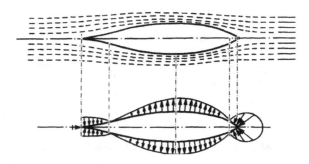

Fig. 3.62. Flow profile and pressure distribution around a moving streamlined body. (After Hertel 1963)

flow up to Re $\approx$ 1000, the critical Reynolds number for a streamlined body is, of course, higher. The body shape of fast swimming animals really does appear to be optimal in this respect. Nevertheless, the Reynolds numbers of fast fishes and dolphins, and also of many birds, lie in the supercritical range. In such cases adaptions can be found that, at least, impede the disruption of the turbulent flow. This, for example, is achieved by the early induction of microturbulences at the surface of the body. Surface structures such as feathers or scales bring this about.

Much has been written in papers on bionics about the specific adaptations of the dolphin which enable it to swim extremely fast. Apparently there is a viscoelastic damping layer of the skin and, furthermore, the ability to induce folds at the body surface by muscular activity, both of which prevent the occurrence of latent instabilities in the flow.

The real drag of a swimming fish is difficult to determine. The simplest way to measure this would be to pull dead or anesthetized fish at a given velocity through water. As a result, the values for the drag determined in these experiments are so high that it would be impossible for the musculature of an actively swimming fish to overcome them. The fact that a fish nevertheless does swim with such a velocity is known as *Grays paradox*. The solution to this apparent contradiction is as follows: in contrast to technical constructions, as for example vessels which consist of a driving component (screw), and a frictional component (body), in case of the fish both elements are integrated. A fish diminishes its friction during active swimming. An actively swimming fish has a lower drag than one that is towed passively through the water.

The part of the body which is involved in propulsion of fishes can be very different. Depending on the relative length and flexibility of the tail, three main types can be differentiated. The *anguilliform type* of movement, for example the eel, involves the whole body for propulsion. Most fishes show the *canrangiform type* of propulsion (Fig. 3.63). In this case tapering tails of medium length allow fast and dexterous swimming. They are able to accelerate quickly, reaching a high speed after a short time. In the case of *ostraciiform type* of movement, named after the trunk of coffer fish, only the fins, like propellers, propagate the fish. These fishes are only able to swim slowly.

In contrast to the mechanism of swimming, where we considered mechanisms of drag reduction, in the biomechanics of flying, lift is of central interest.

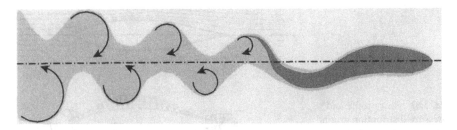

Fig. 3.63. Carangiform mechanism of propulsion of a fish with the vortex street behind

First, we must distinguish between passive gliding and active flying. Various insects (especially butterflies), birds and mammals can glide, as can a number of plant seeds and fruits. Many types of aerodynamic mechanisms are employed by these vegetable objects, ranging from simple adaptations to reduce the rate of descent by means of rotating wings, for example the propeller seed of the maple tree, up to the stable gliding flying-wing of the seeds of the climbing pumpkin species *Zanonia macrocarpa*.

Active flying by animals is achieved through a most varied assortment of mechanisms. Those animals which can rise into the air without some initial help and then move forwards or backwards from this position can be considered as ideal fliers. These are the properties of a helicopter and are only possible in animals which have a large wing area relative to their body weight. It can be deduced from Fig. 3.64 that the wing area does not increase in direct proportion to body mass, but rather to the power of 2/3. The wing loading of heavier animals is usually greater than $1 \text{ kg} \cdot \text{m}^{-2}$ but lower than in many insects.

The wing loading, i.e. the relation between body weight and wing area, is crucial for the power which is required to hover. Table 3.6 shows calculated values for the power outputs which would be required for stationary hovering flight for some animals. In fact, due to the real efficiency of that mechanism which may be lower, as proposed in the calculation, the required muscular power might be even greater than these figures indicate. The upper limit of the specific muscular performance has already been reached in the case of humming

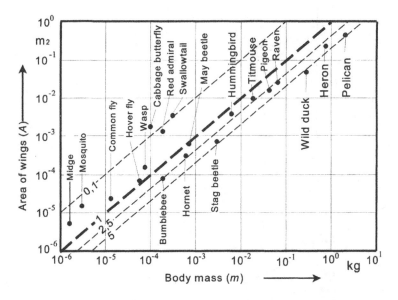

Fig. 3.64. Area of wings (A, in m^2), body mass (m, in kg), and wing loading (*dashed lines* with parameters in $\text{kg} \cdot \text{m}^{-2}$) of various insects and birds. The points enable us to draw an allometric function $A = 1.5 \, m^{2/3}$. (Data from Hertel 1963)

Table 3.6. Wing loading and the required specific power for stationary hovering flight in various animals

Animal	Wing loading (kg m^{-2})	Specific power (in relation to the body mass) (W kg^{-1})
Cabbage butterfly	0.04	2.6
Dragon-fly	0.08	3.7
Bee	0.71	1.0
Humming bird	1.01	3.8

birds. Larger birds are not able to hover on their own although some of them, like the kestrel can do so with the help of the wind.

Figure 3.65 shows schematically the wing movement of a hovering humming bird. The wings move horizontally and each covers about a third of the horizontal area around the bird. The angle of attack is continuously adjusted so that lift is generated as the wing moves forwards as well as backwards. In contrast to humming birds, the wings of flies and bees move in a vertical

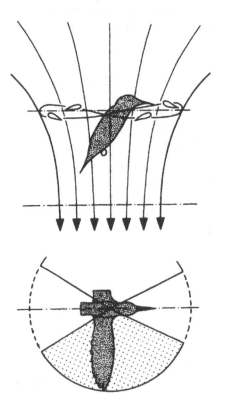

Fig. 3.65. Wing position and air flow of a hovering humming bird. (After Hertel 1963)

Fig. 3.66. Airstream and forces of a bird's flight during a stroke forwards and downwards. (After Hertel 1963)

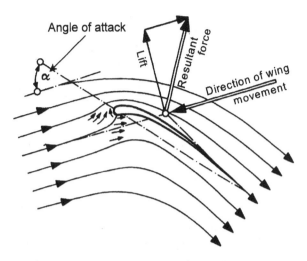

direction. The wings are also twisted but only the downstroke is used to generate lift. In case of birds flying forward, the wings are moved up and down. The angle of attack is also regulated, but additionally the wings are bent or partly folded on the upstroke.

In contrast to the humming bird, all other birds require some assistance to get off the ground. Just as an aircraft must reach a take-off speed to ensure that the wings are generating sufficient lift, so must birds. This is achieved by running, jumping or dropping. Nevertheless, the take-off speed of birds is remarkable low. This means that maximum lift at minimal Reynolds number is required. Figure 3.66 shows the force diagram for a wing that is being thrust forwards and downwards. If the incident flow remains constant (v_0) then the lift generated by the wing increases with the degree of curvature of the wing section as well as with the angel of attack (α). The amount of increase of both these parameters is, however, limited. If they exceed a critical value, the flow is disrupted and turbulence develops. This will not only cause a loss of thrust but will at the same time destabilize the entire system. Such an occurrence would cause an airplane to crash. There are a number of biological adaptions preventing such a disaster and at the same time ensuring maximum lift at low speeds.

Further reading: Alexander 1983; Azuma 1992; Hoppe et al. 1984; Leyton 1975; Videler 1993; Vogel 1994; Webb and Weihs 1983

CHAPTER 4

Physical Factors of the Environment 4

The subject of this section, namely the influence of physical factors on biological systems belongs to environmental biophysics on the one hand and, on the other hand, is important for applications in medicine and biotechnology.

The input of energy from the environment into biological systems can lead to different consequences:

- It can directly interact with the energetics of the organism. This can occur in a specific way, as in the case of photosynthesis, or nonspecifically, for example, just by heating the organism.
- Very small amounts of energy, occasionally being absorbed at particular targets, can significantly disturb biological mechanisms of control. The place of interaction in this case is not a specific receptor, but rather a crucial point in the network of reactions. An example of this is the induction of mutations by ionizing radiation.
- Environmental energy can act as a carrier of information. In this case specific receptors are activated by negligible amounts of energy input.

According to the so-called *Grotthus-Draper principle* it is not the energy penetrating the organism which is effective, but that part of the energy which is absorbed in the system. Investigations into the mechanism of interaction, therefore, start with the question: how, and where is this kind of energy absorbed? Eventually it must be clarified whether alterations are induced by the absorbed energy, and how these influence the biological system. In some cases, biological cascades of amplification exist. The theory of biological amplification was first developed in radiation biology. It is, however, also crucial for a number of other physical influences.

When discussing possible influences of weak physical energies on biological systems, its relation to thermic noise (kT, Sect. 2.3.4) is frequently stressed. This is correct from the physical point of view, assuming the following circumstances are taken into account:

- To relate the energy input to the energy of thermic noise, the primary mechanism must be known which leads to the absorption of this kind of energy in the biological system.
- It must be considered that cooperative effects may be included in these processes. Such interactions are quite probable in more highly organized

molecular systems. This sort of cooperation can significantly promote low energetic influences.

- It must be considered that the organism has developed a multitude of filter systems against thermic noise along the path of evolution. The best example of this is the human ear.

In this context it should be pointed out that the organism has and is adapting through evolution to all physical parameters of its environment. So, for example, the double strand of the DNA is an adaptation to overcome single strand breaks by ionizing radiation. Additionally, a large number of mechanisms have been developed to repair these kinds of damage. On the other hand, during evolution an astonishing number of mechanisms have been developed to sense minimal amounts of energy, in the form of receptors for vibration, for temperature, for light and for other kinds of radiation.

4.1
Temperature

In Section 2.3.5 the Arrhenius equation was introduced to describe the temperature dependence of chemical reactions. We learned that every elementary reaction step depends on an activation energy and will be accelerated by an increase in temperature. In this way, even concentrations of metabolites in a steady-state system can be changed. The degree of these changes depends on the extent to which two opposing fluxes are affected by temperature alterations. If, for example, in an open system the temperature dependence of the rate of decomposition or of the efflux of a certain component is predominant, then with rising temperature the corresponding steady-state concentration decreases, and vice versa. It follows that in complex systems, an increase in temperature can lead to an increase, as well as to a decrease in the overall activity. In biological systems the denaturation of proteins is mainly responsible for the reduction of their activity at elevated temperatures. The complete organism as well as isolated enzyme systems show optimal temperatures of operation. In homoiothermic animals the optimal temperature is regulated carefully. Recently, differences in the temperature response between normal versus cancerous tissues has been used in hyperthermy therapy of cancer. The temperature regulation of an organism is closely connected to its entropy balance [see Eq. (3.1.63)].

In Fig. 4.1 different paths of heat transport between an organism and its environment are illustrated. In general, a distinction has to be made between conduction, convection, radiation and heat loss due to evaporation of water. It is possible to define a thermal flux (J_Q) which is the thermal energy that is exchanged per unit of time across a particular area of the surface.

Thermal conductivity is governed by the same kind of equations which we have used to calculate the diffusion of a substance. In this way the thermal flux (J_{Qc}) resulting from conductivity can be included into the flux matrix [Eq. (3.1.58)], and consequently it can be coupled with transport of matter (so-called

Fig. 4.1. Various paths of heat transport between the body and its environment (After Precht 1955)

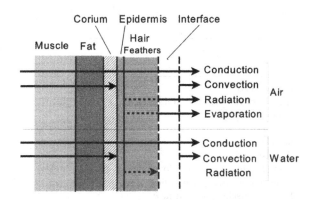

thermodiffusion, see Sect. 3.4.1). For a temperature gradient in the x-direction, the following linear flux equation can be written

$$J_{Qc} = -\lambda \frac{\mathrm{d}T}{\mathrm{d}x}.$$

(4.1.1)

The negative sign indicates the direction of the thermal flux from higher to lower temperature. The factor λ in this equation is the thermal conductivity, also called the thermal coefficient. It has the units $J\ m^{-1}\ s^{-1}\ K^{-1}$. In Table 4.1 some examples of thermal conductivity are listed, showing the good thermal insulating properties of a layer of fat.

Convection allows a very intensive heat transport at physiological temperatures. As shown in Fig. 4.1, convection in homoiothermic animals is mediated internally through blood circulation, and externally through the flow of air or water. At constant flow, the amount of heat flow transported by convection is proportional to the temperature differences between the two phases. The flow rate on the one hand, and the geometric conditions of the boundary layer on the other hand are crucial for the effectiveness of this kind of heat transfer. The level of convection depends on the thickness of unstirred layers of air or water near the body, and on the flow behavior at the boundary layer (see Sect. 3.7.1). This insulating layer can be improved artificially by hairs or feathers.

Thermal radiation is a kind of electromagnetic radiation as illustrated in Fig. 4.13. To investigate thermal radiation, a so-called black body is used. This is a physical body that will completely absorb all of the infrared radiant energy that falls on it. The black body is also suitable for measuring the radiation which is emitted by an object at a given temperature. In contrast to the other kinds of heat transfer, thermal radiation does not depend on the temperature difference between the body and the environment but rather on the absolute temperature of the emitting body itself, whereas the temperature (T) occurs in the fourth power. The amount of heat emitted by a black body (J_{QR}) obeys the Stefan-Boltzmann law:

$$J_{QR} = \sigma T^4.$$

(4.1.2)

Table 4.1. Thermal conductivity for various tissues and other biological materials. (After Precht et al. 1955)

	Thermal conductivity $J\ m^{-1}\ s^{-1}\ K^{-1}$
Tissue in vitro	
Human skin: epidermis and corium	0.34
Epidermis (pig)	0.21
Fat (pig)	0.17
Muscle (pig)	0.46
Living tissues in situ	
Human skin, poor blood flow	0.31–0.33
Human skin, strong blood flow	1.5
Human muscle, no blood flow	0.46
Human muscle, normal blood flow	0.53
Human muscle, strong blood flow	0.63
Hair and feathers	
Wool (loose)	0.024
Feathers	0.024
Rabbit fur	0.025
Other substances	
Air	0.023
Water (20 °C)	0.59
Silver	420.0

The proportionality constant is $\sigma = 5.67 \cdot 10^{-8}\ W\ m^{-2}\ K^{-4}$. Additionally, a correction factor is used if the surface of real bodies does not behave like a black body. For the surface of biological objects it lies between 0.9 and 1. The value depends on the pigmentation of the organism.

Evaporation of water is an important mechanism of temperature regulation of animals living in air. The evaporation of 1 g of water will cause the body to lose about 2.4 kJ. The rate of evaporation depends on the surface structure, on the wind velocity, and on the difference in the vapor pressure between the surface of the body and the surroundings. Because the vapor pressure is temperature dependent, evaporation can still take place when the air is saturated with moisture provided that the ambient temperature is below the body temperature.

The properties of biological systems at extremely low temperatures is of special interest for the techniques of cryopreservation. In this respect, two processes are critical for the survival of cells: the structural damage of the cell by microscopic ice crystals, and the alteration of the osmotic equilibrium of the cell during processes of freezing and the retrieval of them from frozen storage. Similar to other processes of crystallization, the formation of ice crystals starts with certain crystal nuclei in the cell. The rate of cooling determines the number and the size of the developing crystals which can be crucial for the survival of the cell. To allow water to flow quickly out of the cell, cryoprotective agents are used, such as for example glycerol or dimethyl sulfoxide (DMSO). They depress the freezing point in parallel, so that ice crystals begin to form only at about

−5 °C. If, at −5 °C to −15 °C the cell is sufficiently dehydrated, ice crystals will form in the surrounding medium but not in the cell interior. Dehydration of the cells means shrinkage and concentration of the internal solutes. This must be restored in the period of thawing. The size of biological objects which are suitable for cryopreservation is determined by the ability to control the rate of heat exchange during freezing as well as during warming.

Further reading: For temperature regulation: Stanier et al. 1984; for cryobiology: Franks 1985

4.2
Pressure

When aquatic animals move in a vertical direction they are exposed to greatly differing pressure. For an increase in depth of 10 m, the pressure increases by about 0.1 MPa. Deep-sea organisms live under conditions where the hydrostatic pressure can be as high as 100 MPa. Seventy percent of the surface of the earth is covered by oceans with an average depth of 3800 m. Fifty percent of the surface of our planet is covered by water at least 4000 m deep. In fact this is the depth with the greatest number of different species.

Pressure an important physiological parameter not only for aquatic organisms. Humans can also be exposed to changes in environmental pressure. This particularly concerns the field of occupational medicine. It can involve conditions where the pressure is either increased or reduced.

Basically, the following mechanisms should be taken into account when considering the effects of hydrostatic pressure on biological systems:

- volume changes in gas-filled cavities,
- changes in the solubility of gases in blood and tissue water,
- changes in the structure of water and corresponding consequences on hydrophobic bonds.

These mechanisms become important at very different levels of hydrostatic pressure. While the first two mechanisms concern low pressures, changes in water structure, and consequently hydrophobic bonds, only occur at extremely high levels of hydrostatic pressure.

The reduced oxygen solubility in the blood which occurs at great heights, is particularly important for alpinists and pilots. This oxygen deficit can be compensated for by an increase in the partial pressure of oxygen in the inspired air. If humans remain under an increased pressure, for example divers, the reverse effect occurs, namely an increase in the amount of dissolved gas in the blood. This may lead to a dangerous situation during the return to normal atmospheric pressure, especially following a long stay at a great depth. If the decompression is too fast then gas bubbles are formed in the blood (mainly N_2). This can cause aeroembolism also called the bends or caisson disease.

The exposure of biological systems to high pressure can also shift the equilibrium positions of physiological reactions according to the Le Chatelier principle if the reaction leads to alterations of the partial molar volume of the reactants. This will occur mostly in reactions producing a gas. Moreover, it is important in processes associated with modifications of the water structure. We already discussed these circumstances in Section 2.4.2. Although this effect only occurs at pressures greater than 10 MPa, and is therefore important only for deep-sea research, it is particularly useful as a method for studying cellular processes.

Further reading: Péqueux and Gilles 1985

4.3
Mechanical Oscillations

Mechanical oscillations on the one hand can influence biological systems directly by input of considerable energy, and on the other hand they can be recorded by animals with the help of highly specified receptors and serve as physical carriers of information. Oscillations in the air at frequencies between 16 Hz and 20 kHz are perceived by humans as sound. Oscillations with lower frequency, in fact those lower than 20 Hz are called *infrasound*, and those with frequencies higher than 20 kHz are called *ultrasound*. Furthermore, vibrations can influence biological systems. Oscillations of materials of various frequencies which may induce sound, but which influence biological systems mostly by direct contact, are described as vibrations.

4.3.1
Vibration

The effects of vibration are of particular interest in occupational medicine. It concerns induction of oscillations in the human body by various vibrating tools and devices, vibrating seats of tractors, trucks and cars etc. In addition to affecting vegetative functions, leading to kinetosis, or for example to seasickness, a high and long-lasting vibration exposure may contribute to low back pain, or other acute or even latent disorders or injuries.

Generally, the human body becomes vibrated through contact with a vibrating solid object, standing or sitting on it, holding it etc. The intensity of the vibration as well as the efficiency of its transfer to the human body and its own resonance frequency are of interest. From the physical point of view this is the problem of oscillations of resonators consisting of springs and dampers, enforced by external vibrations. In the simplest case, the body will be excited by a simple harmonic oscillation in the form:

$$x = x_{max} \sin \omega t, \tag{4.3.1}$$

where x means the displacement, oscillating with the angular frequency ω and a maximum amplitude x_{max}. Deriving this equation in respect to time (t), the vibration velocity ($\mathbf{v}$) is obtained by:

$$\mathbf{v} = \frac{dx}{dt} = \omega x_{max} \cos \omega t = \mathbf{v}_{max} \cos \omega t. \tag{4.3.2}$$

In practice, man is not influenced by vibrations consisting of simple sinus waves but rather by random vibrations, consisting of a sum of many frequencies differing in amplitude as well as in phase behavior. Using Fourier analysis they can be regarded as a sum of many sinusoidal oscillations.

Investigating the induction of vibrations by contact with vibrating materials, biological systems cannot be considered simply as consisting of rigid materials. We have already discussed viscoelastic properties of biological materials in Section 3.6.3 and demonstrated some viscoelastic models in Fig. 3.53. Such models can be used to construct vibratory biomechanical models of the human body, taking in to account the inertia of masses.

The kinetic properties of similar systems will be discussed later in Section 5.1.2, where we consider the behavior of various systems after a single extension (Fig. 5.3). Here, enforced oscillations of such systems are of interest. Every oscillating body has its own characteristic resonance frequency which is indicated in Figs. 5.3C and 5.4. If the frequency of external vibration inducing these oscillations is much smaller than this resonance frequency, then the system follows exactly the amplitude and phase of the applied oscillations. When the applied frequency is near that of the resonance frequency, or equals it, then the amplitude of the enforced oscillation is greater than that of the applied oscillations. An undamped system in this case would always increase its amplitude until it breaks down (Fig. 5.4, curve 4). Depending on the degree of damping, in the resonance region the maximal amplitude of vibration will be lowered, but nevertheless it can be larger than the amplitude of the applied vibration. A further increase of the external frequency leads to a decrease in the coupling efficiency up to zero, at $\omega \Rightarrow \infty$. In all cases the frequency of the enforced vibration corresponds to the external frequency, but phase shifts may occur.

The oscillation of a mechanical system can be treated as analogous to the electrical oscillations of an RCL circuit (see also Sect. 3.5.3). Complex parameters are used to describe damping and elastic parameters of these mechanical systems. In analogy to electrical systems a *mechanical impedance* of the system can be defined. It is the relation between vibrational force and vibrational velocity.

This coupling of external vibrations to various parts of the body can be measured directly. Figure 4.2 shows the movement of head, shoulders and hips of a person sitting on a vertically vibrating seat. The ordinate shows the relation between the amplitude of the vibration of the designated points of the body, and the amplitude of vibration of the support. It is shown that at frequencies below 2 Hz, the body simply follows the movement of the seat, i.e. the degree of amplification is 1. The fundamental resonance frequency of a human body is

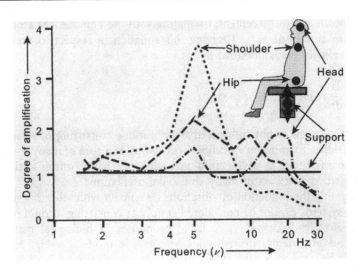

Fig. 4.2. The amplification factor for vertical oscillations at various points of the body of a man sitting on a vibrating support as function of the frequency. (Redrawn according to Dieckmann 1984)

about 5 Hz. This is the frequency of maximal displacement of all three reference points of the body in Fig. 4.2. The head furthermore shows a second maximum at about 20 Hz.

The biological effects of long-lasting vibrations mostly consist of a kind of wear of the joints. The actual forces set up in the joints during vibration can only be measured directly in special cases. They cannot be calculated simply by physical models, using passive mechanical properties of anatomical constituents of the body. It must be taken into account that forces are generated by the muscles trying to compensate the external stress. This, however, in some cases can lead to an opposite effect, namely to an additional strain on the joints. In general, one must consider that the mechanical impedances of parts of the body are influenced by neuro-physiological processes.

In the frequency range below 2 Hz the effects of vibration of the body cannot be simply attributed to mechanical processes. Particularly, the frequency of 0.3 Hz is responsible for sea-sickness, a special kind of so-called kinetosis. All these processes are caused by neuro-physiological influences, and are induced by the vestibular system detecting accelerations of the head. This can lead to alteration in blood circulation with all subsequent physiological consequences.

The mechanisms by which humans and animals can perceive vibrations are a special field of research. Humans, for example, can detect a vibration of 200 Hz with an amplitude of only 0.1 μm! The sensibility of insects and spiders can be several orders greater.

Further reading: Fritz 1998; Qassem et al. 1994; Seidel and Heide 1986; Wilson 1989

4.3.2
Sound

It has been known for a long time that alterations of external signals are perceived more sensitively, the lower the absolute intensity of the signal is. Let us take a sound intensity (I, in W m^{-2}). The intensity difference is dI and a relative difference: dI/I results. This relative alteration of the physical signal, should thus be proportional to the perceived difference of the sound intensity level (dL). This rule which is valid for all kinds of sense organs is called the *Weber-Fechner law*. The integration of the corresponding differential equation leads to the following logarithmic relation:

$$L = 10 \lg \frac{I}{I_0} = 10 \lg \frac{p^2}{p_0^2} = 20 \lg \frac{p}{p_0}. \tag{4.3.3}$$

The dimensionless unit for the sound intensity level (L) is the decibel (dB).

The integration which led to Eq. (4.3.3) needs an initial condition which is included in the parameter I_0. This is the reference sound intensity which is for sound transmitted in air: $I_0 = 10^{-12}$ W m^{-2}. It corresponds to the human hearing threshold level for a reference frequency of 1000 Hz, i.e. for the frequency of maximal sensitivity of the human ear. Furthermore, the sound pressure (p) is introduced, the square of which is proportional to the intensity (I). The reference pressure, correspondingly to I_0 is defined as: $p_0 = 2 \cdot 10^{-5}$ Pa.

Figure 4.3 indicates a frequency diagram of human auditory sense. The sound intensity level (L in dB), and the sound pressure (p in Pa) are plotted against the frequency. Additionally, the unit phon is used. This is a unit for measuring the apparent loudness of a sound, equal in number for a given sound to the sound intensity in dB of a sound having a frequency of 1000 Hz when, in the judgment

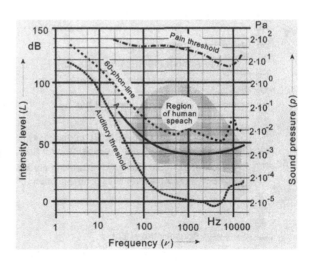

Fig. 4.3. Response of human ear to sound intensity. The curves are extended up to the frequencies of infra-sound ($v < 20$ Hz) according to data from Møller 1984. A-weighting scale according to Fig. 4.5

of a group of listeners, the two sounds are of equal loudness. The 60-phon curve therefore crosses the 60 dB level at the frequency of 1000 Hz. This indicates that a 50 Hz sound, for example, with an objective sound intensity level of 120 dB subjectively seems to have the same loudness as a 1 kHz sound of only 60 dB. The phon unit therefore is a subjective measure which is related to a normal human ear. It is also useful to relate not only various pure tones (sine waves), but also sounds (true musical sounds with overtones) and noise (random mixture of frequencies). This evaluation, however, depends not only on the loudness but also on duration. Sounds with a duration lower than 1 s will be subjectively underestimated.

The use of the unit phon is not sufficient to determine an objective noise exposure of humans in their environment. Figure 4.4 shows some examples of noise production of various sources. It has been known for a long time that in case of permanent exposure, vegetative diseases are possible at 60 dB. The level for dangerous noise at low frequency is at 80 dB, and for high frequencies at 90 dB. Sound intensities which can induce acute injury start above that region.

From the legal point of view, noise is a sound which is able to disturb, to endanger, to cause inconvenience, or to put some one at a significant disadvantage. It is complicated but absolutely necessary to measure this legally definite noise in a physical way.

A simple measurement of an overall sound intensity in dB is too crude to be used for practical environmental protection because of the broad variations of the character of noise. Usually, weighing scales are used to evaluate a particular noise. These scales are determined empirically and fixed by standardization (Fig. 4.5). They are always normalized for 1 kHz and take into account the frequency dependence of the sensibility of the human ear. With the help of these scales it is possible to take into account the lower sensibility of the human ear to frequencies which are lower or higher than 1 kHz. Of course, such scales are

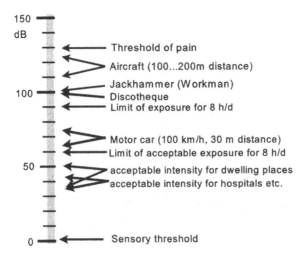

Fig. 4.4. Example of noise production of various sources

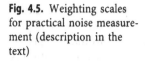

Fig. 4.5. Weighting scales for practical noise measurement (description in the text)

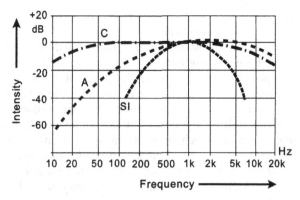

dependent on the real character of the measured noise and on the question being asked. There are differences in the frequency mixture, depending on whether the noise comes from engines, from a discotheque, from traffic or from other sources. It also makes a difference as to whether the noise limit of a working place in a factory, in an apartment or in a hospital is to be determined. Mostly the weighing scale A (Fig. 4.5) is used. It takes into account in particular the low sensibility of the human ear to lower frequencies (Fig. 4.4). A very flat scale B (not drawn) is not used anymore. Instead the C-scale is used for noise in a very broad frequency range. The SI scale particularly evaluates the disturbances of human communication.

Further reading: Harris 1991; Wilson 1989

4.3.3
The Biophysics of Hearing

Hearing on the one hand means the reception and analysis of sound, on the other hand the localization of its source. The process of sound localization is based on the analysis of the time delay for the sound reaching the two ears, and on evaluation of the differences in the corresponding intensities. Humans are able to localize a source of sound with an angular resolution of 3°. This means that the minimal time resolution of the differences of arriving auditory signals must be about 35 µs. This is mainly a question of neurophysiology. In the following text we will for the most part concentrate on the problem of sound perception and frequency analysis.

To differentiate between neurobiological and biomechanical mechanisms of sound analysis, the following experiments are instructive: two pure tones are transmitted by a headphone. In one case, both of them are simultaneously transmitted into both earpieces, in a second case, one tone is transmitted into the right, the other into the left earpiece. If both tones are mistuned, having nearly the same frequency, first order beats occur, i.e. the amplitude of the resulting tone becomes low frequency modulated. This sort of mistuned sound is recognized, if both tones are fed together into both earphones. The resulting first order beat is indicated in the inner ear. In contrast to this, tones with nearly harmonic frequencies, i.e. mistuned fourth, or third etc., produce beats of second order, namely a sort of frequency modulation. In this case, recognition is also possible, if each tone is fed into one of the two earphones. Then no physical interference of the tones can occur. They come together only in the brain.

Figure 4.6 schematically shows the organization of the human ear. Anatomically as well as functionally it can be divided into three parts: outer, middle and inner ear. Each of them participates in the process of sound analysis. Additionally, of course, the sound will be analyzed in the brain as well.

Beside the true nerve impulse from the excited hair cells, the auditory nerve also transmits so-called microphonic potentials. These are obviously generated passively as a result of vibrations of the inner ear as inhomogeneous dielectrics, and are not caused by nerve excitations. It remains unclear as to whether they have any physiological relevance or not.

The outer ear acts as a simple acoustic amplifier with certain frequency characteristics and a particular directional selectivity. The resonance frequency of the human outer ear is between 2 and 3 kHz, i.e. at the maximum of human sound sensitivity (Fig. 4.3).

In the middle ear the auditory ossicles (malleus, incus, stapes) which are suspended by flexible ligaments transfer the vibrations of the eardrum having an area of 65–85 mm^2 to the oval window which is only 3.2 mm^2. This system works passively, without energy supply as a kind of mechanical impedance transformer. It transforms vibration of air into vibration of the lymph fluids, which resemble water as far as density and viscosity are concerned ($\eta = 1.97$ mPa s). During this transduction a lowering of amplitude and an increase in pressure occurs. In the case of the human ear the pressure increases by a factor of about 17, for cats by 60. Recently it was found that regulation of

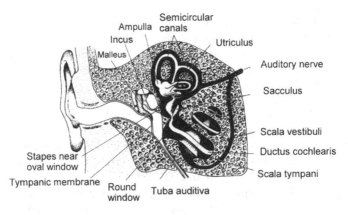

Fig. 4.6. The human ear consisting of the outer ear, the middle ear between the eardrum and the oval window, and the inner ear. (After Penzlin 1991, redrawn)

the properties of this impedance transformation can be controlled by ligaments, protecting the ear against overloading.

The actual sound detection takes place in the inner ear. This is a helical coiled canal that, along its length, is divided by Reissners membrane into two cavities, the scala vestibuli and the scala tympani. The perilymph fluid of the scala vestibuli and scala tympani is separated from the endolymph fluid of the scala media. The sound pressure is transmitted from the oval window to the perilymph fluid and proceeds to the scala tympani (Figs. 4.6, 4.7). At the end of this, the round window buffers the sound waves. There is no reflection of the sound and therefore no standing waves can occur.

Reissners membrane and the basilar membrane transmit the sound pressure of the perilymph fluid everywhere to the endolymph fluid of the scala media. The sensory cells themselves are arranged on the basilar membrane, or better: in the cleft toward the canalis spiralis under the tectorial membrane (Fig. 4.7).

Herrmann von Helmholtz assumed that each point on the basilar membrane along the cochlea is sensitive to a particular pitch. This single-point theory of auditory perception has been confirmed by George von Békésy (Nobel prize winner 1961), but not in the sense of the auditory theory of Helmholtz who postulated a resonance theory, claiming that various places on the basilar membrane indicate resonance properties for various frequencies. There is, however, no experimental evidence for this.

On the other hand, the basilar membrane exhibits very specific viscoelastic properties. In humans the basilar membrane is 35 mm long and gets 10 fold wider over this diameter away from the oval window. At the same time, its elastic properties change. As shown in Fig. 4.8, the stiffness of the basilar membrane is reduced by about 100-fold with increasing distance from the oval window.

Investigations into the vibration characteristics of this system are very complicated. A first approach to this behavior can be made using the model shown in Fig. 4.9. A pendulum transmits its movement to a rotatable rod to

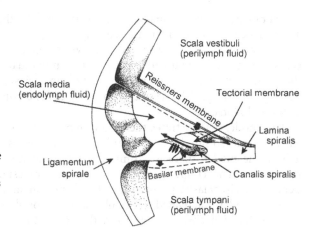

Fig. 4.7. A section through one turn of the cochlea. The *dotted lines* indicate the extension of the membranes induced by local sound pressure. (After Penzlin 1991, redrawn)

Scala vestibuli (perilymph fluid)

Scala media (endolymph fluid)

Reissners membrane

Tectorial membrane

Lamina spiralis

Ligamentum spirale

Basilar membrane

Canalis spiralis

Scala tympani (perilymph fluid)

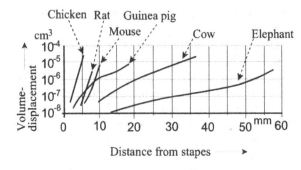

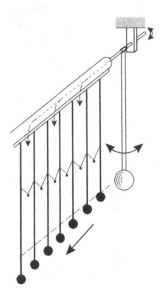

Fig. 4.8. The elasticity of the cochlear partition of various animals as a function of the distance from the stapes. As a measure, the volume displacement is used which is produced in a 1 mm segment by a pressure of 1 cm of water. (After Békésy, in Waterman and Morowitz 1965)

Fig. 4.9. Pendulum resonator model to demonstrate travelling waves in the inner ear. (After Békésy, in Waterman and Morowitz 1965)

which other pendulums, made of balls suspended by threads of various lengths, are attached at equal intervals. These, through the backwards and forwards movements of the axle are caused to move at a given frequency, which in most cases is not the resonant frequency of the individual pendulum. Additionally, the pendulums are coupled to one another in series by threads weighted with small ball bearings. The degree of coupling depends on the point of attachment of these threads.

This model demonstrates the following properties of the inner ear: the sound waves spread so rapidly in the perilymph fluid that, similar to the effect of the axle in the model, there is a simultaneous stimulation along the whole of the membrane. The coupling of the pendulums corresponds with the position-dependent viscoelastic properties of the basilar membrane.

If the model is activated by simple harmonic oscillation, travelling waves are generated which always move from the stiffer to the more compliant region along the row of pendulums. The amplitudes of these waves are not constant. They reach a maximum at certain points of their movement in the direction of the arrow (Fig. 4.10), depending on the frequency of activation. Connecting these maxima, an envelope curve results, the maximum of which depends on the frequency of activation and on the properties of the pendulums, particularly the viscoelastic properties of the basilar membrane and the scala vestibuli. The higher the frequency, the closer the maximum of this envelope curve is to the oval window. Figure 4.11 indicates the positions of maximum stimulation along the cochlear partition of the elephant. In the case of complicated tones, envelope curves occur with several maxima. At the locations of maximal amplitude, stimulations of the sensory cells occur. So in this simple mechanical way a frequency analysis of the sound is already possible.

These sensory cells, the so-called hair cells, are located at the basilar membrane. They contain at their surface bundles of fine hairs, so-called stereocilia in a hexagonal arrangement. These stereocilia differ in length and thickness. Some of them attend the tectorial membrane on the opposite side of the cleft. They are excited by the oscillating flow in the cleft between the scala

Fig. 4.10. Cochlear partition vibrations at two instants within a cycle, and the corresponding envelope curve of a vibration of 200 Hz. (After Békésy, in Waterman and Morowitz 1965)

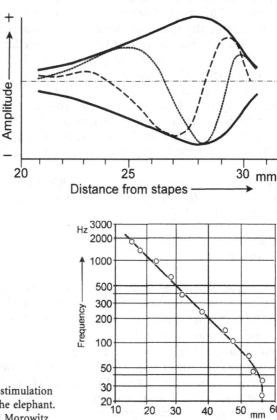

Fig. 4.11. Position of maximum stimulation along the cochlear partition of the elephant. (After Békésy, in Waterman and Morowitz 1965)

media and the canalis spiralis, as well as by oscillations of the distance between the opposite surfaces. The excitation of the hair cells occurs by mechanical coupling between the stereocilia and ion channels of the cell membrane. In this way the mechanical vibration will be transformed into an electrochemical, i.e. an electrical signal.

In this context the enormous spread of the measuring range of our auditory system must be remembered. We learn from Fig. 4.3 that, for example, at 1 kHz the sound pressure for the auditory threshold of pain is about 10^7 times larger than the sound pressure for the auditory threshold. Let us suppose that this pressure is proportional to the maximal amplitude of the displacement of the basilar membrane, or to the amplitude of the oscillating flow in the cleft between the scala media and the canalis spiralis. This means that a variation of this amplitude could occur between about 1 nm = 10^{-10} m near the auditory threshold, and 1 mm = 10^{-3} m near the pain threshold! Near the maximal intensity, damaging the system must be avoided, near the minimal displacement, mechanisms must exist which protect the system against thermic noise.

One of the mechanisms against thermic noise is the mutual coupling of the stereocilia. These hairs of a cell are connected together at their ends by fine filaments. This kind of connection follows a particular pattern. The bending properties of the system of stereocilia in this way become anisotropic. The stereocilia are much more sensitive in the direction of the proposed oscillating flow than to the random bending induced by thermic noise.

Summarizing this chapter we can conclude that the process of auditory analysis of sound frequency is realized in several steps: it starts with the filtration quality of the outer ear and proceeds step by step with the frequency characteristics of the impedance transducer of the middle ear, the frequency characteristics of the travelling waves, the resonances of stereocilia and sensory cells, to the neuronal mechanisms of sound analysis.

Further reading: Hudspeth 1985 and 1989; Markin and Hudspeth 1995; Skalak and Chien 1987; Waterman and Morowitz 1965; especially for biophysics of music: Roederer 1979

4.3.4
Infrasound

Infrasound consists of oscillations of frequencies below 20 Hz, whereby 0.1 Hz is taken as the lower limit. Because of their large wavelength they have some special properties in relation to audible sound. For example they are only poorly damped by stone walls, buildings etc. It is therefore difficult to shield from this kind of noise. Sometimes, it even happens that in spaces with corresponding resonance frequencies amplification occurs. This happens, for example, in cars, or housing spaces with a typical resonant frequency of 2–8 Hz.

There are natural sources of infrasound, like wind, waterfalls, sea surf etc. Technically, infrasound is produced by vibrations of various engines. There is considerable exposure to infrasound in cars if the airstream induces internal resonances. Sometimes the 100 dB limit is exceeded. In recent years special infrasound loudspeakers have been used in discotheques in order to induce psychosomatic resonance of the participants. Because of the above mentioned problems of shielding they may considerably disturb the neighborhood.

Despite the fact that infrasound in general is a non audible sound, it nevertheless leads to unspecific auditory sensations even at frequencies below 16 Hz, if it has sufficient intensity (see Fig. 4.3). Sometimes, even clinking and overtones can be heard. In general, however, infrasound interacts with air-filled cavities in the human body, like lungs, nasal cavities, frontal sinus, middle ear, gut etc. At intensities above 160 dB dramatic damage of the middle ear is to be expected. Infrasound of intensities between 140–155 dB leads to acute psycho-physical injuries like headache, difficulties in breathing and stress sensations. Below 100 dB no physiological influences could be found, but nevertheless psychosomatic reactions occur which are regarded as a disturbance. These occur near the threshold of perception.

Further reading: Møller 1984

4.3.5
Biophysics of Sonar Systems

It has been known since the experiments of the Italian monk and scientist Lazzaro Spallanzani in 1794 that bats have the ability to orientate in complete darkness and to avoid obstacles. It was demonstrated at that time that a bat could still fly safely when blinded, but was quite helpless when its ears were plugged. Only in 1938 was D. R. Griffin able to show by direct measurements that bats are able to emit ultrasound, and detect the echoes reflected from objects in their surroundings to determine their location. In the following years many investigations have been carried out, and it has been found that not only bats, but also rodents, aquatic mammals, birds, fish and insects use echolocation by high frequency sound or ultrasound.

Because bats have been investigated in this respect more than other species, we will start with this example to explain the biophysical basis of echolocation. Figure 4.12 shows an oscillogram of a typical orientation signal of a bat. It is characterized by the following physical parameters: frequency (v), amplitude, duration (Δt), and of course, the distances between this, and the next signal.

The frequency of the emitted sound depends on the size of the object which is to be localized. An effective echo can be expected only from targets which are equal, or even larger in size than the wavelength of the sound. The relation between wavelength (λ) and the frequency (v) is given by:

$$\lambda = \frac{\mathbf{v}}{v},$$

(4.3.4)

where $\mathbf{v}$ is the velocity of sound waves. In air under atmospheric conditions $\mathbf{v} = 331$ m s^{-1}. A sound of 25 kHz therefore has a wavelength $\gamma = 331/25\,000 = 0.01324$ m, i.e. 1.324 cm.

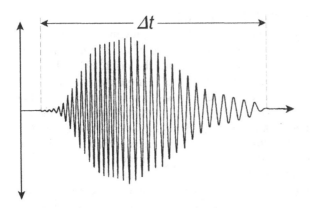

Fig. 4.12. Example of a typical FM modulated signal of a bat

The frequencies emitted by bats differ from species to species. Usually, this amounts to between 25 kHz and 150 kHz. Bats which use their sonar system in order to catch insects must emit signals of higher frequencies than those just orientating in dark caves.

The amplitude of the signal, i.e. its intensity, is determined by the required range of the sonar system. Naturally, the auditory system of the bats must be sufficiently sensitive to receive the corresponding echo. The power of the emitted sound of the big brown bat (*Eptesicus fuscus*), for example, is about 10^{-4} W, whereas as minimum, a power of 10^{-16}–10^{-14} W corresponds to the auditory threshold. At 5–10 cm distance from the head of this bat, the sound pressure amounts to 2–6 Pa. According to Eq. (4.3.3) this means an intensity of 100–110 dB. We learned in Section 4.3.2, Fig. 4.4 that, transformed into audible sound, this would be a considerable noise.

This large difference between the emitted versus the received sound causes problems for the animal when listening during calling. The maximal duration of a signal (Δt), therefore, is more or less limited by the time the sound needs to arrive at the target and come back, i.e. by the distance of the target.

These then are the basic parameters of an echolocation signal. Looking at the sonogram in Fig 4.12 carefully, more details can be detected. The frequency, for example, as well as the amplitude of the signal are time dependent. It is a frequency modulated sound, a so-called *FM signal*. In the case of the little brown bat (*Myotis lucifugus*) the signal starts with a frequency of 100 kHz and goes down by an octave at the end of the 2 ms long signal. In contrast, the horseshoe bats use signals of 50–65 ms with constant frequencies (*CF signals*). Mostly, however, these CF signals have a short frequency modulated final part. The character of the emitted signals depends on the species, on situations, and even on the individual animal.

The frequency of modulation seems to be an important property of the signal, especially for measurement of distances. In some instances it appears that by superimposition of the FM part of the end of the signal with the initial CF part of the returning echo, specific beat frequencies occur which are evaluated by the animal. Furthermore, it has been suggested that a technique known as pulse compression in radar technology, may be utilized by bats. For this, short frequency components of the echo are delayed by a special acoustic filter, a so-called optimal filter which depends on their frequency. The higher frequency parts at the beginning of the signal are delayed to a larger extent than the lower frequency parts at the end of the signal. This process of impulse compression provides an instantaneous cross correlation between the emitted and the received signals. Pulse compression is a way to explain the ability of the animal to measure the distance even in cases when the echo time is less than the duration of the emitted sound. This principle would also minimize disturbances by the Doppler effect. It is unknown which parts of the auditory system could carry out this function or what the molecular basis for this could be.

On the one hand, the Doppler effect disturbs the bat when measuring distances, on the other hand, it is used to measure relative velocities between the animal and its surroundings. For this a Doppler-shift compensation occurs. If a flying bat approaches an object at a certain speed, the echo from that object will be Doppler-shifted upward to a higher frequency than the emitted signal. Subsequently, the bat lowers the frequency of the next emitted CF component by an amount that is nearly equivalent to the Doppler shift in the preceding echo. The amount of the required frequency is used by the bat to calculate its relative velocity.

Many questions on information processing in bat echolocation are still open. These include mechanisms to determine the direction of the object as well as information about the size, shape and material of the target.

Echolocation is also known in some birds. Alexander von Humboldt has already described a South American cave-dwelling bird, the oil-bird or guacharo (*Steatornis caripensis*) which flew around the head of the explorer uttering "screams of fear". Today it is known that these cries serve as signals in an echolocation mechanism. The sound is pitched at frequencies of 6–10 kHz, which means that it can be heard by humans. These birds are herbivores, but they require their sonar system in order to orientate in large dark caves. The frequency employed corresponds to wavelengths between 3–5 cm which means that the sound will be reflected even from small rock projections.

Sonar systems are found not only in airborne species, but also in several aquatic animals. Dolphins, for example, emit short click signals with a duration mostly of only 40–50 ms having a frequency up to 130 kHz with repetition after 20–40 ms. It is still unknown where and how this sound in the animal is produced. Due to the larger velocity of sound in water ($v = 1500$ m s^{-1}), the wavelength of a sound in water is about 4.5 times larger than that of the same frequency in air. This would require a higher frequency for echolocation in water for targets of the same size. Additionally the specific sound distribution in water must be taken into consideration. The transmission loss for a point source of sound in free space in general is:

$$\text{Transmission loss} = 10 \lg \frac{I_0}{I_r} = 10 \lg r^2 + \alpha r, \tag{4.3.5}$$

where I_0 is the intensity at $r = 0$, I_r is the intensity at distance r, and α is the coefficient of absorption.

The first part of the sum describes the simple geometrical distribution of the sound, whereas the second term concerns the absorption of sound in the medium. Especially in sea water, because of its content of magnesium sulfate, the coefficient α is frequency dependent. It amounts 0.002 dB m^{-1} for 20 kHz and rises up to 0.3 dB m^{-1} for 100 kHz. This means that in contrast to airborne animals, the frequency of the emitted signal for sea water animals is additionally a function of the distance.

Further reading: Au 1993; Griffen 1958; Pollak and Casseda 1989; Popper and Fay 1995

4.3.6
The Effects of Ultrasound

Ultrasound consists of longitudinal waves, i.e. periodic alterations of local pressure at a frequency range between 16 kHz and 10^9 Hz = 1 GHz. Sometimes for mechanical waves higher than 1 GHz the term hypersound is used. Technically, ultrasound can be generated by electro-acoustic converters. Because of its short wavelength it can be focused by appropriate reflectors and diffraction lenses. In recent decades ultrasound applications in medical diagnostics and therapy have become increasingly relevant.

The propagation rate of ultrasound in water is 1500 m s^{-1}, in solid bodies about 4000 m s^{-1}, and in water-rich biological tissues it is around 1479 m s^{-1}. Corresponding to Eq. (4.3.4), ultrasound of $v = 10$ MHz in tissues has a wavelength of 0.15 mm. This ensures a satisfactory resolution for sonographic images. It can be increased further by using ultrasound of higher frequencies, however this will decrease their penetration into the body. In fatty tissue, for ultrasound of 0.8 MHz a half-value thickness of penetration of 3.3 cm has been measured; in muscles it amounts to only 2.1 cm. This parameter decreases with increasing frequencies. This is the reason why for prenatal observations and mammographic diagnostics, frequencies of only 1–5 MHz are used. In ophthalmologic diagnosis 10–50 MHz are used because in this case a higher resolution is required. For therapeutic applications, ultrasound of a high power density in the frequency region of 0.2–2 MHz is used.

The image generating techniques of medical diagnostics evaluate differences of the mechanical impedance of various membranes, cells or tissues, leading to reflections, diffractions and refractions of the ultrasound waves. Investigations into blood flow with ultrasound are based additionally on the Doppler shifts of the reflected ultrasound.

The influence of ultrasound on cells and tissues on the one hand is caused by the appearance of local pressures, on the other hand on local accelerations. As a concomitant phenomenon a certain heating occurs. According to the frequency of the ultrasound, alternating positive and negative pressures appear locally, leading to stretch or compression of the material. Homogeneous liquids have a considerable resistance to the disruption effect. In order to pull pure water apart in such a way that two parallel surfaces are formed which are separate from one another, a negative pressure is required of 1500 MPa. Only 150 MPa is necessary to form a spherical cavity. This effect is called *cavitation*. In fact, the local negative pressures which are required for cavitation in real solutions and suspensions are lower. This depends on the concentration of nuclei of cavitation. For short pulses of ultrasound, cavitations in tissue are observed above 10 MPa. This determines the limit of the ultrasound intensity in therapeutic applications.

These caverns are short-lived spaces, filled by the gas which is dissolved in the liquid, and by the vapor of the liquid itself. With increasing pressure, these caverns collapse and vanish completely, or they remain as microscopic crystallization nuclei for the next cavitation. The diameter of these caverns depends on the intensity and frequency of the applied ultrasound. Their mutual distance is also frequency and wavelength dependent. For a frequency of 20 kHz in water, it amounts to about 0.15 mm and becomes lower if the frequency increases. Cavitation can lead to audible noise, finally to "boiling" of the liquid. The sound intensity which is required to produce cavitations increases with the viscosity of the liquid, and decreases with the amount of dissolved gas.

In biological systems this cavitation leads to the destruction of supramolecular structures and cell organelles. Furthermore, various physicochemical processes, for example diffusion of matter, can be influenced. In cases of high intensity, even water radicals can appear, leading to indirect effects similar to that of ionizing radiation.

For the therapeutic application of ultrasound, its property is used to heat the tissue. This effect is called *diathermy*. It depends on the frequency mechanical impedance of the tissue, also being a function of frequency. Caused by diffractions or reflections in the tissue, the thermic effects can be heterogeneous. It is possible that hot spots occur in the tissue, i.e. microscopic locations of increased temperature.

Although ultrasound by definition is non-audible sound, it was found that the human ear may sense tones even up to 40 kHz, although with very low sensibility. Nevertheless, there have been discussions as to whether the weighing scales, discussed in Section 3.4.2 (Fig. 4.5) must be extended up to this frequency.

To prevent risks from ultrasound radiation, the increase of local temperature in tissue must not exceed the limit of 1.5 °C, and local pressures must not exceed 8 MPa. WHO recommend a limit of a mean intensity of 3.10^4 W m^{-2}. The real values applied in diagnostic methods are far below this limit. Therefore one may exclude any risk for health, especially since X-ray methods are being replaced by ultrasound.

Further reading: Nygborg 1985

4.4
Static and Electromagnetic Fields

In this section we will first discuss static fields, i.e. fields with zero frequencies, leading to AC fields of low frequency, and finally to the region of microwaves. Thermic radiation has already been mentioned briefly in Section 4.1, and visible light in context with excitation processes and photosynthesis in Section 2.2. Effects of ionizing radiation will be explained later.

This chapter is based on the previous explanations about cells in static and AC electric fields (Sect. 3.5), but instead of strong electric fields, used for cell manipulations (Sect. 3.5.6) we will concentrate here on fields in the natural and technical environment and on their use in therapeutic treatments.

Figure 4.13 shows the whole spectrum of electromagnetic radiation. Wavelength (λ), frequency (v) and quantum energy (E) are connected together by the usual relations: $c_0 = v \cdot \lambda$ and $E = h \cdot v$ (c_0 – speed of light in vacuum, h – Planck's constant). The energy of ionization of water: $E = 12.56$ eV usually is taken as the beginning of the region of ionizing radiation.

4.4.1
The Static Magnetic Field

A magnetic field with the *magnetic field strength* **H** (in A m^{-1}) causes a *magnetic induction*, or *magnetic flux density B* in a body (in Telsa: T = V s m^{-2}) of:

$$B = \mu\mu_0\mathbf{H}, \tag{4.4.1}$$

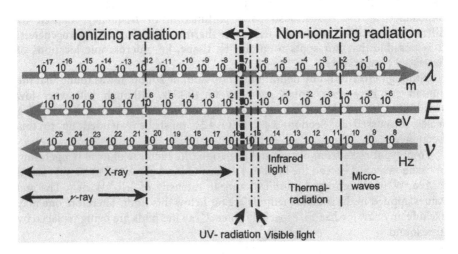

Fig. 4.13. Spectrum of electromagnetic radiation with wavelength (λ), quantum energy (E), and frequency (v)

where $\mu_0 = 1.256 \cdot 10^{-6}$ V s A^{-1} m^{-1} is the *magnetic permeability of vacuum*, and μ is the *magnetic permeability number* – a dimensionless number in relation to vacuum, where $\mu = 1$. The deviations of the magnetic permeability for non-ferromagnetic materials from 1 are very small. Therefore the *magnetic susceptibility* $\chi = \mu - 1$ is introduced.

Magnetic properties of various materials are classified according to their properties as being:

$$\begin{aligned} diamagnetic&: \quad \mu < 1;\ \chi < 0, \\ paramagnetic&: \quad \mu > 1;\ \chi > 0, \\ ferromagnetic&: \quad \mu \gg 1;\ \chi \gg 0. \end{aligned} \qquad (4.4.2)$$

As shown in Table 4.2, cells and tissue in general have diamagnetic properties. An exception are desoxygenized erythrocytes, which are paramagnetic due to the property of the central Fe-atom in the hemoglobin molecule. If it binds oxygen, it changes from a paramagnetic to a diamagnetic state.

Highly organized structures, like the cell membrane, show magnetic anisotropy. Its magnetic susceptibility depends on the direction in which it is measured. This anisotropy, $\Delta\chi$, is defined as follows:

$$\Delta\chi = \chi_\| - \chi_\perp, \qquad (4.4.3)$$

where $\chi_\|$ and $\chi_\perp$ are the corresponding susceptibilities of directions parallel and perpendicular to characteristic directions of the structure. Anisotropic magnetic structures, like for example cell membranes or chromosomes, can be orientated in strong magnetic fields.

In special magnetostatic bacteria, and in low amounts also in animal and human tissue, ferromagnetic particles, so-called *magnetites* have been found.

Table 4.2. The magnetic susceptibility of air, water and various biological materials

	After Maniewski (1991) $*10^{-6}$	After Khenia et al., in: Maret et al. (1986) $*10^{-6}$
Air	+0.34	+0.264
Water	−9.05	−9.04
Arterial blood	−9.1	−9.3
Oxygenized erythrocytes		−9.03
Venous blood	−8.4	−7.8
Desoxygenized erythrocytes		+3.88
Lungs (breathed in)	−3.9	
Lungs (breathed out)	−4.1	
Lungs (15% air content)		−4.2
Muscle	−9.0	−9.03
Liver	−8.8	−8.26
Bone	−10	−10

Magnetites of magnetostatic bacteria can be obtained in sufficient concentrations to be analyzed chemically and by crystallography. These are cuboidal crystals of Fe_3O_4 (a mixture of FeO and Fe_2O_3) with a size of 50–200 nm. The largest of them, at least, are surrounded by a membrane forming *magnetosomes*. The density of these particles is 5.1 g cm^{-3}. Because of their extremely small size, they consist of only one single magnetic domain which means that they consist of a single elementary magnet. These magnetite crystals are synthesized by the cell from iron bound in a chelate complex. From the crystallographic point of view, they are different from artificially produced ferrite particles or from ferrite particles absorbed from a polluted external medium. The magnetosomes of tissue cells are apparently similar to those of bacteria.

In these magnetostatic bacteria magnetic particles are arranged in a 3 μm-long chain which is fixed along the axis of motility of the bacterium. The magnetic momentum of this chain is sufficient to orient the bacterium in the geomagnetic field. Magnetostatic bacteria are usually anaerobic inhabitants of the sediment. Cells with appropriate polarization, and in regions away from the equator, swim along the geomagnetic vectors into the anaerobic depths. This orientated movement is called *magnetotaxis*. Bacteria with wrong magnetic polarizations swim in the wrong direction and die off in oxygen-containing surface water. During the division of the cells by mitosis, the magnetic chains will also be divided into two parts, and the daughter cells get a part of the correctly polarized chain. Adding further magnetosomes after division, to complete the chain, these newly synthesized particles automatically become polarized according to the inherited part of the chain. It is possible to reverse the polarity of these chains in artificial magnetic fields of at least 10 mT.

The magnetic field of the earth is demonstrated in Fig. 4.14. The vector of this field can be subdivided into a vertical component, which has its maximum at the poles, coming up to a field strength of 48–56 A m^{-1}, corresponding to a magnetic flux density of 60–70 μT. The horizontal component has its maximum near the equator, with 24–32 A m^{-1}, i.e. 34–40 μT. The intensity of the geomagnetic field shows various alterations, which are caused chiefly by changes of sun-spot activity. Furthermore, small variations during the day, during the

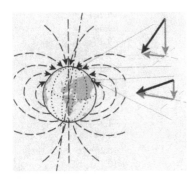

Fig. 4.14. Geomagnetic field lines, and their inclinations at two particular locations

month, and during the year occur. Through history the magnetic pole of the earth has shifted a considerable distance. Additionally, at some locations significant geologically caused inhomogeneities of the geomagnetic field can be measured.

The question as to whether animals can orientate in geomagnetic fields has long been discussed. Strictly speaking, the passive orientation of magnetostatic bacteria cannot be considered as the result of geomagnetic receptors. It seems sufficiently established, however, that various animals really do have the ability of magnetic orientation. In this respect, one must distinguish between several kinds of use of the magnetic sense: compass-orientation (migrant birds, rodents living in the earth), magnetic mapping (bees, pigeons, wales), and probably also magnetic timing. At least in birds the magnetic orientation seems to be connected with optical orientation. Whether human behavior can be correlated with the occurrence of magnetic storms, as is sometimes claimed, is quite questionable.

Although behavioral experiments clearly indicate that many animals are able to orientate with the help of the geomagnetic field, it remains absolutely unclear which mechanisms may be responsible for the reception of these extremely weak magnetic fields. The findings that in some cases light is necessary to allow birds to orientate in the geomagnetic field have led to speculations as to whether transitions of the electron spin could occur during photoactivation processes of rhodopsin in the retina. A further hypothesis claims that this reception is somehow connected with magnetosomes in the tissue. Long magnetite chains which could overcome thermic noise, however, were found only in bacteria. In bees large magnetite crystals were found near nerve endings of particular abdominal hairs. It has been speculated that this ferromagnetic material could focus the magnetic field and amplify it in this way locally. In the case of marine animals, geomagnetic orientation could be realized using the effect of electrical induction. This requires a relative high speed of the animals and a considerable size. In this case a conducting loop occurs between the body of the animal and the surrounding seawater. Moving in the geomagnetic field, a current could be induced in this loop, and electric potential differences could be measured by electroreceptors (see Sect. 4.4.2). It does not seem so evident that similar mechanisms would be possible in terrestrial animals. Sometimes, the vestibular system in mammals and even humans has been discussed in this respect. This, of course, remains quite speculative because of the small size of the induction loop.

Recently, detection of biogenic magnetic fields on the one hand, and applications of strong magnetic fields on the other have become more and more important for medical diagnoses. Using sensitive measuring instruments with superconducting materials (SQUIDs, superconductive quantum interference devices) it is possible to measure magnetic fields which are induced by extracellular electric currents around excited tissue, as well as minimal alterations in the magnetic susceptibility and magnetic anisotropy in the body.

To record magneto-cardiograms, -encephalograms, and -myograms, magnetic fields weaker than 1 pT must be measured. This is seven orders of magnitude lower than the geomagnetic field! These methods are more costly than the corresponding records of ECG or EEG, but they are much more instructive. As we already pointed out, the magnetic permeability (μ) of various materials does not deviate very much from $\mu = 1$. This means that the magnetic field in the body will not become disordered significantly, in contrast to the electric field. Therefore it is much easier to localize the oscillating electric dipole in nerves and muscles, inducing the magnetic field (see also the explanation of the ECG in Sect. 3.5.2, Fig. 3.33).

Furthermore, the application of highly sensitive measuring instruments for magnetic fields can be used in occupational medicine to analyze accumulation of iron dust in the lungs. The measurement of the magnetic susceptibility of the lungs can also be used in lung diagnostics. After inhalation of a small amount of a ferromagnetic label, the rate of magnetic reorganization of these particles can be recorded, which allows conclusions concerning the activity of the ciliated epithelium of the lung.

The construction of superconducting magnets with bores large enough to accommodate human bodies have made it possible to apply *nuclear magnetic resonance* (NMR) in medical diagnostics. This enables one to measure ionic composition and metabolic activity of various tissues and organs. The method of whole-body tomography, or nuclear magnetic imaging (NMI) makes possible detailed pictures with great diagnostic value, for example in cancer diagnostics. During these measurements, the patient is placed in a magnetic field with a flux density of 0.1-3 T, and field gradients in the order of 0.01 T m^{-1}. Signals from nuclei, such as protons differing in their relaxation times are used to create these images.

Furthermore, during the last decade the method of *transcranial magnetic stimulation* has become available. In this case, one activates the primary motor cortex and the motor roots in conscious man by magnetic millisecond pulses of an intensity of 1.5-2 T. This technique has proved sensitive enough to illustrate early abnormalities of central motor conduction in various neurological diseases such as multiple sclerosis, amyotrophic lateral sclerosis, cervical spondylotic myelopathy, degenerative ataxias or hereditary spastic paraplegias.

These kinds of enforced application of strong magnetic fields in medical diagnostics and therapy, as well as increased occupational exposure to magnetic fields raises the question of mechanisms of interaction between these fields and the biological system. In general, the following effects are possible:

- Orientation of particles in static magnetic fields,
- Translation of particles in inhomogeneous static magnetic fields,
- Action of Lorentz forces on moving charged particles in magnetic fields (e.g. ions, molecules, cells in moving blood),
- Induction of electric currents in the electrolyte milieu of the body in case of movements in a magnetic field or in gradients of a magnetic field.

In inhomogeneous magnetic fields, diamagnetic bodies move in the direction of decreasing field strength, whereas para-, and especially ferromagnetic particles are driven in the opposite direction. In context with the fixation of safety standards, these mechanisms are important only in relation to ferromagnetic implants. The orientation of molecules against the Brownian movement requires extremely high magnetic fields. So, for example, the orientation force which is induced by a magnetic field of 13 T on a magnetically anisotropic DNA molecule at room temperature amounts only to about 1% of the energy of thermic noise. Therefore it cannot be expected that field strengths as used in medical diagnostics and therapy would lead to translations of cells based on differences of their magnetic susceptibility (see Table 4.2, page 257).

On the other hand, it is possible to induce cellular movements in special arrangements. The passive shift of cells in strong inhomogeneous magnetic fields is called *magnetophoresis*. This, for example, can be applied to separate paramagnetic desoxygenated erythrocytes from diamagnetic lymphocytes (Table 4.2). For this, a net made of thin iron wire is used, which is placed in a polyethylene tube. Applying an external magnetic field of about 2 T, strong inhomogeneities are induced near the wires, attracting the erythrocytes. Lymphocytes, in contrast to this, pass the net.

The method of *immunomagnetic cell separation* is much more effective. For this, latex particles are used, containing supraparamagnetic iron which indicates ferromagnetic properties only during the presence of an external static magnetic field. These particles are loaded with antibodies against cells which are needed to be separated. After the attachment of these particles to the cells, they can be separated easily even at moderate magnetic gradients.

Charges moving in a magnetic field are mechanically influenced by a so-called Lorentz force. This force also acts on ions and other charged particles under the influence of strong magnetic fields in the flowing blood. In fact, this influence is slight at magnetic field strengths as applied in medicine. The inverse of this effect is the induction of an electric current if a conductor crosses lines of a magnetic field. This can occur in limbs moving in the magnetic field of an NMI engine. The result is painful nerve excitations.

Magnetic fields of 10 mT, oscillating with a frequency of 10 Hz, can lead to excitations of the optical system. As a result, so-called *magnetophosphenes* appear, i.e. the production of visual sensations. The mechanism is the same as in the above-mentioned method of magnetostimulation.

Magnetic fields with flux densities much higher than one Tesla can influence biochemical reactions in isolated enzyme systems and consequently various cellular processes, like proliferation or differentiation. Cellular reactions, which are connected with phase transitions of membrane lipids are apparently influenced most. This is caused by the above-mentioned anisotropy of lipids.

Based on established results on the influence of static magnetic fields, safety standards have been recommended. According to this, the average magnetic flux density of 200 mT must not be exceeded by research personnel or others working in high magnetic fields during an 8-h working day. The limit of possible permanent exposure of the population is 40 mT. A short-lasting exposure of humans without ferromagnetic implants and without cardiac pacemakers is possible in a field of maximum 2 T. Exposure of extremities is allowed up to 5 T. This is the limit for NMI equipment. Subjects with implants or cardiac pacemakers, therefore, are excluded from NMI diagnosis or other magnetic treatments.

Further reading: Influence of magnetic fields in general: Polk and Postow 1996; Maret et al. 1986; magnetic orientation of animals: Able 1994; Wiltschko and Wiltschko 1995; magnetosomes: Dunin-Borkowski et al. 1998; Kirschvink et al. 1985; Vainshtein et al. 1998; magnetophoreses: Higashi et al. 1993; immuno-magnetic separation: Moore et al. 1998; Olsvik et al. 1994; magnetic stimulation: George et al. 1998; deNoordhou 1998; Wassermann 1998

4.4.2
The Electrostatic Field

Aspects of bioelectricity have already been discussed in context with various other circumstances in previous sections of this book: After some basic parameters and laws of electrostatics (Sect. 2.4.1), principles of membrane electrostatics (Sect. 2.5.5) have been explained. Subsequently, we came to transmembrane potentials (Sects. 3.4.2, 3.4.3) and to discussions of electric fields in cells and organisms (Sects. 3.5.1, 3.5.2). Finally, passive electric properties of cells and tissues have been considered (Sects. 3.5.3, 3.5.4), as well as various techniques of cell manipulation using electric fields (Sect. 3.5.5). Now the question arises: how do electric fields in the environment influence cells and organisms? In fact, this can be considered as an interference in the sophisticated electrical structure of the organism, as discussed in Section 3.5.1.

In this respect it must be pointed out that a strong division between electrostatic fields and oscillating AC fields is artificial. In fact, there is a continuous transition between static fields changing for example in a circadian rhythm or in periods of hours or minutes, and very low frequency AC fields. The same holds for static fields switched on and off after some time, and pulsed electric fields (Fig. 4.20). Using Fourier analysis, every kind of pulse can be considered as a sum of sine waves of various frequencies. Many aspects discussed in this chapter, therefore, are important also for low frequency AC fields.

Terrestrial animals live in the electrostatic field of the earth. Its field strength near the earth's surface at sea-level amounts to $100-200$ V m^{-1}. Below a thundercloud it can rise up to $20\,000$ V m^{-1}. The electric field of the earth changes during the day and night and also through the year.

DC fields in the vicinity of technical devices are not very important from the point of view of health protection. They can, however, lead to considerable charging of objects in their vicinities. When touched, such electrically charged surfaces can lead to painful discharges.

The specific conductivity of dry air near the earth's surface about $2.5 \cdot 10^{-14}$ S m^{-1}. Therefore, at a field strength of 200 V m^{-1} a current density of $5 \cdot 10^{-12}$ A m^{-2} results. The conductivity of air depends on the presence of charged particles. These particles have the misleading name *air ions*. In fact, these are particles of quite heterogeneous nature and size, beginning with true ions of nitrogen or sulfate compounds, up to various microscopic dust particles containing surface charges. The content of such "ions" in air in high altitudes amounts to 10^9 ions per m^3. It varies across a broad range depending on many conditions. In cities it may increase up to 10^{11} ions per m^3. There are many speculations about a possible biological role of these charges, but in fact no correlation with any kind of disorder has been established yet.

The biological influence of environment fields depends on their transfer into the body. The electrical conductivity of the human body is about 10^{14} times larger than that of the air. The dielectric constant exceeds that of the air by a factor of 50. This leads to considerable distortion of the field lines around the body (Fig. 4.15). The electric field strength near the surface of the body can significantly increase, especially at the head or at the end of a raised hand.

The external electric field leads to a displacement of internal charges in the body. An external electric field E_e in air ($\varepsilon_e = 1$) induces in a spherical, non conducting body ($g_i = 0$) the following internal field E_i':

$$E_i' = \frac{3E_e}{2 + \varepsilon_i} \quad \text{for} \quad g_i = 0. \tag{4.4.4}$$

If the body is conducting ($g_i \neq 0$), this internal field will immediately be neutralized by an internal current. Taking this process into account, one gets the real field strength E_i as a function of time (t): 0

$$E_i = \frac{3E_e}{2 + \varepsilon_i} e^{-\frac{g_i t}{\varepsilon_i \varepsilon_0}} = \frac{3E_e}{2 + \varepsilon_i} e^{-kt}, \quad \text{where} \quad k \equiv \frac{g_i}{\varepsilon_i \varepsilon_0}. \tag{4.4.5}$$

Using appropriate parameters ($g_i = 0.6 \text{ S m}^{-1}$, $\varepsilon_i = 50$, $\varepsilon_0 = 8.84 \cdot 10^{-12}$ $\text{C V}^{-1} \text{ m}^{-1}$), a corresponding rate constant $k = 1.35 \cdot 10^9 \text{ s}^{-1}$ can be obtained. This means that an internal electric field which is induced by a single rectangular pulse will vanish by a half life time of $\ln 2/1.35 \cdot 10^9 = 5.13 \cdot 10^{-10}$ s. A significant field in the body can therefore be induced only in the case of higher frequency AC, or pulsed fields. We will come back to this point in Section 4.4.3.

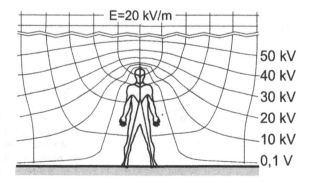

Fig. 4.15. Distortion of an electrostatic field near a standing man. (After Leitgeb 1990, redrawn)

This consideration indicates that an electrostatic field in the air environment may influence only the surface of the body. This chiefly concerns electrostatic erection of hairs. In the case of low frequency AC fields, a vibration of hairs can be sensed. The threshold where humans sense electrostatic fields in the environment is near 10 kV m^{-1}.

In contrast to the conditions in air, electric fields in water are transferred to biological organism to a much higher degree. Vice versa, electric fields of the body are transferred directly into the aquatic environment. This is the basis of the effectiveness of electrical organs of fishes in defense and orientation (see Sect. 3.5.2). Water-living animals have developed electroreceptors which apparently make it possible to localize living prey, or are used for mutual communication and for electro-orientation, or finally, as explained in Section 4.4.1 in some cases they could also be responsible for orientation in the geomagnetic field.

The electroreceptors of some aquatic animals have enormous sensibility. Behavioral evaluation of the lowest field perceived by rays was found to be 0.1–0.2 mV m^{-1}. In most other fishes, however, it is higher by several orders of magnitude. In the same way as for electrical organs, and in the case of electroreceptors the different conductivity of sea water and fresh water must be taken into account. In the case of fresh water fishes (Fig. 4.16 above) because of the low external conductivity, the field strength in the environment can be measured directly as a potential drop across the skin. Marine fishes use so-called *ampullae of Lorenzini* as amplifier systems (Fig. 4.16 below). These are long, well-insulated canals in the body, filled with a highly conductive jelly. At one end these ampullae are in electrical contact with sea water. In this way they transduce the electrical potential from particular points of the surface of the fish almost without loss to the point of their mutual contact. The external field E_e, therefore, will be amplified to an internal field E_i at the distance between the endings of the ampullae, where the electroreceptors are localized.

The molecular mechanism of them is not clear. It seems that there are similarities to thermoreceptors.

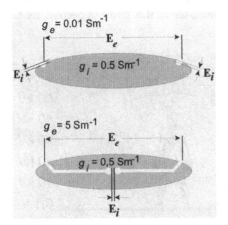

Fig. 4.16. Electroreceptors in freshwater fish (*above*), and seawater fish (*below*). The canals in the lower picture indicate the ampullae of Lorenzini (*g*-specific conductivity of water and tissue). The external electric field E_e becomes amplified to the inner electric field E_i near the receptors

Various calculations indicate that extremely high sensibility of the electro-receptors is possible, even considering thermic and electrical noise. As mentioned in Section 4.4.1 these receptors are probably also used to measure fields which are induced in the electrical loop between fish and environment, if the geomagnetic field lines are crossed.

We already pointed out that there are various biogenic electrical fields in the human body which are probably not simple passive, concomitant phenomena of some electrophysiological processes, but which may have functional relevance. In this context in Section 3.5.2 we mentioned wound potentials, which can create local electric fields of hundreds of V m^{-1}, or electric fields in loaded bones. This is the basis for various attempts of electrotherapy to control biological processes like cell motion, cell differentiation or cell growth etc. by applied electric fields.

One of these cellular properties is *galvanotaxis*. This means an orientation of the motility of living cells by static electric fields with field strengths of about 100–500 V m^{-1}. These kinds of orientation have been observed not only in various protozoans, sperms, and zoospores, but also in many kinds of cells moving slowly on the substratum, like granulocytes, fibroblasts, osteoblasts, nerve cells etc. This property must not be confused with electrophoresis, which is a passive movement of charged particles in an electric field (see Sect. 2.5.4). In contrast to electrophoresis, galvanotaxis predicts an active movement of the cells which usually will not be accelerated by the applied field, but simply orientated. In contrast to electrophoresis, galvanotaxis, being an active biological process, needs a much longer lag phase. Cells may move by galvanotaxis for many seconds or even minutes in one direction even if the polarity of the applied field has already been changed. Most cells, even when charged negatively, move by galvanotaxis toward the cathode.

The mechanism leading to galvanotaxis of tissue cells is rather unclear. Whereas in ciliata the orientation of their motion is apparently controlled by the transmembrane potential, for other cells other mechanisms are proposed. Using fluorescent labels it is possible to indicate that membrane-bound receptors shift under the influence of an external field laterally in the cell membrane, and concentrate on one side of the cell. This does not seem to be the result of individual electrophoresis of these molecules but rather a reaction to local electro-osmotic flow, induced by the interaction of the external field with all charges of the membrane surface.

A further property of cells in electrostatic fields is *galvanotropism*. This means the influence of an external DC field on the direction of cell growth. So, for example, nerve cells in electric fields of 0.1–1 kV m^{-1} form dendrites preferentially at the side of the cell which is directed toward the cathode. In this case, local modifications of the membrane potential are probably responsible for these reactions, but the above-mentioned lateral translation of membrane proteins can also be responsible for this. This property has been applied in therapy to try to stimulate nerve regeneration.

Further reading: Electric fields in the environment: Reiter 1992; Polk and Postow 1996; air ions: Chary and Kavet 1987; electroreception: Kalmijn 1997;

Petracchi and Cercignani 1998; Atema et al. 1987; galvanotaxis: Gruler 1995; Machemer-Röhnisch et al. 1996; Zhao et al. 1999; galvanotropism: Rajnicek et al. 1988

4.4.3
Electromagnetic Fields in the Human Environment

As the result of technical and industrial development, an increasing number of man-made electromagnetic fields have appeared in the human environment. The classification for the frequency ranges, as listed in Fig. 4.17 (ELF, ULF, VLF, LF, MF, HF, VHF, UHF, SHF, EHF) corresponds to the terminology of electronic engineers. Sometimes fields in the GHz range are summarized as "microwaves" (MW), or the generalized term "radiofrequency" (RF) is used for a broad frequency range. Some authors call frequencies below 3 Hz "ultra low frequencies".

According to Maxwell's equations, electromagnetic fields have a magnetic and an electric vector. With increasing frequencies, it is more and more difficult to separate experimentally the magnetic from the electric components of the fields. At ELF and ULF fields, it depends on the techniques of field application, whether mostly an electric, or preferentially a magnetic interaction will occur. To illustrate this, three methods are shown in Fig. 4.18 to apply low frequency fields as examples. If, for example, the object is positioned between the poles of an iron core, it is influenced mainly by the magnetic component of the field, induced in the iron by the current in the coil (Fig. 4.18A). If the object is between two plates of a capacitor in air (Fig. 4.18c), not being in touch with

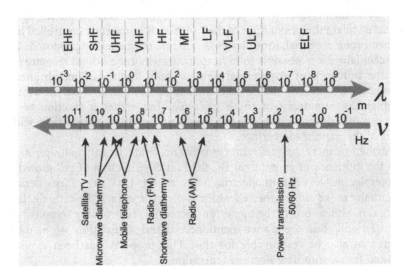

Fig. 4.17. The spectrum of technically used frequencies (HF = *high frequency*, LF = *low frequency*, E = *extremely*, S = *super*, U = *ultra*, V = *very*, M = mean)

Fig. 4.18. Examples of three techniques of experimental application of electromagnetic fields (explanation in text)

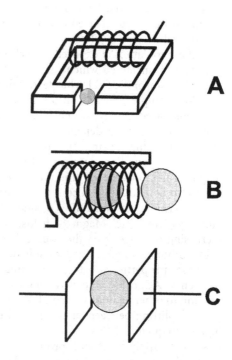

them, almost only electric influences occur. In the same way, AC fields can also be applied, using electrodes connected via physiological solutions with biological objects, or which are directly inserted in them. Using coils (Fig. 4.18B), the magnetic, as well as the electric component of the field interacts with the object. There is no difference whether the object is positioned inside the coil or in its vicinity. This situation resembles the general kind of exposure of humans near technical devices.

These techniques of field application are used in research as well as in therapy. In the case of short wave diathermy (Table 4.3), the VLF fields are applied by electrodes with direct electric contact with the tissue. In contrast, microwave diathermy in the UHF and VHF range uses coils to transmit the

Table 4.3. Frequencies of electromagnetic fields used in medical treatment and their penetration (δ) in biological tissue with specific conductivity $g = 0.6$ S m^{-1}

	Frequency v (MHz)	Wave length λ (m)	Skin depth δ (m)
Shortwave diathermy	13.56	22.05	0.1765
	27.12	11.03	0.1248
	40.68	7.35	0.1019
Microwave diathermy	433.92	0.69	0.0312
	915	0.33	0.0215
	2450	0.12	0.0131

energy into the body. Sometimes additional iron cores are applied to increase the magnetic output. The eddy currents induced in this way in the body lead to the desired diathermy effects.

If the population is exposed to technical facilities, additional circumstances have to be considered which may lead to a dominance of the magnetic or the electric vector of the field. The electric field of a particular position near a power line, for example, depends on the construction of the line, i.e. on the voltage of the transported energy, which is constant. The magnetic field emitted by this line, however, varies depending on the current flowing through it, which is always changing during the day. Furthermore, usually several wires are combined in one line, carrying different currents with phase differences. The magnetic field of a power line therefore can vary considerably. Similar aspects must also be taken into account in other electric devices.

Furthermore, it must be considered that electric fields are shielded by buildings, trees etc. Magnetic fields, in contrast to this diminish only with increasing distance from the source. This leads to the dominance of magnetic fields in buildings and at places shielded in other ways.

These properties of the electromagnetic fields are important for low frequency AC fields, the wavelengths of which (see Fig. 4.17) are large in relation to other geometrical conditions such as size of the objects, or distances. Such a splitting between magnetic and electric field regions is not possible in higher frequency fields. Although up to UHF frequencies it is technically possible to apply fields with pronounced electrical or magnetic components, in practical use of technical devices, both components are connected together.

At higher frequencies, the so-called *skin effect* becomes important. This means that the current density in a conductor induced by a high frequency electromagnetic field falls off rapidly below the surface. This effect determines the depth of penetration of high frequency electromagnetic fields in the body. This is characterized by the *skin depth* (δ) which can be calculated for practical use by:

$$\delta = \sqrt{\frac{1}{\pi \mu \mu_0 g \nu}}. \qquad (4.4.6)$$

This skin depth (δ) is the distance over which the field decreases to $1/e = 0.368$. Inserting into Eq. (4.4.6) the magnetic permeability of vacuum: $\mu_0 = 1.256 \cdot 10^{-6}$ V s A^{-1} m^{-1}, the magnetic permeability number $\mu = 1$, and a mean conductivity of the tissue: $g = 0.6$ S m^{-1} (see Fig. 3.38), one understands that this effect will become important only at high frequencies (ν). As listed in Table 4.3, the skin depth plays an important role in therapeutic use of electromagnetic fields. The values shown in this table should only be taken as examples because of the dielectric heterogeneity of the human body.

These differences in the physical properties of low versus high frequency fields are also reflected in the use of measuring units for dosimetry. In the case of low frequency fields the magnetic component is expressed in A m^{-1}, or mostly as magnetic flux density in T, and the electric field strength is measured in V m^{-1}. Multiplying the electric field strength (E) by the specific conductivity (g) of the medium, one gets the current density (j) in A m^{-2}. This parameter is important for evaluating possible excitation phenomena in the body.

At higher frequencies the parameters: *plane power density* (W m^{-2}), and the *specific absorption rate* (SAR) in W kg^{-1}, or W m^{-3} are used. In contrast to the plane power density, which just depends on technical parameters of the field-emitting device, the SAR is the energy which is absorbed in a unit of mass or volume of the body per unit of time. For low frequency fields, using Ohm's law, this means:

$$P = jE = \frac{j^2}{g} = E^2 g. \tag{4.4.7}$$

Knowing the SAR and the heat capacity of the tissue, thermic effects of absorbed electromagnetic fields can be calculated. In fact, the calculation of SAR distribution in a heterogeneous dielectric body, like for example the human head, is very complicated.

As shown in Fig. 4.17 for radiation protection, two frequency regions are of particular interest: the low frequency electromagnetic fields of power transmission and the high frequency fields of various kinds of electronic communication. The worldwide discussions about possible hazards of these fields have led to various technical modifications to minimize exposure. In the low frequency range, one can show that at common distances from electrical devices, a whole body exposure of 30 µT will never be exceeded for the general public. In most cases exposure is much lower. Directly beneath a power line, the magnetic flux densities even in worst cases usually do not exceed the 10 µT limit. At a distance of 50 m it is already lower than 0.3 µT. In some industrial appliances, as for example inductive heaters with very high current density, magnetic flux densities of 10 mT may occur.

From the physical point of view, low frequency fields can interact in different ways with the human body. As already pointed out, the magnetic component of a low frequency field fully penetrates the body. In general, it is possible that the magnetic field can interact with the biological system directly. On the other hand, it can act via the induced eddy currents. The current density produced in this way depends on the frequency of the field, on its intensity, and on the geometrical parameters of the induction loop. This is determined by the size or diameter of the body in a projection plane perpendicular to the magnetic field vector, but also by anatomy and the impedance of the corresponding tissue. Differences in the conductivity of the organs in the body can lead to preferential conductivity loops. Such loops may occur via special arrangements of blood vessels or even between electrically communicating cells. The cell itself as a space of high conductivity, surrounded by an isolating membrane, can act as a conducting loop, whose diameter is, however, very small. This in fact complicates the calculation of the real field conditions in the body even if the environmental field parameters are known.

In Section 4.4.2 we considered the possibility of an external electric field interacting directly with a body, inducing charge displacements [Eqs. (4.4.4) and (4.4.5)]. The field strength induced in the body in this way depends on the frequency of the AC field, the size of the body, its conductivity and its dielectric constant.

In the practice of environmental and occupational protection of humans against electromagnetic fields, various physical approaches are used to calculate the resulting current densities in the human body. The difficulties of these calculations and their uncertainties are easy to understand, considering the discussion above. For this reason, two types of parameters are used in safety standards: *basic restrictions*, these are current density in the body, or SAR at high frequency fields and *reference levels* describing the corresponding parameters (electric or magnetic field strength or the plane power density introduced above) in the environment. If a measured value in the human environment exceeds the corresponding reference level, it does not necessarily follow that the basic restriction will also be exceeded. However, an investigation should be carried out whenever a reference level is exceeded.

Besides technically induced fields, humans are also exposed to natural electromagnetic fields in the environment. These fields have wide-spread frequencies, starting with variations of the electrostatic field of the earth, as discussed in Section 4.4.2, in a region far below 1 Hz. Pulses induced by lightning are above 10 Hz. These worldwide discharges induce permanent low frequency electromagnetic oscillations in the atmosphere which can be measured even in clear weather conditions. In this case the so-called *Schumann resonance* is important, a global electrical resonance frequency of 7.8 Hz with corresponding harmonics at 13.8, 19.7, 25.7 and 31.7 Hz. At higher frequencies, so-called *atmospherics* (or *spherics*) can be measured. These are pulses induced by various discharges in the atmosphere, containing electromagnetic frequencies up to tenths of kHz. Their magnetic flux density does not exceed 20 nT. The intensity of these atmospherics depends on various atmospheric conditions. It is speculated that the weather sensitivity of humans could be based on this influence. Recently, experiments have been carried out to test volunteers with artificial spherics simulation systems.

Further reading: Dosimetry and safety standards: Polk and Postow 1996; therapeutic use of fields: Low and Reed 1992; natural fields and spherics: König et al. 1981; Reiter 1992; Schienle et al. 1995

4.4.4
Biological Effects of Electromagnetic Fields

The question as to how electromagnetic fields may affect biological systems is of particular interest for problems of health protection and at the same time it opens new possibilities of therapeutic use. Health protection studies are focused mainly on two frequency regions with quite different kinds of biological interaction: the low frequency range of the power frequency (50 or 60 Hz) and high frequency ranges which are used for devices of radiocommunication and diathermic heating (Fig. 4.17).

Applications of electromagnetic fields for medical purposes already have a long tradition. From a scientific point of view, however, it must be considered as an area still in development. Various excitation methods are now scientifically established which work with low frequency pulses, for example transcutaneous electric nerve stimulation (TENS), as well as high frequency field applications for diathermy, as listed in Table 4.3 (page 267). Many other attempts have been made to use electromagnetic fields in bone and wound healing, nerve regeneration etc., but up to now mostly without sufficient scientific understanding.

As explained already in Section 3.5.4, low frequency AC fields and field pulses modify the transmembrane potential proportional to the external field strength and the length of the cell in the direction of the field vectors [Eq. (3.5.19)]. In the case of muscle or nerve cells this can lead to excitation processes. The excitation of long nerve or muscle cells by a given external electrical field strength is easier, if the effective length of the cell in the direction of the electric field vector is larger. The word "effective" means that it is not simply the length of the cell which is responsible for this potential alteration, but rather the length of its conductive area surrounded by the isolating membrane. If two cells, for example, are connected by gap junctions with high electric conductivity, then the "effective" length means the extension of this whole system. Because of the time constant of the excitation process, i.e. the chronaxy, the threshold of cells is lower for low frequency, rather than for high frequency fields (Fig. 4.19).

Using the parameters of Fig. 4.19 and inserting them in Eq. (4.4.7) one can easily see that for low frequency fields the SAR, even at the highest threshold in this figure, namely absolute danger to life, amounts only to 0.001 W kg^{-1}. This value is very small in relation to the basic metabolic rate, i.e. the heat production of the resting human body of 1 W kg^{-1} which increases during active movement by a factor of 10. It can be related directly to the dissipation function (Φ) which

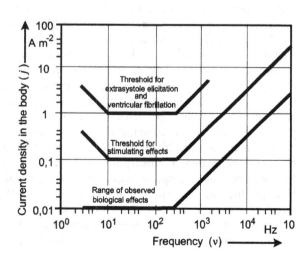

Fig. 4.19. Threshold values of the electric current density in the human body in the low frequency region. (After Bernhardt 1988 simplified)

we discussed in Section 3.1.4 [Eq. (3.1.64)] which for small mammals is even larger (see Fig. 3.7).

This means that in the frequency range up to 100 kHz excitations of nerves and muscles are the dominating mechanisms of interaction, but not thermic effects. The excitation properties shown in Fig. 4.19 lead to recommendations that in the frequency range between 4 Hz and 1 kHz occupational exposure should be limited to fields that induce current densities of less than 10 mA m^{-2} in the body. For the general public an additional factor of 1/5 is applied.

At higher frequencies this situation reverses. As seen in Fig. 4.19 the threshold of nerve excitation increases, corresponding with the decrease in the induced membrane potential ($\Delta\psi_{ind}$) with increasing frequencies, as depicted in Fig. 3.45. In the frequency region $\nu > 10^5$ Hz, diathermic heating becomes dominant. The basic restrictions are therefore determined by the SAR (for definition see Sect. 4.4.3). Established biological and health effects could occur at a rise in body temperature of more than 1 °C. This temperature increase results from exposure of individuals under moderate environmental conditions to a whole body SAR of about 4 W kg^{-1}. This means that we now reach the magnitude of the basic metabolic rate. Of course, it is not the mean metabolic rate of the whole body which must be considered when evaluating health risks, but rather that of the most active organs like muscles or brain which may be higher by one order of magnitude. Using an additional safety factor for whole body irradiation with frequencies between 100 kHz and 10 GHz an average whole body SAR limit of 0.08 W kg^{-1} is recommended.

These safety factors are important chiefly for two reasons: firstly, the questions arise: are there low frequency effects below the level of nerve and muscle excitation? And: are their high frequency reactions below the level of diathermic heating? The second reason concerns the problems of microdosimetry.

Let us start with this second question. In the previous section we discussed the problems of local conducting loops in the body in case of eddy currents induced by low frequency fields. But microdosimetry is also complicated for high frequency fields. As we learned in Section 3.5.3, the biological system must be considered as a heterogeneous dielectric. This means that even for a constant irradiation with high frequency fields, because of the skin effect, and differences in impedance of various cells and various tissues, and partly because of reflections of the waves by bone surfaces etc. a heterogeneous spatial energy absorption occurs. So-called "hot spots" appear, i.e. small or even microscopic regions of heating. This means not only local temperature differences, but also permanent temperature gradients occur. Furthermore, one must consider that at points of increased physiological heat dissipation, for example in muscles or in the brain, an enforced blood circulation occurs to control the temperature. The hot spots generated by absorption of high frequency electromagnetic fields, however, do not correspond necessarily to the locations of high physiological heat production. Therefore it might be possible that in organs with lower circulation a particular SAR may lead to a stronger temperature increase than in other parts of the body with stronger blood flow. A typical example for this is the lens of the eye, a small part without any blood supply.

Biological effects of low frequency electromagnetic fields below the threshold of excitation, as well as high frequency fields below thermic alterations has been the subject of many investigations. Nevertheless, knowledge in this field is quite incomplete. Most investigations have been undertaken in a pragmatic way, limited to the particular frequencies of power supplies, or other technical situations. Because of the political relevance of safety standards on the one hand, and of commercial aspects of therapeutic applications on the other, many investigations unfortunately lack scientific seriousness. Many results of this kind of research indicating some field effects on protein expression, on cellular metabolism, on neuronal reactions, on proliferation and differentiation, up to processes of carcinogenesis must be reproduced and verified. Epidemiological investigations do not show any significant effect and mostly fail because of the limited number of people exposed in a definite way. In fact it seems clear that this kind of non-ionizing electromagnetic radiation, in contrast to X-rays or other types of ionizing radiation, does not induce any kind of mutation. So far, there has been no serious reason to decrease the safety level as discussed above.

As already mentioned, further research is necessary not only for health protection, but also in the direction of possible medical applications of electromagnetic fields. For use in therapy much stronger fields can be applied than limited by the guidelines for radiation protection. Apart from the already established methods of electric and magnetic nerve and muscle stimulation, as well as methods of shortwave and microwave diathermy (see Table 4.3, page 267), a large number of more or less scientifically based methods are used to promote bone healing, nerve regeneration or other therapies. Research in this region is quite divergent, using various techniques of field applications (Fig. 4.18) and various kinds of pulses and sine fields (Fig. 4.20). These range from application of continuous sine fields (Fig. 4.20A), pulsed sine fields (PEMF; Fig. 4.20B), modulated sine fields up to various types of unipolar and bipolar electric pulses (Fig. 4.20C, D). Sometimes coils are used (Fig. 4.18B) in order to induce eddy currents in the tissue. The analysis of the real electrical conditions near the cells is mostly unknown, especially if rectangular magnetic pulses are used.

Mean theoretical approaches of biophysical mechanisms of biological field effects can be summarized briefly as follows:

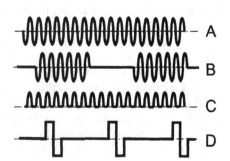

Fig. 4.20. Some examples of AC-fields used in research and therapy

- It is unlikely that one single primary reaction of electromagnetic fields will be found in analogy to the ionization process of X-rays. Probably non-ionizing electromagnetic fields interact with biological systems in many different ways depending on their intensity, frequency and modulation, as well as on physiological properties of the biological system.
- The basic question, especially in the low frequency region, is whether the magnetic, or the electric component of the field is responsible for the primary reaction. There are experiments supporting both kinds of interactions. Clarification of this question is of fundamental importance for radiation protection and dosimetry, as explained in Section 4.4.3.
- Possible targets of magnetic interactions are radicals occurring during various biochemical processes as intermediate products. The recombination of these radicals will generally take place only if the valence electrons in the two radicals are in a singlet state. The effect of the magnetic field consists in the influence of the spin precessions that affect these processes of recombination. The question is whether weak magnetic fields are able to produce this reaction with sufficient probability. Another kind of magnetic hypothesis to explain possible effects is connected with the occurrence of magnetites in the tissue (see Sect. 4.4.1).
- There are various hypotheses of possible electrical interaction mechanisms. In low frequency regions oscillations of the membrane potential occur which could come into resonance with transport processes or oscillating binding reactions of ions with receptors or transport proteins. In the same way, an influence of the electrical field on electrochemical processes at the membrane surface is possible.
- High frequency fields can interact with proteins by a kind of quantum-mechanical resonance. This furthermore requires a coherent behavior of proteins, organized in the strong electrostatic field of the membrane. The theory of this mechanism predicts sharp frequency windows in the field effects. Recently the role of bound water near macromolecular surfaces has been discussed. In contrast to free water, the dipoles of which are oscillating at the frequency of 18.7 GHz (Sect. 3.5.3), the resonance frequency of bound water can be shifted down to frequencies in the order of 10^7 Hz. Alteration of the membrane structure can affect various macromolecular structures and reactions (Sect. 2.4.2).

Research in this field is in fact interdisciplinary to an extremely high degree: it starts with the techniques of generating a well-defined electromagnetic field in the biological system. Medical physics must then help to calculate the field distribution in the body. Biophysics is responsible for explanation of primary effects on molecules and membranes. Subsequently, molecular and cell biology should explain possible reactions of the cells to these effects, and finally, medicine can investigate whether these effects could lead to diseases or at least health risks, or whether they could be used in therapy.

These effects obviously require a systemic rather than a molecular approach. One the one hand, field distribution can be calculated knowing only the dielectric properties of anatomical as well as of cellular organization. On the other hand, the effects of weak fields, in particular, can be understood only by taking into account cooperative processes and mechanisms of biological amplification.

Further reading: General aspects of field effects: Bernhardt et al. 1997; Matthes et al. 1997; Polk and Postow 1996; Reilly 1998; Repacholi 1998; Repacholi et al. 1999; Wilson et al. 1990; aspects of health protection: International Commission on Non-Ionizing Radiation Protection 1988; history of medical applications: Rowbottom and Susskind 1984

4.5
Ionizing Radiation

Studies on effects of ionizing radiation on living organisms have become necessary following the introduction of X-rays in medical therapy and diagnostics at the beginning of the twentieth century, and have acquired a special relevance in the so-called nuclear age. In this book only a few important aspects of radiation biophysics can be focused on.

In the spectrum of electromagnetic waves (Fig. 4.13), ionizing radiation begins at a quantum energy in the order of 10 eV. Usually the energy of ionization of the water molecule, which amounts to 12.46 eV, is taken as a borderline. This corresponds to a range of wavelengths in the upper limit of ultraviolet light. UV-radiation per definition starts with a wavelength of 380 nm which corresponds to a quantum energy of 3.29 eV. This region is divided into three subregions. Ultraviolet-A (UVA) with a wavelength up to 315 nm, corresponding to 3.94 eV, followed by ultraviolet-B (UVB) up to 280 nm, corresponding to 4.42 eV, and finally the region of ultraviolet C (UVC) lasting up to the quantum energy of water ionization. Because the quantum energy of ultraviolet radiation is lower than the energy of ionization of water, it is usually considered as a kind of non-ionizing radiation. Nevertheless, mechanisms of interaction occur which are similar to those of ionizing radiation.

Corpuscular radiations, such as α-, β-radiation, as well as accelerated elementary particles and ions of various mass numbers are also considered as ionizing radiation.

4.5.1
Nature, Properties and Dosimetry of Radiation

Ionizing radiation in our environment is emitted by extraterrestrial sources, by various technical devices, and by spontaneous decay of certain atomic nuclei. Even if the quality and quantity of human exposure to ionizing radiation has been altered by technical development, a natural exposure of all biological organisms has always existed and has been a part of environmental conditions for evolution.

X-rays and γ-radiation are explained as a part of electromagnetic waves in Fig. 4.13. Let us now characterize briefly some properties of corpuscular radiations.

α-rays are of fast-moving helium nuclei with a mass number = 4 and an atomic number = 2 ($^4_2$He). The helium nucleus consists of two neutrons and two protons and thus carries two positive charges. Because of this strong charge, there is a correspondingly large interaction between α-rays and the elementary particles of matter. The energy of α-rays emitted by radionuclides is generally quite high, namely of the order of several millions of electron volts (MeV). As the α-rays pass through matter, this energy is dissipated as a result of ionizing processes and the particles are slowed down. For example, the energy which is required to generate one pair of ions in air is 34.7 eV. It can be easily calculated that a radiation energy of several 10^6 eV is sufficient to induce about 10^5 ionizations. Consequently, α-particles will leave behind a straight-line track of limited length in irradiated matter which consists of ionization products. This can be visualized by special methods.

β-rays consist of fast moving electrons. These particles carry a single negative charge and an extremely small mass. Although it is possible to generate β-radiation of very high energy by means of particle accelerators, the energy of β-radiation from radioactive nuclides on average is lower than that of the α-radiation. In contrast to α-radiation, the energies of electrons of β-emitting isotopes are not at all equal to each other but are spread across a specific range. Therefore the terms *mean energy* and *maximum energy* are used when referring to the β-emission of radionuclides. The distance to which a β-particle penetrates matter will, naturally, correspond with its energy. The ionization tracks of β-particles in contrast to those of α-particles, do not follow a straight line but their path becomes increasingly curved towards the end. In addition, the density of ionization increases as the energy of the particle decreases so that ionization is more densely packed with low energy radiation than with high energy radiation and, for the same reason, this effect can also be seen towards the ends of the tracks.

Neutron radiation occupies a rather special position in the classification of corpuscular radiation. Its particles are electrically neutral and possess a considerable mass when compared with β-particles. For this reason they are able to penetrate an atom and reach the nucleus where they can cause nuclear transformations. This is technically how radionuclides are made. The irradiated substance itself becomes radioactive. Neutron emission does not occur when a radionuclide decays spontaneously but takes place during nuclear fission or in the wake of other, externally induced reactions.

Recently, various kinds of particles, such as π-mesons, accelerated ions and small nuclides have been used in cancer therapy.

The process of energy dissipation that occurs during the absorption of electromagnetic radiation by matter follows the same pattern as that during the absorption of corpuscular radiation in that it is not continuous but takes place in steps i.e. in quantum leaps. The size of energy quanta absorbed in this way depends on the type of interaction. In general, three types of absorption processes may be distinguished according to the relation between the required ionization energy and the available quantum energy:

1. With low quantum energies, the *photo effect* is produced. In this, a γ-quantum (also called a *photon*) is absorbed and this causes a displacement of an orbiting electron from the shell of an atom. The excess energy, above that which is required for the ionization process, serves to accelerate this so called *photoelectron*.
2. The *Compton effect* occurs at a quantum energy of about 10^5 eV. In this case a *Compton electron* is ejected from the atom and is accompanied by the scattering of a secondary γ-radiation. However, this latter radiation has a quantum energy lower than that which was originally absorbed.
3. If the quantum energy of the γ-radiation is above 1.02 MeV then *electron-pair formation* can occur. The quantum energy disappears producing a negative and a positive electron. Such formation of an ion pair can only take place close to the nucleus of an atom i.e. it can only occur in an absorbing material and not spontaneously in a vacuum.

The radiation energy that is absorbed per unit mass is measured as energy dose in Gray units (abbreviated to Gy):

$$1\,\text{Gy} = 1\,\text{J kg}^{-1}.$$

In earlier papers the units Roentgen (R) and rad (rd) can be found. These units can be converted as follows:

$$1\,\text{rd} = 10^{-2}\,\text{Gy}.$$

In water and tissues:

$$1\,\text{R} = 0.93\text{--}0.98\,\text{rd}.$$

In addition, as measure of the dose rate, the unit Gy s^{-1} is used.

For considerations of radiation protection a further parameter is used which additionally takes into account a factor which marks the relative biological effectiveness of the radiation. So, for example, irradiation by neutrons or positrons are 10–20 times as effective as γ-rays or X-rays. For α-rays this factor amounts 20. This parameter is the dose equivalent with the unit "Sievert" (Sv). This means that the following relations occur: for X-rays: 1 Gy = 1 Sv, for α-rays, however: 1 G = 20 Sv.

4.5.2
Primary Processes of Radiation Chemistry

The action of ionizing radiation of biological systems initiates a complicated sequence of reactions as depicted in Fig. 4.21. The biological reaction starts with

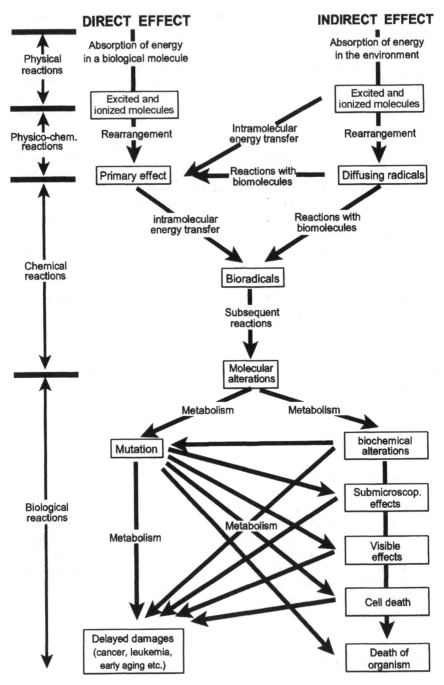

Fig. 4.21. Schematic representation of various steps of the interaction of ionizing radiation on biological systems. (After Dertinger and Jung 1969, redrawn)

an alteration of a biological molecule such as an enzyme or a DNA. Such a change may be induced in two ways:

- By a direct interaction of the energy quantum with the appropriate molecule (*direct effect*);
- Through an interaction of a biologically important molecule with any other product of radiolysis (*indirect effect*).

Water is by far the most common molecule in biological systems and as such plays a special role in indirect effects of radio-biological reactions. In Fig. 4.22 some steps of radiolysis of water are illustrated schematically. In a first step within a time span of 10^{-18}–10^{-16} s an electron is pulled out of the molecular binding by an energy input of 12.56 eV. This primary process of absorption of the energy quantum hv can be expressed by the following reaction:

$$H_2O \xrightarrow{\text{h}v} H_2O^* \rightarrow H^\bullet + OH^\bullet.$$

Both of these primary reaction products undergo further reactions. Greatly simplified, the H_2O^+ ion breaks up as follows:

$$H_2O^+ \rightarrow H^+ + OH^\bullet.$$

The radical $OH^\bullet$ has oxidizing properties and can be regarded as one of the most important of the radiolysis products that give rise to secondary reactions. It can act through electron transfer, by addition of OH or by H-transfer.

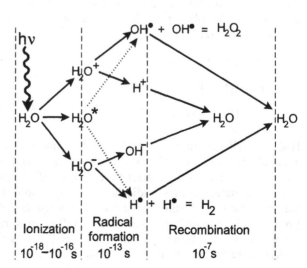

Fig. 4.22. Schematic illustration of the most important reactions of water radiolysis ($*$ – excited states, $\bullet$ – radicals)

The released electron (e_{aq}) becomes hydrated, similar to ions in aqueous solutions (Sect. 2.4.2). For this, in Fig. 4.22 it is depicted as H_2O^-. Because of this shell of bound water, its life span, depending on the pH in the solution, can achieve 600 µs. It can interact with various other molecules. Furthermore, reactions with other water molecules in the hydration shell are possible in the following way:

$$e_{aq} + H_2O \rightarrow H_2O^- \rightarrow OH^- + H^\bullet.$$

In this way a new product of radiolysis appears, the H-radical which indicates strongly reducing properties.

It can also be shown that an excited water molecule can be formed, a process that requires an energy uptake of 7 eV only. This excited molecule can directly break up into its free radicals:

$$H_2O \xrightarrow{h\nu} H_2O^* \rightarrow H^\bullet + OH^\bullet.$$

Summarizing this, three direct products of water radiolysis are obtained: e_{aq}, $H^\bullet$, and $OH^\bullet$. The probability of these products appearing can be expressed by a so-called *G value*. It is defined as the number of molecules which become altered per 100 eV absorbed energy. At a pH of the solution in the range between 3–10, the following G values are obtained:

$$G(e_{aq}) = 2.65; \quad G(H^\bullet) = 0.55; \quad G(OH^\bullet) = 2.7.$$

It has been known for a long time that as the dose rate of the radiation and the ionization density are increased, hydrogen peroxide and molecular hydrogen are formed in growing amounts. The following recombination reactions are assumed to take place provided that locally, for a sufficient time a sufficient number of free radicals are present.

$$OH^\bullet + OH^\bullet \rightarrow H_2O_2$$

$$H^\bullet + H^\bullet \rightarrow H_2$$

$$H^\bullet + OH^\bullet \rightarrow H_2O$$

H_2O_2 is well known to be toxic but the amount generated radiolytically is generally not very great. On the other hand, these reactions eliminate aggressive radicals, thus the indirect effect of radiation will be lessened.

Recombination of these radicals with oxygen, solved in the aqueous phase, is very important for further reactions (*oxygen effect*). This oxygen forms various products with the hydrated electrons as well as with hydrogen radicals, leading eventually to additional production of H_2O_2. Oxygen-rich tissue therefore suffers more from irradiation than that with low oxygen content.

The most important reactions between H• and OH• radicals and organic molecules (MH) can be summarized as follows:

$$MH + H^\bullet \rightarrow MH_2^\bullet$$

$$MH + OH^\bullet \rightarrow MHOH^\bullet$$

$$MH + H^\bullet \rightarrow M^\bullet + H_2$$

$$MH + OH^\bullet \rightarrow M^\bullet + H_2O.$$

The hydrated electron reacts primarily with those parts of an organic molecule that have high affinity for electrons such as SH-, CO-, NO$_2$- or NO-groups. Either it simply becomes attached to the molecule, thus imparting a negative charge to it, or a dissociative electron capture takes place which, similar to the instances described above, leads again to the formation of a free radical.

In addition to the indirect effects of radiation initiated by the radiolysis products of water, biologically important molecules can also become ionized directly. Amino acids generally become deaminated. Aromatic amino acids such as tyrosine, phenylalanine and tryptophan can be destroyed by cleavage of the benzene ring. Amino acids bearing an -SH group (cysteine) and their derivatives (for example glutathione) are particularly sensitive. These groups are oxidized to become disulfide groups. When aqueous solutions of amino acids or proteins are irradiated then both the direct and the indirect mechanisms are involved in the effect that ensues.

The particular sensitivity of SH-bearing amino acids to ionizing radiation is retained even after they have been incorporated into large protein molecules. In such molecules energy transfer is of considerable importance because energy can pass from one monomer to the next until a sensitive group is reached.

Radiolysis of DNA is possible via sequence of many reaction steps, whereas direct as well as indirect effects via water radicals can occur. Single strand breaks which do not necessarily lead to the break of the whole molecule appear, as well as direct breaks of the double helix. Furthermore, the loss of single bases is possible. Irradiation can also lead to formation of whole denatured zones as well as to intermolecular cross-links with neighboring macromolecules. The double strand helix can be considered as a construction of higher stability against irradiation in relation to the single stranded DNA helix. If the hydrogen bridges between the two molecular filaments are broken by placing the DNA in a solution of urea, then fragmentation of the DNA by irradiation is easier.

Using the example of radiolysis of DNA the particular property of UV-radiation must be pointed out. Even in the region of frequencies in which the quantum energy is not sufficient to lead to radiolysis of water, alterations in the base structure of DNA can be observed. Irradiation with UV-light with wavelength of about 260 nm with a quantum energy of 4.8 eV leads to formation of base dimers and some other products of photolysis. This is the region on the borderline of UVB and UVC. This process mostly concerns dimers of pyrimidine, thymine and cytosine, and purine to a lesser extent. In these reactions two neighboring bases of the same strand are interconnected, via opening of the 5-6-double bounds and formation of a cyclobutyl ring structure. Dimer formation of two different bases is possible too. This process is reversible.

Further reading: Farhataziz and Rodgers 1987; Kiefer 1990

4.5.3
Radiobiological Reactions

In this section we leave the narrow region of biophysics and come to the biological reactions according to Fig. 4.21. It can be easily understood how the processes of radiolysis as described in the previous section may interact with biological processes in various ways. As shown in Fig. 4.21, this kind of interaction is marked by a multitude of different paths.

A significant starting point for radiobiological reactions, of course, are alterations in the genome. This kind of reaction is considered as a *stochastic effect* of irradiation. Although DNA molecules play a dominant role in primary reactions of irradiation, this does not mean that every damage of an DNA molecule must lead to the manifestation of a mutation. Thanks to a multitude of repair mechanisms, only a very low percentage of damaged DNA molecules will finally lead to biologically important alterations. Of course the repair process itself can be influenced by irradiation. This, by the way, is a mechanism which is hypothesized for the possible carcinogenic influence of non-ionizing irradiation. Furthermore, these repair processes themselves can lead to incorrectly repaired molecules.

To determine the genetic defects caused by ionizing radiation, the dose of irradiation must be integrated over the whole life span of an individual or even over the whole population. The reason for this is not a biophysical, but a genetic mechanism of accumulation. Recessive mutations especially, i.e. such which become effective only when two of them occasionally meet by arbitrary combinations, become dangerous if they become accumulated in the population. The same holds for radiation-induced cancer as result of somatic mutations.

Mutations caused by irradiation are of particular importance for those cells that are proliferating even in the adult body at a fast rate. These are found, for example, in the blood forming organs and in the gonads. It is therefore understandable that following irradiation, drastic changes in the blood picture can be seen. In this context the reduction of the number of lymphocytes is particularly dangerous.

The different phases in the life cycle of cells show different sensitivities towards radiation. The phase when DNA is being synthesized, the S-phase of the cell cycle, is particularly sensitive. Even the process of mitosis can be disturbed by several effects. As a result of irradiation of cells during metaphase, chromosome aberrations may occur. Disturbances that arise during the ana- and telophase of mitosis are mainly attributable to the effects of radiation on the spindle apparatus. The result is an incomplete separation of the chromosomes. The chromosomes that are left behind as well as some parts of chromosomes form para-nuclei that are not viable.

When tissues are irradiated while they are proliferating intensively, then changes in the incidence of mitosis can be observed over a period of time. Immediately after irradiation by a sub-lethal dose, the number of dividing cells is reduced. After a certain recovery time it returns to its normal value. This is obviously due to the fact that the further development of cells that are at a particular stage in their life cycle, is inhibited. Later, these cells undergo mitosis together with the undamaged cells, sometimes leading to a small overshoot of the number of proliferating cells.

Considering the facts that have been outlined up to now it is understandable that juvenile organisms, particularly those in the embryonic phase, are especially sensitive to irradiation. Damage to the genetic and mitotic apparatus can lead to severe malformations or even death of the individual. There are some phases during the development of an organism that are particularly critical in relation to radiation damage. One such phase is during the early stages of cleavage of the ovum up to the morula formation. Radiation damage in these states, however, does not usually lead to malformations because the injured cells either die or are successful in repairing the damage. Irradiation during the stage when the organs are being laid down is substantially more critical. Exposure during these stages frequently causes malformations. In this context, the development of the central nervous system is particularly sensitive, in contrast to the completely developed brain which is one of the most resistant organs in relation to ionizing radiation.

In addition to the influence on the genetic or epigenetic systems of the cell, other effects of radiation may occur, caused by other mechanisms. A particular role is played by radiochemical alterations of enzymes. Their sensibilities to radiation show great variations. Among the most sensitive enzymes are ATPases and catalases. This sensitivity of proteins in general depends on their composition. We already mentioned that proteins with a high percentage of sulfur-containing amino acids are particularly sensitive to radiation. These proteins which are to be found in substantial amounts in the epidermis as well as its products (hairs, nails, horn) are characterized by a high content of cysteine. Radiolysis of these proteins leads to loss of hair, to skin damage and some other reactions.

One process that can be attributed to an alteration of protein structures as a consequence of irradiation is an increase in the permeability of biological membranes. This reaction, however, occurs only at relatively high doses of radiation, but may have a profound influence on physiological processes. These alterations, for example, are responsible for radiation effects in the central nervous system.

These circumstances help us to understand the course of acute radiation sickness. Figure 4.23 shows the average life span of some animals. All of the curves first show a decline at doses below 10 Gy, then there is a plateau followed by a second decrease at about 80–100 Gy. The shape of these curves reflects two different types of radiation sickness. At low doses the radiation sickness is mainly caused by pathologic alterations in blood-forming organs, which is reflected in dramatic changes in the blood picture. A considerable deficiency of lymphocytes occurs which reduces the resistance of the body to bacterial

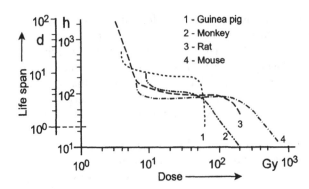

Fig. 4.23. Average life span of various mammals as a function of the dose of a single whole-body irradiation with X-rays. (After Grodzenskij 1966, redrawn)

infections. This can eventually lead to a situation where the intestinal bacteria become pathogenic. This type of radiation sickness lasts from several days up to weeks and can lead to the death of the organism.

The pathological course of this type of radiation sickness is maximal at a dose of about 10 Gy. A further increase in radiation intensity does not accelerate this process. If the animals are irradiated with doses over 80 Gy, the central nervous system becomes damaged directly. In this case the animals die in a much shorter period of time.

Because of the complexity of radio-biological reactions it is difficult to find bench mark figures as a measure of the resistance of different organisms to irradiation. Similar to the characterizations of poisonous drugs, the parameter LD_{50}^{30} (*mean lethal doses*) is used. This means the dose of radiation that causes 50% of the irradiated animals to die within 30 days.

From the values shown in Table 4.4 the following conclusions can be drawn: the sensitivity of radiation varies over several orders of magnitude for different species. A comparison of the sensitivity at different ages and stages in the development of a given species clearly shows the embryonic organism to be more sensitive than the adult one.

Further reading: Kiefer 1990; Nias 1990

4.5.4
Some Aspects of Radiation Protection

Radiation protection means on the one hand the protection against irradiation from the environment, and on the other hand possible irradiation emitted by incorporated radionuclides must be considered. Besides naturally occurring radionuclides in the body, mostly ^{40}K, and to some extent ^{14}C and some others beginning with the nuclear age, many technically produced radioisotopes are incorporated. The protection of man against incorporation of radionuclides from the environment is the task of *radioecology*. This discipline concerns the distribution of radionuclides in the ecosystem, its accumulation in various products of the food chain and finally their degree of resorption, accumulation

Table 4.4. Examples for the mean lethal dose (LD_{50}^{30}) of various organisms after single irradiation with X-rays. (After Grajevski and Šapiro 1957)

Organism	LD_{50}^{30} (Gy)	Organism	LD_{50}^{30} (Gy)
Algae		Low vertebrata	
Chlorella spec.	180	Goldfish	6.7
Chroococcus spec.	80	Frog	7.0
Ancistrodesmus spec.	110	Turtle	15
Synechococcus spec.	1000	Serpend	820
Protozoa		Birds	
Amoeba spec.	1000	Chicken	10
Colpidium colpoda	330	Mammals	
Paramecium caud.	350	Mouse	4–6.5
Crustacea		Rat	6–10
Calliopius laeviusc	10	Guinea pig	1.7–4.1
Artemia salina (egs)	800	Rabbit	7.5–8.3
Insects		Hamster	7.25
Drosophila (adults)	950	Dog	2.75
Drosophila (larvs)	1.3	Donkey	6.5
Drosophila (eggs)	1.5	Monkey	5.0
Calliphora (eggs)	0.4	Man	4.0–5.0

and excretion by man. In fact it is much more complicated to calculate the dose or irradiation caused by accumulated radionuclides, than to calculate intensity of irradiation coming from external sources. Radionuclides emitting α-, or weak β-rays are most dangerous. Empirically obtained coefficients are listed indicating concentration limits of these nuclides in air, water and various foods, as well as in humans themselves.

The protection of man against ionizing radiation is based on recommendations of the International Commission of Radiological Protection (ICRP), a non-governmental organization founded in 1928. The philosophy of these recommendations is the exclusion of any non-stochastic radiation effects and the minimization of inevitable stochastic effects. These stochastic effects really are the critical point of radiation damage. It includes possible genetic aberrations or induction of cancer. In general, one can assume that any increase of irradiation exceeding that of the natural dose of radiation will lead to an increase of the mutation rate. There is no threshold below which no influence of the mutation rate can be expected. It has been evaluated that the doubling of the mutation rate of man occurs at a radiation dose below 1 Sv. The probability of carcinogeneses, evaluated in Japan, is considerably lower than that.

The everyday irradiation of man is caused by natural, as well as by technical sources as products of our civilization. As shown in Table 4.5 as an example, the largest radiation dose of about 2.4 mSv yearly comes from natural sources. Of all technical sources, probably the irradiation caused by medical treatments seems to be dominant. Of course, this is only an example, the numbers may differ depending on cultural, geological as well as individual conditions.

Because of the possibility of stochastic effects becoming integrated over the whole life span, the limits of maximal exposure are formulated over longer periods of time. People with occupational exposure are under special supervision. For occupational exposure, a maximum yearly irradiation dose of 50 mSv is laid down. For the general population 1/30 of this value is considered to be acceptable. As seen in Table 4.5 this is below the level of natural irradiation.

The recommendations of the ICRP states that all exposure should be kept "as low as reasonably achievable", economic and social factors taken into account. This philosophy is commonly known as ALARA-principle.

Further reading: International Commission of Radiological Protection (1991)

4.5.5
Mathematical Models of Primary Radiobiological Effects

Since the first investigations of biological effects of ionizing radiation as early as the 1920s, efforts were made to formulate theoretical approaches, helping to understand the observed phenomena. This was before anything about molecular genetics or details of primary molecular effects of water radiolysis etc. were known. Even though of limited relevance in modern radiation research, these approaches nevertheless helped to understand various quantum biological processes in general, such as photosynthesis, or scotopic vision.

The starting point of these approaches was the realization that radiation effects are characterized by the following particularities:

Table 4.5. Average effective dose of the population in Germany in 1993 (in mSv). (According to the report of the Federal Office for Radiation Protection)

1. Natural sources	
– Cosmic radiation	approx. 0.3
– External terrestrial radiation	approx. 0.4
– By inhalation of radon-decay products	approx. 1.4
– By ingestion of natural radioactive substances	approx. 0.3
	approx. 2.4
2. Technical sources	
– Nuclear plants	<0.01
– By medical applications of radionuclides and ionizing radiation (rough evaluation)	approx. 1.5
– By application of radionuclides and ionizing radiation in research, techniques and in daily life	<0.01
– Occupational exposition (contribution to average exposure of the population)	<0.01
– By fall-out products of atomic bomb tests	<0.02
	approx. 1.55
3. Exposure, caused by Tschernobyl accident	
	approx. 0.02

- The effects not only depend on the total amount of the absorbed energy, but additionally on the quality of the applied radiation.
- When comparing biological effects obtained by irradiation with rays of the same quality, a dose dependence was found, but not a proportionality between effects and irradiation dose.

Furthermore, the total amount of energy absorbed in a biological system, even at irradiation with the mean lethal dose (LD_{50}^{35}), is negligible in relation to the thermic energy produced in it under physiological conditions.

These observations have led to the following conclusion: apparently, there are molecules in the living organisms, which must be considered as "neuralgic" points in the system, the destruction of which would produce a sequence of reactions, finally leading to the death of the individual. This sort of primary event which is caused by a reaction with a single energy quantum of radiation was called a "hit" and the corresponding location where the process occurs, a "target". This leads to the name of the first approach, called *target theory*, firstly applied to a population of single cells.

Let us assume that a population consists of n identical individuals, each of them organized physiologically in such a way that a single "hit" would set off a sequence of reactions, leading to death (*single hit process*). According to the laws of probability, with increasing time of irradiation, an increasing number of individuals would become killed by this *single hit* process. The rate at which the number of living individuals decreases in this way can be expressed by the differential quotient $-dn/dt$. It should be proportional to the number of still living individuals (n), the time of irradiation (t), and an irradiation constant (σ) which depends on quantity and quality of irradiation:

$$\frac{dn}{dt} = -\sigma n. \tag{4.5.1}$$

Using the initial conditions, whereby: $t = 0$; $n = n_0$ it results in:

$$n = n_0 e^{-\sigma t}. \tag{4.5.2}$$

The reduction of the number of living cells therefore follows a simple exponential function.

In the same way, the number of killed cells (n') can be calculated. Because of the relation: $n' + n = n_0$, we obtain:

$$n' = n_0(1 - e^{-\sigma t}), \tag{4.5.3}$$

or:

$$\frac{n'}{n_0} = 1 - e^{-\sigma t}. \tag{4.5.4}$$

In Fig. 4.24 the curve for $m = 1$ represents this function. At $t = 0$ is $n' = 0$ and therefore $n'/n_0 = 0$. With increasing irradiation time and doses (σt) the value $n'/n_0 = 1$ will be approached.

In a similar way, another model can be calculated. In this case it is assumed that the sequence of processes leading to the death of the individual is only triggered if the organism has received a minimum number (m) of hits. Blau and Altenburger have calculated the following equation describing this *multi-hit process*:

$$n' = n_0 \left\{ 1 - e^{-\sigma t} \left[1 + \sigma t + \frac{(\sigma t)^2}{2!} + \frac{(\sigma t)^3}{3!} + \cdots + \frac{(\sigma t)^{m-1}}{(m-1)!} \right] \right\}. \qquad (4.5.5)$$

Some of the curves corresponding to this equation are shown in Fig. 4.24. These functions help to explain experimentally obtained dose dependent effects.

This theory was extended by J. A. Crowther by introducing a *target area* into this concept. This means he transformed this formal target into a concrete area where the "hits" occur. For this he used results of experiments with radiation of different wavelengths. As already stated, radiation leaves an ionization trail as it passes through an object. The density of ionizations along this path depends on the character of the radiation, i.e. of its quantum energy. Suppose a "hit" means an ionization of a quite specific molecule or a molecular configuration, then a certain volume must be sensitive to the radiation, which is called the *target area*.

Figure 4.25 schematically illustrates two experiments where a biological object with a given target area is irradiated by two kinds of ionizing radiation. In case A a radiation is used which induces a lower ionization density than in case B. Let us suppose that the biological reaction in this case is triggered by a single hit process, then in case B the energy will be squandered because the diameter of the target area is larger than the mutual distance between two points of ionization. This theoretical prediction has been verified by experiments, where particular dose-effect curves were obtained when different types of radiation were applied.

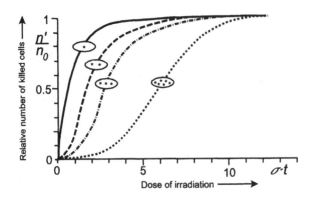

Fig. 4.24. Theoretical curves for one- and multi-hit processes (m corresponds to the number of stars in the "cell"). (After Dessauer 1965)

Fig. 4.25. Schematic illustration of the irradiation of a single target area (*circular area*) with a radiation of lower (**A**), and higher (**B**) ionization density. The points mark the locations of ionizations

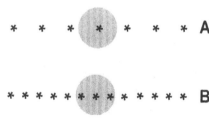

In a similar way the target theory has been modified step by step in adaptation to experimental results. Meanwhile, a considerable number of additional factors have been taken into account influencing the course of radiobiological reactions. For example, it is possible to calculate a multi-hit process, the targets of which have different sensibility. Furthermore, it is also possible to take into consideration some processes of repair. For example, one can assume that in the course of a multi-hit process the efficiency of subsequent "hits" depends on the time period in between. Processes of energy transfer, e.g. indirect effects, can be included. Of course any extension of the model leads to the introduction of additional parameters. This finally leads to ambiguous interpretation of experimental results. This seems to be the limit of the extension of these sort of calculations.

Nevertheless, the target theory has achieved a certain importance in the analysis of various kinds of radiobiolgical effects. So, for example, mortality curves for bacteria have been characterized as multi-hit processes. Induction of mutations as a result of irradiation has also been analyzed by this formalism. Some mutations, even those of chromosomes, could be characterized as single-hit processes.

These considerations have also been extended to the influence of non-ionizing radiation. So, for example, scotopic vision at dawn could be explained by curves of three to seven hit processes. The process of photosynthesis has also been investigated in the same way.

Further reading: Dessauer 1964; Hug and Kellerer 1969; Kiefer 1990; Zimmer 1961

CHAPTER 5

The Kinetics of Biological Systems

This section provides an introduction to systems theory as applied to living organisms, which is an important part of theoretical biophysics. Originally, thermodynamics as a universal theory of the energetic basis of all movement in nature on the one hand, and kinetics, as a theory to predict the time course of processes on the other hand, were considered two separate directions in science. Thermodynamics gives an answer to the questions: what is the cause, the driving force, and the direction of a movement, and finally what is the equilibrium situation which will be eventually arrived at? Kinetics, on the other hand, just studies the course of a given reaction, its time constants and rates, its possible stationary states and oscillations.

After introducing time as a new variable into the framework of thermodynamics, starting with the thermodynamics of irreversible processes, i.e. nonequilibrium thermodynamics, these two directions became more and more linked. A new name was even proposed to characterize this connection: it was called *synergetics* (see Hagen 1983).

The idea of *isomorphism* still forms the basis of system analysis. This means that quite different processes in nature, technology or society can formally be described by similar mathematical approaches and therefore show similarities in their properties. This is also the basic idea of cybernetics.

5.1
Some Foundations of Systems Theory

The system analysis concerns consequences of movements, and interplay of various processes determining the behavior of the system as a whole. Therefore, we will follow the definition of the term "system" which was given in Section 2.3.3 and the general discussion on system properties in Section 3.1.1. The definitions of stationary states and stability, which have been discussed from the thermodynamic point of view in Section 3.1.4 are important here as well.

5.1.1
Problems and Approaches of System Analysis

The main concern of systems theory is to investigate the behavior of a system and the interplay between its elements. Formally, one considers all kinds of parameters influencing the system as *input*. This leads to a response of the system, which is named the *output* of the system. The transformation of the input into an output signal is performed by the *input-output law* or: *law of system behavior* (see Fig. 5.1). This law in simple systems can be expressed by a mathematical equation.

In general, there are three kinds of questions which are directed to system analysis:

1. The input and the input-output law of the system are known. It is necessary to find the output (*direct problem*). This is the problem that the practicing doctor has to solve before deciding on a therapeutic method.
2. The input-output law and the output of the system are known. The input has to be determined (*indirect problem*). This is the problem of finding the cause for an alteration in a known system. To give another medical example: it is the problem of diagnosing a disease.
3. The input and the output of the system are known. What is input-output law of the system (*black-box problem*)? This third question is typical of the problems facing the scientist and is, at the same time, the most difficult of the three. The "black box" must be converted into a "white box", the law of which should be clarified.

The cognitive process usually follows an iterative course, continuously setting up hypotheses and testing them by special experiments on the system. The more complex the system is, the more tedious and ambiguous this problem will be. Studies on such systems often require that they be broken down into smaller subsystems that will still allow individual measurements to be made. Then the laws governing the behavior of these subsystems must be analyzed. Subsequently, an attempt must be made to reintegrate the whole system. A functional diagram can be drawn indicating the interaction of the individual components (Fig. 5.2C), each of which is similar to that shown in Fig. 5.1. Even if one knows the rules governing the function of each of the subsystems, it is not possible to determine the behavior of the system as a whole without more information because usually the arrangement of the elements in the system is not known.

The approaches used to analyze the system behavior are based on differential equations containing time derivatives of the corresponding parameters. To describe complex systems, usually a system of many differential equations is required. It is usually complicated to find sensible solutions for these equations. It is therefore reasonable to attempt to simplify the system, taking into account the time hierarchy of the individual processes. This approach has already been

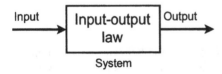

Fig. 5.1. Schematic illustration of behavior of a simple system

Fig. 5.2. Schematic representation of various
types of systems: A system without feedback,
B system with feedback, C directed graph;
the points in C mean subsystems of type A
or B with corresponding inputs and outputs

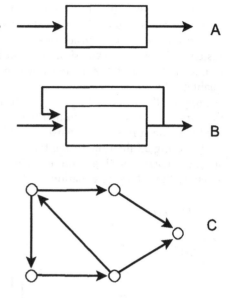

considered in Section 3.1.4 in connection with Fig. 3.8. We will discuss this
problem in more detail in Section 5.2.2 in the context of biochemical reactions.

Schemes of reactions such as those shown in Fig. 5.2C are called *directed
graphs*. *Theory of graphs,* or *network theory* which was established by Leonhard
Euler in the eighteenth century enables general laws to be derived to explain
their behavior and their possibilities of transformations.

Further reading: Doucet and Sloep 1992; Rashevsky 1960; Haken 1983

5.1.2
General Features of System Behavior

Systems theory is concerned with the behavior of the parameters of a system as a
function of time. The theoretical approaches therefore describe the time
dependence of these parameters in the form of ordinary differential quotients. In
some instances, such as for example the propagation of organisms as a function
of time in an ecological model, or where there are problems of growth, or
changes in the distribution in space taking place at the same time, approaches
are necessary which contain partial differential equations.

In general, the system behavior will be characterized by the order and the
degree of the corresponding differential equations. The *order* of a differential
equation is the order of its highest derivative. The *degree* of an equation is
determined by the highest power to which one of its variables or their
derivatives is raised, provided that the equation has been so arranged that only
positive powers of the dependent variables or their derivatives occur. The
product of two variables has to be considered as a kind of second power.

The thermodynamic characteristics of non-linear processes have already been discussed in Section 3.1. We will complete these considerations by introducing kinetics. To illustrate this, let us consider again a simple mechanical system, namely the extension of a spring. We will now calculate the time characteristics of these systems which we discussed in Section 3.6.3 only qualitatively.

Suppose there is a spring hanging from a rigid support. At the lower end of the spring is a pointer which marks the length (l) of the spring. A certain force (F) holds the spring under tension and is in equilibrium with the pulling force of the spring all the time. If the force is suddenly changed, the extension of the spring is correspondingly altered. The relation between these two variables is given by the following equation:

$$F = k_1 l, \tag{5.1.1}$$

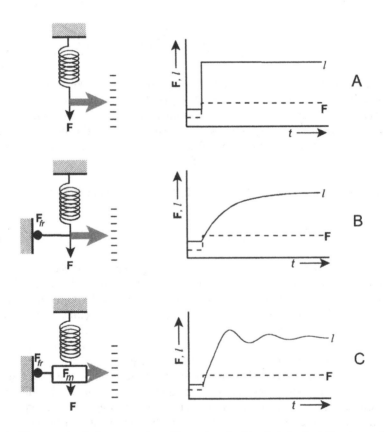

Fig. 5.3. Three mechanical constructions illustrating the kinetic behavior of systems of zero (A), first (B), and second (C) order as response to a rectangular function of the extending force F

where k_1 is a constant containing the Young's modulus. Equation (5.1.1) is a simple algebraic equation. If it were to be included in the category of differential equations, it could formally be classified as of zero order, because it does not contain a derivative of the variable l. The curve $l(t)$ shows a sudden jump corresponding to the alteration of the force $(\mathbf{F})$.

The example in Fig. 5.3B has a brake incorporated into the system. The frictional force of which $(\mathbf{F}_{fr})$ is taken to be proportional to the rate of displacement (dl/dt):

$$\mathbf{F}_{fr} = k_2 \frac{dl}{dt}, \tag{5.1.2}$$

where k_2, again, is a proportionality factor. In this case the action of the force $(\mathbf{F})$ is compensated by the frictional force $(\mathbf{F}_{fr})$ as well as by the force of extension:

$$\mathbf{F} = k_1 l + k_2 \frac{dl}{dt}. \tag{5.1.3}$$

This is already a first order differential equation. The behavior of this system shows a correspondingly complex time course. A sudden alteration of the force $(\mathbf{F})$ leads to an extension of the spring according to a simple exponential function.

The third example (Fig. 5.3C) additionally has a mass (m) built into it, which imparts a certain amount of inertia to the system. The inertial force $(\mathbf{F}_m)$ is, according to Newton's law, proportional to the acceleration. Acceleration, however, is the second time derivative of a distance:

$$\mathbf{F}_m = k_3 \frac{d^2 l}{dt^2}. \tag{5.1.4}$$

Considering this additional force, we get a differential equation of the second order:

$$\mathbf{F} = k_1 l + k_2 \frac{dl}{dt} + k_3 \frac{d^2 l}{dt^2}. \tag{5.1.5}$$

It is easy to imagine that a second order system like this will show oscillations as it settles down, like pairs of scales. After practical acquaintance with such scales, we know that these oscillations can be made to slow down at different rates by use of various damping devices.

Non-linear systems have further possibilities for kinetic behavior. Figure 5.4 shows in principle how a given parameter c in a system can change when it is displaced by an amount Δc from its stationary value. When the system is stable it then returns to its original state. This can occur without oscillations increasing the time constant up to an aperiodic limiting condition, or by oscillations that die away asymptotically. Under certain conditions non-damped

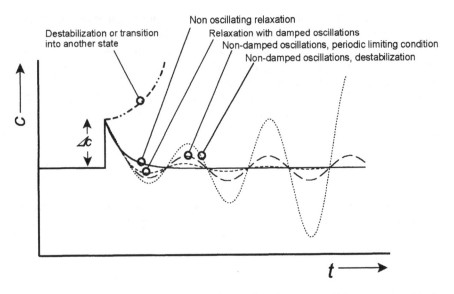

Fig. 5.4. Possible kinds of system behavior, illustrated as time course of the parameter $c(t)$ of a system after a particular displacement Δc

oscillations can occur (periodic limiting condition). Disturbances of the system, however, may also lead to instabilities, possibly leading to another stationary state.

Each state of a system with n independent parameters can be characterized as a point in an n-dimensional phase space. We will illustrate this for example in the biochemical reaction model of Fig. 5.15, or in the mixed culture of two species of paramecium (Fig. 5.18). Let this system be determined by the two variables c_1 and c_2. It is possible to characterize the state of this system by plotting both curves $c_1(t)$, and $c_2(t)$ separately (Fig. 5.5, left part). At any moment (t) a value of the parameter c_1 corresponds to a particular value of c_2. The state of the system therefore can also be represented by a graph, where c_2 is plotted against c_1. At the time t_B, for example, the state of the system corresponds to the point B on the right graph. Following the functions $c_1(t)$, and $c_2(t)$ on the left plot from the moments t_A, t_B, t_C, t_D, t_E ... it is possible to construct a trajectory following the points A, B, C, D, E ... in the graph at the right side of Fig. 5.5. When there are only two variables, the trajectory is just a line in the plane. The arrow indicates where the system proceeds. In systems with more parameters, this is a function in an n-dimensional space.

Figure 5.6 shows such trajectories in a two dimensional parameter space. These curves were obtained in the same way as the trajectory in the right part of Fig. 5.5. This illustration is a kinetic completion of the energetic schemes in Figs. 3.5 and 3.6. At first we can distinguish between stable systems (curves A, D, F) and unstable ones (curves B, C, E), whereby the curve F represents a system with stable permanent oscillations. The stability of a system in this plot is

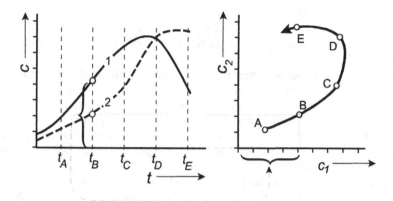

Fig. 5.5. Plot of the functions $c_1(t)$ (curve 1), and $c_2(t)$ (curve 2) on the *left*, and the corresponding trajectory of these values $c_2(c_1)$ on the *right* side

Fig. 5.6. Trajectories in a two-dimensional space to characterize various states of the system: A stable node, B unstable node, C saddle point, D stable focus, E unstable focus, F stable limit cycle

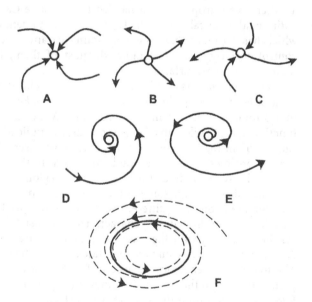

characterized by a trajectory leading to a fixed point or a limit circle, whereas in case of instability, the arrows lead away. Metastable states (see Fig. 3.5), which are not plotted in Fig. 5.6, would be characterized by an arrow leading from one point to another.

Furthermore, in this figure different ways are indicated for systems used in order to attend the stable states. The aperiodic approximation to this state, as shown in Fig. 5.4, corresponds to the case A in Fig. 5.6. The situation with damped oscillations in Fig. 5.4 is represented in case D of Fig. 5.6, and the undamped oscillation corresponds to case F in Fig. 5.6. The distortion of the circle into an ellipse (case F) simply depends on the chosen coordinates. In the case of oscillations superimposed by several frequencies, this ellipse will be deformed further.

In this way, the set of differential equations describing the behavior of the whole system determines a vector space, with the n variables as coordinates. This means that at each point in this n-dimensional vector space a signpost is located, indicating the direction of further system development. There are some "watersheds", or borderlines where the directions become bifurcated, like the point in case C (Fig. 5.6). This would correspond to the energy maximum between two metastable states (Fig. 3.5).

System trajectories can be plotted in 2-dimensional coordinates only if the systems are simple. Generally, multidimensional spaces are required to describe alterations in the state of more complex systems in mathematical terms. In 1969 the French mathematician, Rene Thom developed a comprehensive topological theory of dynamic systems which primarily enables the occurrence of multi-stationary states to be systematized. He called a sudden change from one state to another a catastrophe, which has led to the name Catastrophe Theory for this entire mathematical approach. Thom was able to demonstrate that for a system which has no more than four controlling parameters, seven different types of control surfaces can be formulated, most of them, however, being 3- or 4-dimensional hypersurfaces.

The basic concepts of this theory will now be elucidated by referring to one of these seven functions, the so-called *cusp catastrophe* which can be demonstrated in a 3-dimensional system of coordinates. Above a control surface (x, y), the function $z(x, y)$ builds up a behavioral surface, indicating a state parameter z for each value in the x, y-control surface. In most cases, there is only one point in the z-coordinates that corresponds to a point in the x, y-plane. The behavioral surface, however, is folded at a particular region. As a result, there are three possible values of z for some points of the x, y-control surface. The system can, therefore, take on three different states for some variables. The value of z in the middle of those three points characterizes an unstable state (shaded area). This can be compared with the thermodynamic picture of a meta-stable state (Fig. 3.5) where there is also an unstable state between the two stable stationary states. The fold in the behavioral surface in Fig. 5.7 starts at the point P which is the point from which the bifurcation lines diverge and can be seen clearly when the fold is projected onto the x, y-control surface.

This picture of a cusp catastrophe can be used to demonstrate the isomophism, namely the fact that quite different processes can be treated mathematically in a similar way. Take z to be a parameter that characterizes the viscosity or the structure of water. It will depend on pressure (x) and on temperature (y). The value of z at the same time will be an indicator for the corresponding aggregation state of the water. When the temperature falls (trajectory A in Fig. 5.7) the system will reach a bifurcation point at 0 °C at normal pressure. It can either freeze (i.e. move on the lower surface), or remain as a super-cooled liquid (upper surface). The super-cooled liquid, however, will

Fig. 5.7. Functional space of the state function (z) depending on two independent state parameters (x, y) as an example of a cusp catastrophe. A trajectory with bifurcation point (P); B, C trajectories with a phase jump ("catastrophe") near the borderline of stability; *shaded area* – region of instable states

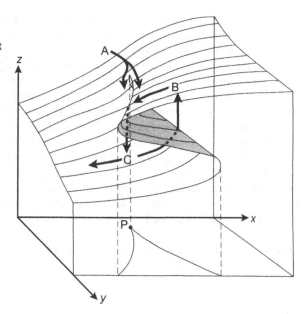

spontaneously freeze (trajectory B) if there is any disturbance of the system (alteration of pressure, vibration etc.). This reaction in the trajectory B would be a "catastrophe" in the terminology of this theory.

In Section 5.2.2 we will characterize the behavior of a biochemical reaction chain with such a function (Fig. 5.15B). Furthermore, the electrical potential of a nerve cell was explained in this way. In this case the upper and the lower part of the folded behavioral surface are considered as non-activated and activated nerve membrane, corresponding to its resting potential and action potential. The "catastrophe" in this case was the process of excitation (trajectory B). In this case a transition corresponding to the trajectory C is also possible as the restoring reaction of the nerve.

A quite different example can be taken from behavioral biology: let z be a parameter that characterizes the behavior of a dog, and x and y are parameters corresponding to its disturbance. When these reach a certain size (point P) the dog will be critically irritated, with the result that it either runs away (i.e. the upper surface), or it attacks (lower surface). Either position, once taken, will be maintained and can even be intensified if the degree of irritation is increased. If the flight reaction is suddenly limited (e.g. by an obstruction of some sort), or if the dog is convinced that the attack is senseless, then the dog's behavior may be reversed. The running away in panic changes to an enraged attack or vice versa.

The above example indicates on the one hand the broad field of possible applications of this theory and at the same time the inherent danger of its overuse. Of course, the example from behavioral science is quite graphic, and similar examples can be stressed taken from human behavior, from some social responses, and even from politics. But of course it is impossible to perform real calculations in these situations. Therefore it remains just an illustration without any new, real insights into the mechanism. It remains a game with words. On the other hand, in cases of simpler systems, in biochemical network theory etc. catastrophe theory really can lead to important results.

In recent decades a kind of system kinetics was investigated which is called *chaotic*. This means an obviously irregular and non predictable behavior, which nevertheless is the result of interaction of deterministic processes. Non-linear systems with several stationary states can jump from one state to another. Sometimes, oscillations are the result, sometimes these oscillations degenerate into chaotic movements. The turbulent flow, which we discussed in Section 3.7, may be taken as an example of a chaotic process. In medicine, the disturbance of the cardiac cycle is analysed by approaches using chaos theory.

Further reading: Doucet and Sloep 1992; Harrison 1993; Heinrich and Schuster 1996; Thom 1969; Zeeman 1976

5.1.3
Cybernetic Approaches to System Analysis

In 1948 Norbert Wiener's book *Cybernetics – or Control and Communication in the Animal and the Machine* was published. This was the signal for a rapid development in bio-cybernetics which, when treated seriously, has made substantial contributions to the understanding of biological regulation and control processes. This theory is largely based on experiences with systems in mechanical technology which are made up of discrete functional units (subsystems) that have a flow of information and energy between them. The analogy with biological systems can be drawn, especially if neuronal processes are involved. In the case of thermal regulation of the body, for example, nervous and hormonal effects are essential, controlling blood flow, muscle contraction etc. This is an area where the application of cybernetic approaches is reasonable. On the other hand, in biochemical and cellular or ecological systems, where the elements of the control system are not clearly separated from each other, direct approaches using the theory of non-linear systems seems to be more appropriate.

The title of Wiener's book indicates two major problems which are included in cybernetics: the control of animals and machines, and the processes of communication. In this section we will discuss only the first of these. Some aspects of information theory have already been dealt with in Section 2.3.2.

The module of cybernetic modeling is the control system (Fig. 5.8). In its simplest form it consists of two subsystems, the *controlled system* itself, and the *controller*. Both of these subsystems process one or more input signals to produce one or more output signals depending on their input-output law (see Fig. 5.1). Cybernetics is mainly concerned with systems that are made up of

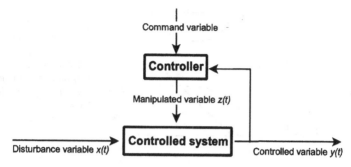

Fig. 5.8. Elements of an automatic control system

linear subsystems in which the system behavior conforms to a linear differential equation of the nth order.

The purpose of a control circuit is to adjust a *controlled variable* $y(t)$ to a given standard which is specified by a *command variable*, in spite of a supplementary effect of a *disturbance variable* $x(t)$. If the command variable remains constant then the network is called a *retaining control circuit*. However, if the command variable is a function of time, which means that the system has an additional control on it, this is a *follow-up control circuit*. The command variable, the controlled variable, and the disturbance variable may differ from each other in their physical properties both in quality and quantity. The regulation of blood pressure (controlled variable), for example, can be controlled through a sequence of nerve impulses acting as the manipulating variable. It could just as well be regulated and controlled by the concentration of a hormone in the blood. Hence, the control signals can be classified as either discrete (sequence of nerve impulses) or analog (hormone concentration).

The difference between the command variable and the controlled variable is called *control deviation* $[\Delta y(t)]$. Depending on the type of the controller, the *manipulated variable* $z(t)$ is calculated in a particular way. According to this, let us consider two types of controllers:

– The proportional controller (*P-controller*):

$$z(t) = k_P \Delta y(t); \tag{5.1.6}$$

– The integral controller (*I-controller*):

$$z(t) = k_I \int \Delta y(t) \, \mathrm{d}t. \tag{5.1.7}$$

In case of the *P*-controller, the modified variable $[z(t)]$ is proportional to the control deviation $[\Delta y(t)]$. This type of controller is not able to completely compensate for an induced disturbance of the system. As illustrated in Fig. 5.9,

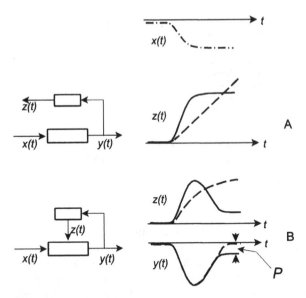

Fig. 5.9. Characteristic responses of a closed (**B**) and an open (**A**) control loop with P-controller (*solid line*) and I-controller (*dashed line*) on a disturbance variable $x(t)$. P-proportional deviation

there always remains a proportional deviation (P) because, if $\Delta y = 0$, then according to Eq. (5.1.6), $z = 0$.

This problem does not arise with the *I*-controller. In this case the modified variable is proportional to the time integral of the control deviation. If the system is brought back to its starting positions then the parameter z in Eq. (5.1.7) becomes constant, which means that an optimal compensation for the disturbance variable is achieved.

In nature, true *P*- or *I*-controllers are hardly ever found. Mostly, the two types exist in combination with each other or are combined with a differential controller (*D-controller*).

For a *D*-controller the following equation holds:

$$z(t) = k_D \frac{d[\Delta y(t)]}{dt}.$$ (5.1.8)

This type of controller does not respond to the control deviation but to the rate at which it is changing.

Neither the controller nor the controlled system respond without inertia. In Eqs. (5.1.6)–(5.1.8) terms have to be included that take into account the time dependence of the adjustments. This will result in differential equations of the nth order. The system behaves therefore like those demonstrated as mechanical examples in Fig. 5.3. This means that the curves representing the functions $y(t)$ and $z(t)$ in Fig. 5.9 are, in fact, more complex than shown. Curves like those illustrated in Fig. 5.4 will more likely result. At least there are no multistationary states to be expected with linear systems. The amplification depends on the factors k_P, k_I and k_D in Eqs. (5.1.6)–(5.1.8) whereas the attenuation correlates with the time constants of the rates at which the systems make necessary adjustments.

The sooner and stronger a system reacts to a disturbance, the faster it will achieve the required correction. However, because control circuits are systems of higher order, the compensation can be overshot. If a system is deflected by a step-wise function then the following types of system behavior can occur as the amplification is increased (see also Fig. 5.4):

- At a low degree of amplification, up to the aperiodic limiting condition, the control variable approaches the target correction value asymptotically without an overshoot.
- As the amplification increases, exceeding the aperiodic limiting condition, the system starts to oscillate. It reaches the target correction value faster but overshoots it and reaches it eventually by damped oscillations. As a result, it may take longer to achieve a stable value than in the first case.
- A further increase in amplification leads to the periodic limiting condition whereby undamped oscillations around the correct parameter occur.
- If the periodic limiting condition is exceeded because of a further increase in the amplification then the system becomes unstable. The oscillations get bigger and bigger and cause the system to become increasingly more deflected.

Such instabilities can lead to the system destroying itself unless the appropriate safety measures are provided. The deflection of the system is usually limited by the energy available. However, these observations illustrate the factors involved in optimizing a control circuit. The fastest possible rate of attaining the target correction value, of course, should be considered as an optimal condition. However, as has already been discussed, this cannot be done by simply increasing the amplification but is best achieved by a combination of different control systems.

In order to illustrate this and, at the same time, to demonstrate the application of cybernetic methods to physiology, the example of target-orientated movement of the hand will be discussed. The person to be tested is asked to move a pointer from position A to position B immediately on being given a command. This movement can be recorded as a function of time (Fig. 5.10). After a lag period of about 145 ms, the hand starts to move and reaches the target 500 ms after the signal, without oscillating. This behavior cannot be achieved by a simple I-controller. The aperiodic limiting condition, i.e. the fastest possible reaction of this controller without oscillations is shown in the diagram as well. The advantage of the faster movement of the hand in relation to that of the I-controller can be explained in the following way: the target-orientated movement of the hand is controlled by two processes. First, the distance the hand has to travel is estimated through the eyes, and a rapid movement with a built-in limit is programmed. This takes up the observed lag period of 145 ms. Only during the last phase of the target-orientated movement, do the eyes correct the hand, a process that can be regarded as being performed

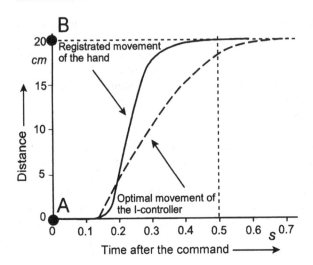

Fig. 5.10. The course of moving the hand (*solid line*) after an order, given at time $t = 0$, to move to a point with a distance of 20 cm from the starting-point. The *dashed line* indicates an optimal reaction of an *I*-controller with a lag period of 145 ms and an amplification up to the aperiodic limit. (Data from Hassenstein 1966)

by an *I*-controller with a time constant of 47 ms. This model is not only mathematically consistent with observations but can also be verified by temporarily preventing the subject from seeing what they are doing during the experiment.

Further reading: Goldman 1960; Morecki et al. 1984

5.2
Systems of Metabolism and Transport

In Section 3.3 flux was introduced as a generalized parameter to explain the transport of matter, of energy and of entropy, and also to calculate the formation of a substance in the course of a chemical reaction. The: "παντα ρει" (panta rhei = everything is flowing), which was formulated by Aristotle to summarize the dialectics of Heraclitus are related to all levels of biological organization. Propagation of organisms, or their movements in ecological reservoirs can be described mathematically by similar kinetic models, as can biochemical reactions or transport processes across a membrane. In this section we will explain these approaches first for simple examples of fluxes of matter and for biochemical reaction rates. Subsequently, we will extend these applications to more complex processes, such as population dynamics, as well as growth and evolution.

As stated before, systems theory concerns the consequences, but not the energetic reasons for movement. Therefore, in case of kinetic approaches it is not so important whether the considered system is in a true steady state with entropy-producing fluxes (Fig. 3.4B, C), or in a thermodynamic equilibrium with thermic fluctuations (Fig. 3.4A; see also scheme in Fig. 3.6). In Section 3.3.2 (Fig. 3.22) we introduced the net-flux as the difference between two unidirectional fluxes. The following kinetic approaches will not distinguish whether a unidirectional flux is part of an entropy-producing net flux or whether it just results from thermic fluctuations.

5.2.1
Introduction to Compartmental Analysis

The aim of compartmental analysis is to classify the kinetic terms of exchange processes in various systems. If we consider the flux as a generalized parameter, i.e. as the rate of a transport process, or of a chemical reaction, or as some kind of proliferation rate etc. then we also need a term which formally characterizes the considered system. This term must be applicable to a phase with spatial borders as well as to a component of a chemical reaction and even to populations of cells or organisms.

For this, the concept of the compartment has to be introduced. In principle, a compartment could be a vessel in the hydraulic models of Fig. 3.4. The general definition is as follows: *a compartment is an element of an open system that, with respect to the matter under investigation, has limits which are defined by a specific process and contains no gradients within itself.* The term "matter" is used in a generalized form. It can represent a substance in a reaction system, a cell in a proliferation system, an organism in an ecosystem etc. The "specific process" depends on the type of the considered flux or movement. It could be membrane transport, a biochemical reaction, mitosis of a cell or death of an organism. The absence of gradients inside the compartment means that it is to be considered as homogeneous, or "well mixed". Otherwise the compartment must be subdivided into smaller compartments, and a multi-compartment system appears. A compartment model is always defined for a special question and is only valid "with respect to the matter under investigation".

Let us illustrate the polymorphic use of this approach with three examples:

- Investigating the exchange of sodium ions in an organism, the compartment is a space and the process is a flux. The space means the cell when the exchange between the organelles and the cell plasma is much faster than the transport across the cell membrane. In this instance the interior of the cell can be regarded as having no gradients inside. However, the space can also be an organ, such as a muscle, provided that the rate-limiting step in the exchange is the flux through the blood vessel wall and not the transport across the membranes of the individual muscle cells.
- In the compartmental model of radioactive phosphorus (^{32}P) in yeast cells the exchange of P between phosphorus-containing compounds must be taken into account as well as its spatial distribution. In this instance the term "compartment" not only means a spatially limited phase but at the same time it includes the chemical components of biochemical reactions. As a result, the compartments are separated from each other by chemical reaction steps as well as by transport processes.

- Considering a population of cells, for example various blood cells in a suspension, the step defining the borderline of the compartment will probably be the process of cell differentiation, as for example the differentiation of reticulocytes into erythrocytes, or the elimination of erythrocytes by phagocytosis. In this case various states of cell maturation are considered to be the compartments. This example is similar to that used in population kinetics.

It must be mentioned that sometimes in the biochemical literature the term *compartmentation* is used in quite a narrow sense. It means just a spatial separation of components of a chemical reaction leading to additional processes of diffusion.

As in all kinds of systems theory, compartment analysis is realized by an interplay between hypothesis and experiments. The following general questions must be solved:

- What is the number of compartments in the system?
- In which way are these compartments connected to each other?
- What are the rate constants of the individual net fluxes?

Compartmental analysis requires the use of specific markers. This is necessary in order to measure not only net fluxes, but also unidirectional fluxes (see Fig. 3.22). It is very important to take care that the use of a marker does not alter the behavior of the system. Only the observer, but not the system should "see" the markers. The labelled components of the system must behave identically to the unlabelled.

The techniques of labelling depend on the kind of investigation. Frequently, chemical elements are labelled using their radioactive isotopes. In case of elements with low atomic mass, an *isotope effect* must be considered. This means that the properties of various isotopes may differ from each other. Tritium (^3H), for example, differs in its binding properties in relation to hydrogen (^1H). This difference can be rectified by a correction factor. The isotope effect becomes smaller for elements of higher atomic mass. It is already negligible when using the isotopes of carbon. Of course the isotope effect can be neglected if, for example, tritium is used to label an organic molecule and if reactions of this molecule which are to be studied do not include any modification of this particular binding.

The setting up of a compartmental model starts with the design indicating single compartments and their mutual interactions as well as their connections with the environment. The simplest model is the one-compartment system in exchange with its environment (Fig. 5.11A). Such a model, for example, could represent the exchange of ions in a non-nucleated blood cell, the exchange of phosphorus by a protein etc.

If a suspension of cells is to be analyzed containing two different types of cells, or cells in two different physiological states, then this must be regarded as a two-compartment system where the compartments are arranged in parallel

Fig. 5.11. Various compartmental models: A one-compartment system, B two-compartment system in series, C two-compartment system in parallel, D three-compartment system

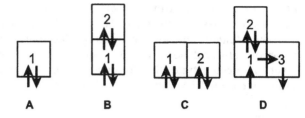

A   B   C   D

without direct connection between them (Fig. 5.11C). In contrast, for cells containing organelles with significant barriers of diffusion or transport, as for example mitochondria, vacuoles etc. then in the simplest case a two-compartment system in series must be applied (Fig. 5.11B). In this case one step is the exchange of matter between the environment and the cytoplasm, another one, the fluxes between cytoplasm and the organelles.

A simple physiological three-compartment system is shown in Fig. 5.11D. This could represent, for example, the transport of a drug in the body: its uptake in the blood (compartment 1), the subsequent accumulation in muscles (compartment 2) and its excretion via the kidneys (compartment 3). We will discuss this model in detail in Section 5.2.3 in the context of problems of pharmacodynamics (see Fig. 5.16).

To test a proposed compartmental model it is necessary to calculate its behavior and to relate the results to measured points of performed experiments. The mathematical representation of such a model starts with the formulation of a system of differential equations using the following basic approach:

$$\frac{dn_m}{dt} = \sum_i J_{im} - \sum_i J_{mi}, \quad \text{where:} \; i \neq m. \tag{5.2.1}$$

The change of the amount of the substance (n_m) in the compartment m as a function of time (dn_m/dt), therefore, is obtained from the sum of all the unidirectional fluxes (J), where the fluxes with the direction outside-in (J_{im}) carry a positive sign, and the fluxes in the opposite direction (J_{mi}) carry a negative sign. In this equation the symbols J are used instead of $\mathbf{J}$, because these are unidirectional fluxes and not fluxes according to the thermodynamic definitions (see Sect. 3.1.3). Furthermore, these fluxes do not necessarily describe transport per unit of membrane area. This corresponds with the generalized approach discussed above. If necessary, these kinetic fluxes can be converted into fluxes in the thermodynamic sense.

To calculate these fluxes, in the simplest case a linear relationship can be applied:

$$J_{im} = k_{im} n_i. \tag{5.2.2}$$

We used a corresponding approach in Section 3.3.2 [Eq. (3.3.39)]. There we characterized the parameter k_{im} as a rate constant with a unit s^{-1} determining the flux from the compartment i into the compartment m. Of course, there are other ways to describe fluxes. In Section 5.2.2 we shall use, for example, the Michaelis-Menten equation introducing biochemical reaction rates into this framework of calculations [Eq. (5.2.19)].

Let us demonstrate the formulation of the equations for the three-compartment model as illustrated in Fig. 5.11D:

$$\frac{dn_1}{dt} = k_{01}n_0 + k_{21}n_2 - k_{12}n_1 - k_{13}n_1,$$

$$\frac{dn_2}{dt} = k_{12}n_1 - k_{21}n_2, \tag{5.2.3}$$

$$\frac{dn_3}{dt} = k_{13}n_1 - k_{30}n_3.$$

The change of the content of the substance n_i in all the three compartments is expressed by the corresponding differential quotient dn_i/dt. The expressions on the right side of these equations correspond to the sums of Eq. (5.2.1), using the linear approaches of Eq. (5.2.2). The environment is characterized by the subscript "0". The parameter n_0 can be considered as constant.

This approach allows interesting conclusions about the character of the stationary state of this system. In Section 3.1.4 we defined the stationary state as a time independent state. This means for the system of Eqs. (5.2.3) the following conditions:

$$\frac{dn_1}{dt} = 0; \quad \frac{dn_2}{dt} = 0; \quad \frac{dn_3}{dt} = 0. \tag{5.2.4}$$

Introducing this in Eqs. (5.2.3) we get a system of simple linear algebraic equations. This makes it possible to calculate the parameters n_1, n_2, n_3 for the case of the stationary state without further conditions. As can easily be seen, solutions for the parameters n_1, n_2, n_3 can be calculated depending only on the rate constants (k_{nm}). Furthermore, because all equations are linear, only one single solution for these three parameters will be obtained.

This leads to the following general conclusions:

- This system has only one single stationary state.
- The parameters n_i of this stationary state depend only on the rate constants (k_{nm}) and not on the initial conditions of the system (Fig. 5.12).

This property which holds for all linear systems is called *equifinality*. Using the phase diagram illustration which in this case is built up by the three coordinates n_1, n_2 and n_3, this means there is only one single point characterizing a stable stationary state. The system approaches this state starting from any point of the phase space. It would correspond to the cases A or D in Fig. 5.6. This, by the way, is the same conclusion as drawn in Fig. 3.6 for the linear steady state and

Fig. 5.12. Illustration of curves of the general solution of the differential equation, Eq. (5.2.5) (- - - -) and its particular solution [Eq. (5.2.9)] (———)

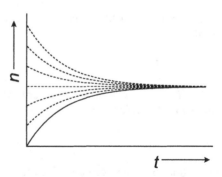

which is expressed thermodynamically by the Prigogine principle of minimal entropy production.

If it is the kinetics, rather than the simple stationary state which is of interest, then it is necessary to solve the differential equations [Eqs. (5.2.3)]. This requires the prediction of initial conditions. Graphically, this means that the point in the n-dimensional phase space where the trajectory starts must be declared. In our case, in a system of three differential equations of the first order, leading in the course of transformation to differential equations of the third order, three initial conditions are required, according also to the 3-dimensional phase space of the coordinates n_1, n_2 and n_3.

We will demonstrate this kind of calculation for the simplest example, the behavior of the one-compartment system (Fig. 5.11A). Corresponding to Eq. (5.2.1) one can write:

$$\frac{dn_1}{dt} = J_{01} - J_{10} = J_{01} - k_{10}n_1. \tag{5.2.5}$$

In contrast to the flux J_{10} where the linear approach [Eq. (5.2.2)] is used, flux J_{01} is considered as constant. To integrate this first order differential equation, only a single initial condition is required.

Let us first consider the case of a decorporation experiment, i.e. an experiment where the extrusion of a previously incorporated substance is investigated. This means that at $t = 0$ only the compartment is labelled in contrast to its environment, which is free of label. Considering furthermore that the environment is very large in relation to the volume of the compartment, in this case the influx of labelled substance in the system can be neglected ($J_{01} = 0$). In this case Eq. (5.2.5) becomes simpler:

$$\frac{dn_1}{dt} = -k_{10}n_1. \tag{5.2.6}$$

Supposing that n_1 and k_{10} were always positive, this equation, looked at from a purely geometrical point of view means that the function $n_1(t)$ is negatively inclined ($-dn_1/dt$) and, the larger the value of n_1, the steeper this becomes.

Equation (5.2.6) can be easily integrated, leading to the following general solution:

$$n_1 = Ke^{-k_{10}t},\qquad\qquad(5.2.7)$$

where K is an integration constant which can be determined by the initial condition. Provided that at $t = 0$, $n_1 = n_{10}$ then it results from Eq. (5.2.7):

$$n_1 = n_{10}e^{-k_{10}t}.\qquad\qquad(5.2.8)$$

This equation, therefore describes the decrease of the amount of labels (n_1) in the compartment as function of time, if at time $t = 0$ the amount n_{10} had been incorporated.

In a similar way it is possible to calculate the kinetics of incorporation, i.e. experiments where the accumulation of a label from the compartment was investigated. In this case the initial conditions are: $t = 0$, $n_1 = 0$. Integrating Eq. (5.2.5) for this case, the following particular solution will be obtained:

$$n_1 = \frac{J_{01}}{k_{10}}(1 - e^{-k_{10}t}).\qquad\qquad(5.2.9)$$

It can easily be seen that for $t \Rightarrow \infty$ the parameter n_1 will attain the value J_{01}/k_{10}:

$$\frac{J_{01}}{k_{10}} = n_\infty.\qquad\qquad(5.2.10)$$

In Fig. 5.12 a group of general solutions of Eq. (5.2.5) as well as the special course of Eq. (5.2.9) is depicted. At the same time this figure illustrates the term equifinality introduced previously.

The analysis of a multi-compartment system is of course more complicated. Here, we will discuss only some general properties of the corresponding equations. Computer models are mainly used for this kind of analysis. In Section 5.2.3 we will describe a multi-compartment system in context with problems of pharmacodynamics (Fig. 5.16).

The integration of the system of differential equations describing a multi-compartment system allows calculation of a time function of the sum content of all compartments together (n_s), which in fact can be measured in experiments. In the case of a decorporation experiment the following type of exponential function results:

$$n_s = A_1e^{-\alpha_1 t} + A_2e^{-\alpha_2 t} + A_3e^{-\alpha_3 t} + \cdots + A_me^{-\alpha_m t},\qquad\qquad(5.2.11)$$

where m is the number of compartments. Only when all compartments are arranged in parallel, do the exponents α_1, α_2, α_3, ... , α_m correspond directly to the rate constants of the individual net fluxes. The same holds for the factors A_1,

A_2, A_3, ..., A_m. In fact, at $t = 0$ it must be: $A_1 + A_2 + A_3 + \cdots + A_m = 0$, but these parameters correspond to the initial contents of the single compartments (n_{10}, n_{20}, n_{30}, ..., n_{m0}) only in the case of a parallel arrangement of all compartments without direct mutual connections. Otherwise, these exponents and factors are more or less complicated functions of all rate constants of the system and furthermore depend on the initial conditions of the system. These particularities can be used to analyze the arrangement of the compartments in the system.

Usually, computer simulations or methods of non-linear regression are used to analyze the results of these kinds of experiments. A better illustration of the character of these curves is given by graphical analysis. Plotting the measured points (n_s) against the time (t), for a one-compartment system, a curve should result according to the functions of Eqs. (5.2.8) or (5.2.9) (Fig. 5.13 left side, above). In the case of a half-logarithmic plot i.e., plotting $\ln n_s$ versus t, the function of Eq. (5.2.8) will show a straight line (Fig. 5.13, right side, above). The crossing point of this line with the ordinate gives the value n_0. Furthermore, there is the following connection between half-life time ($t_{1/2}$) and rate constant (k):

$$k = \frac{\ln 2}{t_{1/2}} = \frac{0.693}{t_{1/2}}. \tag{5.2.12}$$

If it is impossible to fit the measured parameters in a half-logarithmic plot by using a straight line, then at least a two-compartmental model must be supposed. Frequently, such curves indicate a straight line after an initial curved part (Fig. 5.13 right side, below). This is caused by the fact that after sufficient time only one compartment with the largest half-life ($t_{1/2}$) remains.

This is the point at which the graphic analysis of compartmental systems starts. Supposing that at a certain time only one compartment is determining the behavior of the whole system, then the curve ends with a straight line. One can extrapolate this line back to the point $t = 0$ and evaluate the parameters of one of these exponential terms in the sum [Eq. (5.2.11)]. Now, numerically the difference between these extrapolated values and the parameters in the curved part of the measured function can be calculated. The results will show the behavior of a system containing one compartment less. In the case of a two-compartment system, the rest therefore reflects the behavior of a one-compartment system. Thus, it must indicate a straight line in the half logarithmic plot. If not, the procedure must be repeated. In this way all exponential terms in the sum of Eq. (2.5.11) can be obtained successively. In this way this method however, as well as the method of direct fitting of the measured points to corresponding function, will be successful only if there are not too many compartments, and if the rate constants of these parameters differ significantly from each other.

The analysis of multi-compartment systems for an incorporation experiment can be performed in a similar way, if, instead of the function $n_s(t)$, the parameter $[n_\infty - n_s(t)]$ is used. In this way the curved character of the drawn line from Fig. 5.12 will be transformed into that of Fig. 5.13.

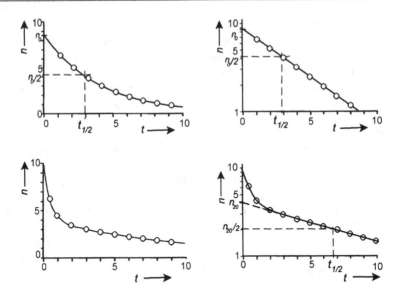

Fig. 5.13. *Above*: illustration of points following the function $n = n_0 e^{-\alpha t}$ (for: $n_0 = 8.6$ and $\alpha = 0.25$); *below*: function $n = n_{10} e^{-\alpha t} + n_{20} e^{-\beta t}$ (for: $n_{10} = 5.8$, $n_{20} = 4.2$, $\alpha = 1$, $\beta = 0.1$); *left*: linear plot; *right*: half-logarithmic plot (all parameters in relative units)

Further reading: Anderson 1983; Shipley and Clark 1972; Solomon 1960

5.2.2
Models of Biochemical Reactions

The rate of chemical reactions can be described by the time derivative of the advancement number (ξ) which was introduced in Section 3.1.5, as well as directly by differential quotients dn_i/dt or dc_i/dt as shown in the previous section. The individual steps in a chain of biochemical reactions can be *mono-molecular*, or *bi-molecular*, depending on how many molecules react together to bring about the reaction. The reaction

$$A \underset{k_2}{\overset{k_1}{\rightleftharpoons}} B + C$$

is a mono-molecular reaction when it runs from left to right and a bi-molecular reaction in the opposite direction. Because the rate of reaction is proportional to the probability of a molecular event leading to the reaction, the unidirectional flux (J_A) for the mono-molecular reaction can be formulated as follows:

$$J_A = k_1 c_A. \tag{5.2.13}$$

The unidirectional fluxes J_B and J_C as a result of bi-molecular reactions are, however,

$$J_B = J_C = k_2 c_B c_C. \tag{5.2.14}$$

Only Eq. (5.2.13) therefore corresponds to a linear approach, similar to the equations which we used in compartmental analysis in Section 5.2.1 [Eqs. (5.2.2) and (5.2.3)]. Furthermore, the rate constants k_1 and k_2 have different units. Eqs. (5.2.13) show that k_1 have the usual unit for a reaction rate: s^{-1}. From Eq. (5.2.14) one can derive the unit of k_2, which is $m^3 \ s^{-1} \ mol^{-1}$ (or $l \ s^{-1} \ mol^{-1}$).

Equation (5.2.14) leads to a non-linear, first order differential equation because it contains the product of two variables ($c_B \cdot c_C$). Correspondingly a non-linear behavior of these systems is to be expected. The chemists call reactions that follow the kinetics described by equations of the type of Eq. (5.2.14) second order reactions. This discrepancy between the nomenclature of systems analysis and chemistry sometimes may lead to misunderstandings.

Bi-molecular reactions do not always lead to non-linear kinetic approaches. If one of the two reactants is present in vast excess, then the rate of the reaction is limited only by the concentration of the other reactant. In this case a bi-molecular reaction behaves like a first order reaction. Such conditions arise, for example, in all biochemical reactions where water is one of the reactants.

Biochemical reactions are usually catalyzed by enzymes. In its simplest form an enzymatic reaction may be described as follows:

$$S + E \rightleftharpoons ES \Rightarrow E + P.$$

A substrate S forms a complex ES with the enzyme E. This complex then breaks down to give the product P and release the enzyme E.

Usually, the rate limiting step of this sequence of reactions is the breakdown of the enzyme-substrate complex. Therefore, the rate of this reaction can be described in the following way:

$$J = k c_{ES}, \tag{5.2.15}$$

where k is the rate constant of this breakdown reaction. Assuming further that the first step of this reaction is always in thermodynamic equilibrium, then by the mass action law:

$$K_M = \frac{c_S c_E}{c_{ES}}. \tag{5.2.16}$$

The total concentration of the enzyme (c_{E0}) must remain constant:

$$c_{E0} = c_E + c_{ES}. \tag{5.2.17}$$

Combining Eqs. (5.2.16) and (5.2.17) one obtains:

$$c_{ES} = \frac{c_S c_{E0}}{K_M + c_S}. \tag{5.2.18}$$

Introducing this into Eq. (5.2.15) one gets the Michaelis-Menten equation which describes the reaction rate (J) of simple enzymatic processes as a function of the substrate concentration (c_S):

$$J = \frac{k c_S c_{E0}}{K_M + c_S}. \tag{5.2.19}$$

K_M, as the dissociation constant of the enzyme-substrate complex, is known as the *Michaelis constant*. It has the units of concentration and corresponds to the concentration of the substrate when the reaction rate is at half of its maximum value (see Fig. 5.14). It usually lies somewhere between 10^{-2} and 10^{-5} M.

As shown in Fig. 5.14 there are two particular regions in this function where some simplifications are allowed. At high substrate concentrations ($c_S \gg K_M$) a maximum reaction rate J_{max} is attained. For this case Eq. (5.2.19) leads to:

$$J_{max} = k c_{E0}. \tag{5.2.20}$$

In this case, the flux is independent from substrate concentration (c_S). J_{max} can be introduced as a constant parameter into the general system of equations [Eq. (5.2.1)].

In contrast, at very low substrate concentrations ($c_S \ll K_M$), Eq. (5.2.19) becomes:

$$J \approx \frac{k c_S c_{E0}}{K_M} = \frac{J_{max}}{K_M} c_S \quad \text{for:} \quad c_S \ll K_M. \tag{5.2.21}$$

In this case a linear flux equation similar to Eqs. (5.2.2) or (5.2.15) can be used. In many particular situations, therefore, simple approaches are possible even in the case of enzymatic reactions.

In all other cases the Michaelis-Menten equation must be directly introduced into the system of flux equations. Obviously, this leads to non-linear equations. This seems to be the typical case for enzymatic reactions and for flux systems where such reactions are included. Furthermore, in many cases additional

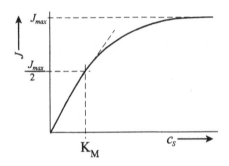

Fig. 5.14. The reaction rate (J) as a function of the substrate concentration (c_S) of an enzymatic reaction with constant enzyme concentration

interactions of enzyme activation or inhibition occur. In the case of allosteric reactions the enzyme contains two binding places, the occupation of which leads to conformational changes in the molecule and alterations of its function. These reactions are governed by equations which are more complicated than the Michaelis-Menten equation.

In Section 5.1.2 we discussed the general behavior of non-linear systems with particular reference to metastable states. We will illustrate this now by using an example of a simple chain of biochemical reactions (Fig. 5.15).

A product (P) is formed by an enzymatic reaction from a substrate (S). The substrate (S) itself is formed by a first order reaction from a substance A. It can, however, break down with the same kind of kinetics to Z. The time dependence of the concentration of the substrate (c_S) can be formulated using Eq. (5.2.1) in the following way:

$$\frac{dc_S}{dt} = J_{AS} - J_{SZ} - J_{SP}. \qquad (5.2.22)$$

Let the flux J_{SP}, which denotes the rate of the enzymatic destruction, be dependent on c_S in a way that corresponds to the Michaelis-Menten equation (substrate inhibition), because of a special kind of regulation. The function $J_{SP}(c_S)$ is illustrated in Fig. 5.15A. The fluxes J_{AS} and J_{SP} shall be determined by

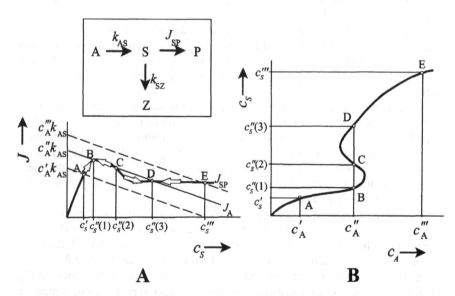

Fig. 5.15. A kinetic model of a simple biochemical reaction chain. **A** Reaction rate (J) as function of the substrate concentration (c_S) for the reaction $S \Rightarrow P$ (J_{SP}), and for the difference of fluxes: $J_A = J_{AS} - J_{SZ}$ at various concentrations of the substance A (c'_A, c''_A, c'''_A). The intersections of the curves (A, B, C, D, E) indicate stationary states. The *arrows* are directed towards stable states. **B** Trajectory of the stationary states of the *left* picture

the linear approaches according to Eq. (5.2.2). In this case Eq. (5.2.22) can be transformed in the following way:

$$\frac{dc_S}{dt} = k_{AS}c_A - k_{SZ}c_S - J_{SP}. \tag{5.2.23}$$

Formally introducing a flux:

$$J_A = J_{AS} - J_{SZ} = k_{AS}c_A - k_{SZ}c_S, \tag{5.2.24}$$

then Eq. (5.2.23) transforms into:

$$\frac{dc_S}{dt} = J_A - J_{SP}. \tag{5.2.25}$$

For the case of a stationary state the following must apply:

$$\frac{dc_S}{dt} = 0 \quad \text{and therefore} \quad J_A = J_{SP}. \tag{5.2.26}$$

The function $J_A(c_S)$, as defined by Eq. (5.2.24), is a straight line with a slope of $-k_{SZ}$, intersecting the ordinate at a point $k_{AS}c_A$ (Fig. 5.15A). The points where the straight line cuts the curve of the function $J_{SP}(c_S)$ denote stationary states according to Eq. (5.2.26).

Let us first consider the case $c_A = c_A''$. This straight line intersects with the function $J_{SP}(c_S)$ at points B, C and D. This means that there are three stationary states for one single value of c_A. However, these points are not equivalent to each other. Only states B and D are stable, C being unstable. This can easily be deduced from the curve.

First consider the point B: a small increase in the substrate concentration will lead to the situation where function $J_{SP} > J_A$. Reference to the balance equation [Eq. (5.2.25)] will show that this will bring about a reduction in c_S. The system therefore is moving towards its initial state. If c_S'' is reduced slightly, then $J_{SP} < J_A$ and this results in an increase in c_S. Small deviations from the stationary value B will thus be corrected. This behavior is indicated by the arrows on the function drawn in Fig. 5.15A.

Near the point C, the system behaves differently. Small reductions in $c_S''(2)$ cause J_{SP} to become larger than J_A. The substrate concentration will continue to fall, until point B is reached. An increase in the value of $c_S''(2)$ by a small value will cause the system to slide in the direction of state D which is stable like the state B.

The stationary states in this system can be summarized by a trajectory. In Fig. 5.15B the function of the substrate concentration (c_S) versus the concentration of the substance A (c_A) is shown. First, the three stationary values for c_A'' shown in Fig. 5.15A are plotted. A change in c_A causes a parallel shift of the curve J_A in Fig. 5.15A (for example the dotted lines). The three stationary values occur only within a relatively narrow range of c_A, or better for the condition: $J_A = c_A k_A$. If all of the stationary states of c_S for different values of c_A are plotted in the coordinates of Fig. 5.15B then an S-shaped trajectory is formed. This corresponds to a section through the behavioral surface of the cusp catastrophe which is shown in Fig. 5.7. This particular example clearly shows that the points on the shaded area of the surface corresponding to the point C in Fig. 5.15, are unstable states.

For the curve Fig. 5.15B the flux J_{SZ} is taken to be constant. Any variation in the rate at which the substrate (S) is used, i.e. a modification of the coefficient k_{SZ} will change the slope of the straight lines representing the function J_A in Fig. 5.15A [according to Eq. (5.2.24)]. It can easily be seen that if the slope of this curve exceeds a critical value, a situation will develop where, independent of c_A, there will only be a single possible intercept. When compared with the functional surface in Fig. 5.7, this situation corresponds to a section through the bifurcation point P. In this instance, k_{SZ} (or J_{SZ}) would have to be plotted on the y-axis of the coordinate.

The analysis of metabolic networks of course leads to systems of many differential equations. The integration of them is complicated, not only because most of them are non-linear, but they are also often quite ambiguous. Many possibly occurring stationary states are probably metastable. With small alterations of the parameters of the system, or by changing the initial conditions, the characters of the stationary states may be modified in a not easily predictable way. Also, chaotic situations are possible. These properties not only make calculations difficult, but may also lead to a higher degree of system destabilization.

In Section 3.1.4 we mentioned the time hierarchy of biological systems. In fact, the rate constants of biologically important reactions are distributed over a brought region of magnitudes. In the context of thermodynamic properties of systems we noted that as a consequence of a steady state ($\sigma > 0$) in the whole system, fast reacting equilibria ($\sigma = 0$) of sub-systems can be established. Considering this time hierarchy from the view point of systems analysis, the following conclusion about the behavior of systems can be drawn from mathematical considerations.

Formally, in a system of many differential equations, simplifications occur because differential quotients of fast processes may approach zero. On the one hand, in a non-equilibrium stationary state, some parts of the system, in fact, can be in equilibrium, because their time constants are quite small. Very slow processes, on the other hand, can be taken as non-changing during the considered time scale. Their parameters, therefore, can be taken as constants. These kinds of simplifications do not only help in mathematical modelling, they automatically reduce the number of stationary states and stabilize the system itself. It seems that the established time hierarchy of biological reactions is the result of biological optimization in the course of evolution.

Further reading: Hayashi and Sakamoto 1986; Heinrich et al. 1987; Heinrich and Schuster 1996; Murray 1993; Palmer 1995; Reich and Sel'kov 1981; Segel 1993

5.2.3
Pharmacokinetic Models

Pharmacokinetics, sometimes also called pharmacodynamics is concerned with the time course of the concentration of drugs in the body or at the sites where they are effective. The aim of such studies is to develop dose and application regimes that will result in an optimal therapeutic effect.

In the simplest situation an amount of an intravenously administered drug is rapidly and homogeneously distributed throughout the whole blood volume and is then excreted or metabolized. The decline in the concentration of the drug in the blood as a first approximation corresponds with the release from a one-compartment system as described in Section 5.2.1. This process can therefore be described by the basic equation [Eq. (5.2.5)] and will follow the time function corresponding to Eq. (5.2.7).

The analogous equation in pharmacokinetics is:

$$\frac{dm}{dt} = -C_1 c, \tag{5.2.27}$$

where m is the total amount of the drug administered, c is its concentration in the blood and C_1 is a rate constant, which in contrast to the parameters k which are used in the previous sections, has the unit $m^3 \ s^{-1}$, or more commonly: $l \ h^{-1}$. This constant is known as the *clearance constant*. Equation (5.2.27) allows us to define this parameter:

$$C_1 = -\frac{dm}{dt} \cdot \frac{1}{c}. \tag{5.2.28}$$

This definition is also valid when the exchange does not follow the kinetics of a one-compartment system. In this case, of course, the system can no longer be described by a simple exponential function like Eq. (5.2.27) but rather by a sum of several exponential terms [Eq. (5.2.11)]. In this case, the clearance constant according to the definition in Eq. (5.2.28) itself becomes time dependent.

In context with the real compartmental model of the body for a given drug the question arises: what is the volume of distribution of this drug in the body? It can be larger than the volume of the blood; as maximum it can include the whole amount of body water. Therefore, the coefficient of distribution of a drug is calculated as the quotient of the real volume of distribution to the body weight, or body volume.

As a first approximation it can be taken that the concentration of the drug at the receptor side is the same as its concentration in the blood. The amount of drug in the blood which therefore is able to arrive at the receptor side, when expressed as a percentage, is called *bioavailability*. In the case of intravenously administered drugs this has a value of 100%. If the drug is administered orally, the bioavailability is lower and depends on its resorption.

Usually, pharmacokinetics studies are based on linear models. In some instances a concept, known as *capacity-limited elimination* is used. This is the excretion of a drug using saturation kinetics. For the direct binding of the drug to a receptor, non-linear approaches should be used similar to the Michaelis-Menten equation [Eq. (5.2.1)].

In general, however, the linear approaches have proved useful for the overall kinetic analysis of the distribution of drugs in the body. In Section 5.2.1 we depicted a three-compartment model (Fig. 5.11D) which can be regarded as the basic pharmacokinetic model. It can be described by the system of differential equations mentioned there [Eq. (5.2.3)]. In the simplest case it contains the following processes: the incorporation of the drug into the blood stream (compartment 1), its uptake into the tissue (compartment 2) and its excretion by the kidneys (compartment 3). The kinetics of this system depends on the mode of drug administration.

Figure 5.16 shows the concentration curves for different parts of the body following a single oral dose of a drug. It can be seen that the concentration reaches a maximum in the blood and subsequently in the tissues. These curves were obtained by computer simulation of the process. This is a method by which the most promising modes of administration for maintaining the required concentrations of a drug at the desired point of action, for a sufficient time can be determined. It is obvious that repeated oral administration will result in the concentration in the blood showing a series of maxima. The maximum concentration and the subsequent time course will depend on the dose and frequency of administration.

5.3
Model Approaches to Some Complex Biological Processes

In this section biophysical models of processes will be introduced that are among the most complex of their kind. Although rapid progress in this field

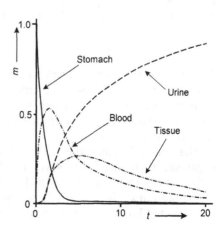

Fig. 5.16. Time course of drug concentration in basic pharmacokinetic model. The normalized parameters according to the system in Fig. 5.11D are: $k_{01} = 1$, $k_{12} = k_{21} = k_{23} = 0.25$. (Data from Knorre 1981)

has been achieved in recent years, not least because of the stormy development of the theory of non-linear systems, such theories have remained at the introductory state. The explanations given in this section should help readers to understand recent approaches and hypotheses.

5.3.1
Models of Propagation and Ecological Interactions

The kinetic models of growth and interactions in biological populations are based on the same kind of mathematical constructions as those of metabolism and transport processes, discussed in the previous section. Let us consider first a population of organisms of a single species in an infinitely large milieu. Let the number of individuals be n, then any changes of this number can be described by the differential quotient dn/dt. Because the external milieu has been taken as infinite, the physiological conditions can be regarded as constant, and thus not influenced by the organisms themselves.

In this case the propagation rate as well as the mortality rate of the organisms can be expressed by time constants which are determined both genetically and physiologically. The rate constant of propagation (k_p) and that of mortality (k_m) can be combined as follows:

$$k = k_p - k_m \quad (k_p, k_m > 0). \tag{5.3.1}$$

In this way the proliferation of a population can be described by the following differential equation:

$$\frac{dn}{dt} = kn. \tag{5.3.2}$$

This expression shows in fact that both the rate of proliferation as well as the rate of mortality are proportional to the actual number of individuals in the population. Considering that at $t = 0$ the initial number of individuals is n_0, then the integration of Eq. (5.3.2) leads to the following equation:

$$n = n_0 e^{kt} \tag{5.3.3}$$

This law of exponential growth was discovered as early as 1798 by the English economist Thomas Robert Malthus. If the propagation coefficient (k_p) preponderates in Eq. (5.3.1), then $k > 0$ and the number of individuals will increase towards infinity. The time course of this function is depicted in Fig. 5.17 (dashed line).

The original assumption that the external milieu is not affected by the organisms themselves is, however, an ideal condition. Only when the number of cells in the suspension is very small, and the time scale is sufficiently short, can this condition apply. In real cell suspensions, this period of growth corresponds to the initial logarithmic period of growth.

Fig. 5.17. The function $n(t)$ according to the Verhulst-Pearl equation [Eq. (5.3.6)] ———, and the Malthus equation [Eq. (5.3.3)] - - - -. Using arbitrary values in ordinate and abscissa, the parameteres are: $n_0 = 0.1$; $k_1 = 1$; $k_2 = 0.14$. For the Malthus curve, $k = k_1$

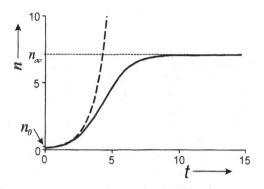

In most cases a mutual influence of the individuals on one another, for example via competition for food, is to be expected. This circumstance can be taken into account by modifying Eq. (5.3.1). It may be postulated that k_p and k_m are not constants but themselves functions of the population density. So, for example, one can assume that the rate constant of proliferation (k_p) will decrease and the rate constant of mortality (k_m) will increase, when n increases, because of shortage of food. The degree of physiological variability of these parameters, however, will be limited for genetic reasons. The simplest approach is the following linear equation:

$$k = k_1 - k_2 n \quad \text{for: } k_1, k_2 \geq 0. \tag{5.3.4}$$

This takes into account an optimal propagation constant k_1, determined genetically, in combination with a further constant k_2 leading to a diminution of k_1 with an increase of the population density (n). If $n \Rightarrow 0$ then $k_1 \Rightarrow k$. Introducing Eq. (5.3.4) into Eq. (5.3.2) one obtains:

$$\frac{dn}{dt} = (k_1 - k_2 n)n = k_1 n - k_2 n^2. \tag{5.3.5}$$

This approach therefore leads to a first order non-linear differential equation. The integration of this equation for the same initial conditions as above ($t = 0$; $n = n_0$) gives an equation which in mathematics is known as the *logistic function*. In population kinetics it is called the *Verhulst-Pearl law*:

$$n = \frac{n_0 k_1 e^{k_1 t}}{k_1 + k_2 n_0 (e^{k_1 t} - 1)}. \tag{5.3.6}$$

In contrast to Eq. (5.3.3), in this function n does not approach infinity at $t \Rightarrow \infty$ but rather tends towards a saturation value (n_∞). This saturation value can easily be determined from Eq. (5.3.6) considering that with increasing t the exponential terms increase up to a value where neither the subtraction of 1, nor the addition of k_1 become important:

$$\lim_{t \to \infty} n = n_\infty = \frac{k_1}{k_2}. \tag{5.3.7}$$

Furthermore, it is easy to show that for small values of t, the Eq. (5.3.6) takes the form of Eq. (5.3.3). Both functions for some arbitrarily chosen parameters are illustrated in Fig. 5.17. An example of experimentally obtained data is shown in Fig. 5.18 (above).

If the system contains individuals of several different species then the inter relation between them must be taken into consideration. This can involve competition for the food supply (Fig. 5.18, below) or a hunter-prey relationship (Fig. 5.19). Systems like this were studied in the 1930s by Volterra and Lotka.

The following set of equations describes the simplest case of a hunter-prey relation between two species:

$$\left.\begin{aligned} \frac{dn_P}{dt} &= (k_{pP} - k_{mP}n_H)n_P \\ \frac{dn_H}{dt} &= (k_{pH}n_P - k_{mH})n_H \end{aligned}\right\}. \tag{5.3.8}$$

The subscripts p and m have the same meaning as in Eq. (5.3.1) and the index P in these equations refers to "prey", whereas H means "hunter". For the prey

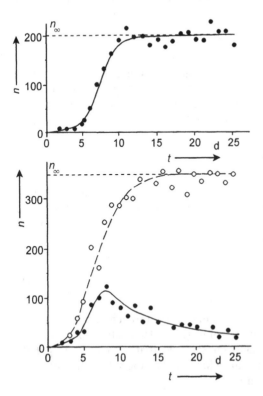

Fig. 5.18. The time dependence of the growth of a pure population of *Paramecium caudatum* (•——• above) and mixed together with a population of *Paramecium aurelia* (o- - -o). The ordinate indicates the number of individuals (n) per 0.5 ml medium. The abscissa gives the time in days (d). (Data from Gause 1935)

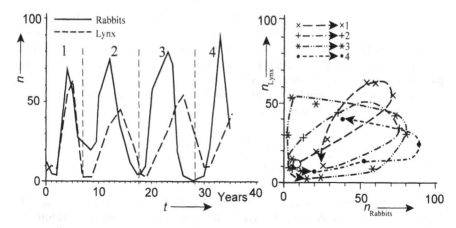

Fig. 5.19. A particular example of population kinetics: the number of pelts, obtained from hunted lynx and snow rabbits in Canada after 1900 (numbers in thousands). On the *left*: time course; on the *right*: the same numbers in a Volterra-plot. The population waves 1–4 of the *left* part of this figure appear as Volterra cycles in the graph on the *right*, starting with the point ○. (Data from Haken 1983)

organisms the mortality ($k_{mP}n_H$) will increase as the number of hunters (n_H) increases. The greater the number of hunters, the shorter the life expectancy of the prey organisms will be. Looked at the other way round, an increasing number of prey organisms means better living conditions for the hunters. For this reason the propagation coefficient of the hunters (k_{pH}) is multiplied by the number of prey organisms (n_P) in the second equation. Removing the brackets shows that these equations resemble those for bimolecular reactions [Eq. (5.2.14)]. In fact, the present "reaction" also depends on the coming together of the two "reaction partners".

This set of Eqs. (5.3.8) represents an extremely simplified situation. The interaction of individuals of the same species, for example, which was already reflected in Eq. (5.3.5), has been ignored here. This example, however, demonstrates that even the simplest ecological models already lead to complicated systems of non-linear equations. The solution of such systems of equations frequently leads to functions indicating oscillations. This corresponds with experiences where ecological systems oscillate even without additional input of external signals, such as daily or annual oscillations of light and temperature.

These oscillations can either be plotted against time, or represented as trajectories. In the case of Eq. (5.3.8), for example, the functions $n_P(t)$, and $n_H(t)$ can be plotted separately, or combined as the trajectory $n_P(n_H)$. Again, a system trajectory would appear like those shown in Figs. 5.5, 5.6, 5.7 or 5.15B. In the case of periodic oscillations closed curves occur, known as *Volterra cycles*.

Figure 5.19 demonstrates a concrete example of such oscillations. It uses data from the hunting lists of the Hudson-Bay Company in Canada which were kept from 1845 to 1935. Suppose the numbers of hunted animals were proportional to the number of those really existing, one can roughly use this as a representation of the ecological situation. In Fig. 5.19 the data from 1900–1935 were used indicating the time course with typical oscillations (left) and the same data in a parameter plot showing Volterra cycles (right).

In a similar way to the hunter-prey system, a kind of symbiotic interrelation of two species can be formulated:

$$\left.\begin{array}{l} \frac{dn_A}{dt} = (k'_{pA} - k''_{pA}n_B - k_{mA})n_A \\ \frac{dn_B}{dt} = (k'_{pB} + k''_{pB}n_A - k_{mB})n_B \end{array}\right\}. \tag{5.3.9}$$

This set of equations takes into account the fact that when $k'_{pA} > k_{mA}$ and $k'_{pB} > k_{mB}$ then one species of this symbiotic community can survive without the other. The symbiotic cooperation is expressed by the product terms $k''_{pA}n_B$ and $k''_{pB}n_A$.

Mathematical approaches of this kind form the basis for modelling of ecological systems. However, other factors must also be considered such as a large number of trophic levels and complex interrelations, time delays caused by processes of developments, annual rhythms, and also ethological and socio-logical structures.

Further reading: Ginzburg and Golenberg 1985; Jeffries 1989; Hallam and Levin 1986; Logofet 1993; Renshaw1990

5.3.2
Models of Growth and Differentiation

The morphology of an organism, its geometric structure and the corresponding morphogenetic mechanisms have provided, time and again, subjects for modelling. In 1917, in his well-known book *On Growth and Form*, D'Arcy Thompson first formulated the mathematical ideas of morphometry. His theory states that different types of body shape, when distorted by geometric projections, can be transformed from one to another. If the shape of an animal or of a part of an animal body is projected onto a system of coordinates, then by means of the mathematically defined distortion, i.e. by a sort of coordinate transformation, one shape can be systematically transformed to another. This method of D'Arcy Thompson has been demonstrated for the shapes of various species of fishes as well as the shape of a crab's, carapace, a mammalian skull and even leaves of plants and many other structures. In this way, for example, phylogenetic changes can be expressed mathematically. Such approaches belong more to the realms of biometry but they do, however, form a starting point for more extensive biophysical modelling.

In the same way, *allometric models* of growth primarily have a purely phenomenological importance. The basic principle of allometry is that both morphological and functional parameters of similar animals of different sizes can be represented by exponential functions of the body mass. Such functions have been empirically determined (see Fig. 3.64).

The basal metabolic rate (J'_Q) of animals (in $J s^{-1}$) can be obtained from the following equation where it is given as a function of the body mass (m, in kg):

$$J'_Q = 3.41 \cdot m^{0.743}. \tag{5.3.10}$$

In this way functions for many parameters of living systems have been formulated based on empirically obtained data. Although such equations rarely explain the reasons for the phenomenon under consideration they do have their practical use. In sport medicine, such relations can be applied to compare the performance parameters of subjects of different sizes; similar information is also useful in anthropology and pharmacology. Furthermore, phenomenological growth curves are important for calculating economic aspects of agriculture.

Biophysical models of morphogenetic processes are based on the following premise. The differentiation of embryonic cells is controlled by growth regulating substances which can be either growth-stimulating, or growth-inhibiting. Such substances, or growth factors, have already been found in many experiments. They are also indicated in classical ligation experiments carried out on embryos and larval stages on animals. It can be assumed that such substances control the morphogenetic movements of cells as well as the activation of the genes that are responsible for differentiation. In this context reference must be made to the discussions on the possible role of electric fields in embryogenesis, leading to electrodiffusion of growth factors (Sect. 3.5.2).

As usual, the models start with the simplest approach. It is supposed that these substances will be formed at particular sites and then distributed by diffusion. In this way complex concentration patterns, with a space-time structure, are set up. If the growth rate, or the mode of differentiation depends on the concentration ratio of these factors, such concentration patterns finally bring about a morphologically visible structure.

There are many ways in which such processes can be modelled. Consider the equation for a simple unidimensional case where the concentration of an activator (c_A) and an inhibitor (c_I) are distributed only in the x-direction. For this case the following set of equations can be formulated:

$$\left. \begin{array}{l} \frac{\partial c_A}{\partial t} = k_A \frac{c_A^2}{c_I} - k'_A c_A + D_A \frac{\partial^2 c_A}{\partial x^2} \\ \frac{\partial c_I}{\partial t} = k_I c_I^2 - k'_I c_I + D_I \frac{\partial^2 c_I}{\partial x^2} \end{array} \right\}. \tag{5.3.11}$$

This is a system of partial, non-linear differential equations, because the variables c_A and c_I are differentiated both with respect to time and position. These equations correspond in principle to the scheme of the flux balance given in Eq. (5.2.1).

In particular, the following assumptions are made in this model. The rate of formation of the activating factor A is determined by a process that is autocatalytically promoted by c_A. This corresponds to the term: $k_A(c_A^2/c_I)$. The inactivation of the activator is expressed by the simple linear term $-k'_A c_A$. The diffusion along the x-axis $D_A(\partial^2 c_A/\partial x^2)$ is determined by the second Ficks law [Eq. (3.3.8)]. The differential equation for the concentration of the inhibitor (I) differs from that for the activator (A) only in the fact that there is not an inhibition but an autocatalytic activation ($k_I c_I^2$).

The behavior of such a system depends to a great extent on the choice of the constants. The relation of the diffusion coefficients D_A and D_I is of particular importance. If the diffusion constant for the inhibitor (D_I) is much greater than that for the activator (D_A) then the distribution of the factor I can be regarded as being homogeneous.

Now consider the initial state of a likewise homogeneously distributed activator A (Fig. 5.20A). If, in such a system, the concentration of A is raised for a short period (R_1) equally through the whole system, the system can compensate this and will attain the previous state. If, however, the excitation is local (R_2), it may become unstable and take on a new quality. A concentration

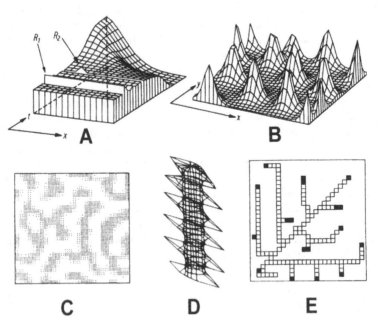

Fig. 5.20. Examples of biophysical models of morphogenesis. **A** Time course of distribution of an activator in a uni-dimensional model (R_1 and R_2 two kinds of excitation); **B** Formation of a 2-dimensional pattern of the activator concentration; **C** Development of a 2-dimensional zebra-pattern, corresponding to a model with stochastic elements; **D** An axi-symmetrical model with local concentrations of an activator, resembling a botanical system; **E** A model representing a bifurcation of vessels; After Meinhardt 1982. **C** Was kindly placed at our disposal by H. Meinhardt

gradient of the activating substance A is set up. Occasionally, local fluctuations are sufficient to induce such changes.

A great number of observed morphogenetic phenomena can be explained on the basis of similar equations. There are systems of equations and particular combinations of parameters that show oscillations with respect to both time and space. If the models are not simply based on a 1-dimensional approach as in the set of Eqs. (5.3.11), but consider diffusion in two or three dimensions then concentration patterns could be modelled on a plane or in a space.

Figure 5.20B shows the genesis of a stationary concentration pattern in a plane of x, y-coordinates. Such a distribution of activators and inhibitors, for example, could result in a distribution pattern of guard cells in the epidermis of a leaf, or distribution of hairs, pigment areas etc. Axial symmetric models may explain the arrangement of leaf nodes on the stem of a plant (Fig. 5.20D). If a term is included in the equation to arrange the mutual catalytic effects of neighbor cells then a model for the development of branching systems of vessels is obtained (Fig. 5.20E). In agreement with the actual situation, the system constants in these cases only determine the principle of branching, whereas the specific configuration of the network is a product of chance.

At the present time, these models still have a predominantly hypothetical character. Nevertheless, there is an increasing number of correlations with the results of experiments in developmental physiology.

Further reading: Britten 1986; Goel 1988; Kaandorp 1994; Meinhardt 1982, 1995

5.3.3
Models of Evolution

The work of Manfred Eigen (1971) stimulated comprehensive investigations on biological evolution on the basis of the theory of non-linear systems. The evolution of life from inorganic matter to its present form can be divided into three, qualitatively different phases:

- *Chemical evolution* – the path from inorganic molecules to the low molecular weight components of biopolymers.
- *Self-organization of biomacromolecules* and the formation of cell-like structures.
- *Evolution of organisms.*

The decisive step in Eigen's theory was his discovery of laws governing pre-biotic evolution. This means that even at the level of simple proteins and nuclear acids, mechanisms of competition and selection occur that reduce the degree of freedom of stochastic combinations. Earlier calculations had indicated that there was only a minute probability that a functional biopolymer could be formed from the chance combination of monomers (Albert Einstein had already compared this probability with the chance of a complete dictionary resulting from an explosion in a printing office!).

A relatively small biopolymer with an information content of 10^3 bits would result in this way, on average, once in the course of $2^{1000} \approx 10^{300}$ chance combinations. This means that out of 10^{300} resulting molecules, one could be considered as being able to initiate life. Such an incredibly large number might be illustrated as follows: there is enough space in the hydrosphere of the earth to contain about 10^{32} molecules, and about 10^{100} molecules can be accommodated in the entire universe. The earth is "only" 10^{17} s old so, even if the reaction of combination and re-combination were very fast, the possibility of a successful combination is minute (such stochastic considerations were also discussed in Sects. 2.3.2 and 2.3.3).

In fact, it must be assumed that already at the beginning of evolution the principles of the struggle for life controlled the development of particular chains of reactions. The formation of a functional sequence of proteins, therefore, is as unlikely to be a product of chance, as is the functional morphology of an organ according to the principles of Charles Darwin.

These principles of molecular selection can be ascribed to processes of catalysis and auto-catalysis. Auto-catalysis can be regarded in this context as a fore-runner of self-reproduction.

The number of monomers in a given region is limited and this will lead to a form of competition between the different kinds of polymerization reactions. However, such a system would rapidly reach an equilibrium state if it were not exposed to a continuous flow of energy and entropy. It will be continually disturbed by geophysical effects and from the thermodynamic point of view, it is a kind of dissipative structure.

Under these conditions, selection of polymers with high catalytic activity will take place. A network of catalytic relations between the individual molecules is built up (Fig. 5.21). If a closed chain occurs in such a network this will form an auto-catalytic cycle. This is a structure with a good chance for survival. Such cycles sustain themselves and in doing so can change their own structure. Side chains, which are formed through the poly-catalytic activity of several molecules are removed and distensible intermediate linkages are eliminated. In this way the cycle becomes smaller. At the same time, it can be shown that heterogeneous cycles, made up of alternating proteins and nucleic acids possess a selection advantage. Eigen called such structures *hypercycles*.

This concept of evolution therefore possesses a dialectic unity of stochastic and deterministic processes. The stochastic component is derived from the stability of bonds and from the stochastics of molecular contacts in the course of the reaction. At higher levels of development, this stochastic element is realized by the mutation rate or the dependability of self-reproduction.

This theory is backed up by mathematical modelling and computer simulations. It is possible to formulate equations for systems of chemical reactions as well as for processes of reproduction and selection at the highest levels of evolution. With regard to the equations that were given in the previous sections, the following simplified approach shall be demonstrated:

Fig. 5.21. The network of catalytic correlations in a mixture of polymers. The points with numbers connected by *thick arrows* indicate an auto-catalytic cycle

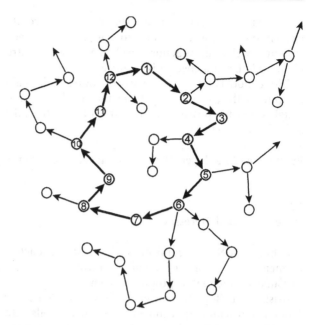

$$\frac{dn_i}{dt} = \left(k_{pi}q_i - k_{mi} - k_i'\right)n_i + \sum_{j\neq i} k_{m,ji}n_j. \tag{5.3.12}$$

This equation for the time course of the number of individuals (n_i) of a species i embodies the following postulates: the individuals propagate at a rate that bears a linear relation to the existing number of individuals and is determined by a time factor k_{pi}. This coefficient is, however, reduced by the factor q_i being a little less than one. In this way mutations which disturb the identical reproduction of individuals are taken into account. The coefficient governing the mortality (k_{mi}) is the same as in Eq. (5.3.1). In addition, another coefficient, k_i' is introduced, expressing the possibility that some individuals will leave the area or will diffuse away. The summing term at the end of the equation reflects the possibility of the emergence of an individual of the species i from another of the species j by mutation. In contrast to the ecological equations in Section 5.3.1, n_i is thus capable of increasing even under the initial conditions, when $t = 0$ and $n_i = 0$. This sum term is a prerequisite for the neogenesis of a species.

This is of course a very primitive approach. It presupposes that propagation is based on linear relations, an assumption having limited validity as has already been discussed in Section 5.3.1. The relations between species have not been considered either. Nevertheless, evolution because of competition between species can be demonstrated on the basis of this approach if the following mass balance is used:

$$\sum_{i=1}^{m} n_i = \text{const} \quad \text{therefore:} \quad \sum_{i=1}^{m} \frac{dn_i}{dt} = 0. \tag{5.3.13}$$

The advantage for selection of a species will then result from the relations of the chosen system constants, one to another. With more complex evolutionary systems, weighing functions will have to be introduced but these must necessarily be of a speculative nature.

These models for evolution are, of course, still of a highly speculative nature. In spite of this, it is due to them that we have some important fundamental insights into the possible mechanisms of evolution, the highest form of motion in biological systems.

Further reading: Eigen 1971, 1992; Haken 1983; Hoppe et al. 1982; Rowe 1994

5.3.4
Models of Neural Processes

The first models of neural networks used approaches of Boolean algebra, the mathematical theory of binary control elements. As early as the 1940s, McCulloch and Pitts proposed such models. They investigated networks consisting of "formal neurons". These are elements performing basic logical functions, resembling in some instance nerve cells with synaptic connections. It could be proved that these kinds of networks are able to perform all operations of Boolean algebra.

Such a formal neuron can accept several digital input signals S_i and transform them into an output signal S_o. Figure 5.22 shows this process schematically. The signals may obtain the values: 0 = "non excited" and 1 = "excited". A weighing factor k_i determines whether the influence is stimulating (+1) or whether it is inhibiting (−1). In the neuron these signals are summarized to get a parameter h. If $h \geq h^*$ this neuron produces an output signal which again has the value 1.

This basic model has been modified in subsequent years. So, for example, the refractory period of nerves has been included. This means that for some milliseconds after an input signal the nerve is not able to accept a further one. This requires a synchronization of all impulses in such a network, and the introduction of a certain sequence of operation. Despite the fact that in the brain neurons actually exist which produce such a sequence, for example cortical cell columns, there is no evidence of an overall synchronization in the brain.

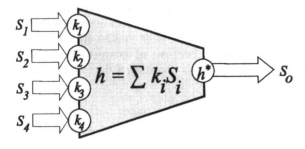

Fig. 5.22. Schematic presentation of the function of a formal neuron according to the theory of McCulloch and Pitts (explanation in text)

Meanwhile, so-called integrate and fire models have been developed to overcome this problem.

Another problem concerns the huge number of possible connections in networks containing a large number of neurons. Even in the case of a ganglion of a nematode with only 300 neurons and about 5000 synapses, the number of possible kinds of interconnections is enormous. The human brain contains about 10^{10} neurons, each of them being connected with others by about 10^4 synapses. The total number of synapses in the brain, therefore, is about 10^{14}. The number of possible permutations of these connections is unimaginably large.

These numbers lead to the following conclusions: a construction of a neuronal network as a copy of the human brain is neither possible nor would it be useful. Furthermore, it is impossible that the pattern of synaptic connections in the brain has been determined genetically. The amount of information which would be required by this exceeds by far the storage capacity of the genome. This means that the problem of neuronal networks is not only a question of its function but additionally of its self organization.

This problem of the immense number of possible realizations, is something which the theory of neuronal networks has in common with other fields of molecular physics. So, for example, approaches of statistical thermodynamics can be used as well as the cooperative behavior of molecules in amorphous materials. In this context, for example, the theory of spin-glasses is of interest.

Spin-glasses are viscous materials containing particles of different magnetic properties. These include ferromagnetic particles, orientating themselves by mutual influences parallel to each other, as well as anti-ferromagnetic particles, orientating anti-parallel (Fig. 5.23). The analogy between spin-glasses and neuronal networks is the direction of the particle and the state of the neuron as well as their mutual interactions. A parallel or anti-parallel orientation of a particle in the spin-glass would correspond to a "silent" or an "excited" cell in a neural network. The ferromagnetic or anti-ferromagnetic interaction of the particles simulates the interaction of the nerve cells.

In this respect, spin-glasses may show a quite complicated pattern of self-organization. Of particular interest are the situations where two opposite kinds of interaction come together at one point. In this case a particle will be influenced at the same time by two forces tending to orientate it in opposite ways. This disturbs the stability of the whole system. This situation is called a *frustrated* state. It can be considered as a sort of metastability and could reflect a kind of memory state of a neuronal network. This is special kind of phase transition which cannot be described like a catastrophe as depicted in Fig. 5.7, but instead shows a more continuous kind of transition. These spin-glass models therefore resemble not only the problem of self-organization but also the function of neuronal networks directly.

The theories of neuronal networks in general are orientated in two directions: on the one hand, neurobiologists are interested in understanding processes of neuronal data processing and memory in the brain, on the other hand there is a branch of modern computer technology with the aim of producing self-organizing systems with so-called artificial intelligence.

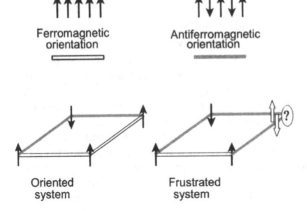

Fig. 5.23. Ferromagnetic and anti-ferromagnetic interactions in spin-glasses and the occurrence of frustrated systems

It is interesting to note that many years ago computer technology started with basic elements resembling a nerve system, like the formal neurons of Fig. 5.22, but so far they are a long way from modelling the living brain. At best, their results are constructions showing some formal similarities with the function of the central nervous system. Despite the importance of the theory of neuronal networks for computer development, it has now become an area which is more or less moving away from neurobiology and biophysics. In some respects research involving big digital networks will support the understanding of neuronal systems, even if there are fundamental differences in the properties of their elements.

The most important difference between brain and computer of course are the time constants of their corresponding elements. In contrast to computers, the elements of which react faster than 10^{-9} s, a nerve cell needs 10^{-3} s for an action potential (Fig. 3.27). Furthermore, the rate of impulse conduction in electronics exceeds that of nerves by about 9 orders of magnitude. The property of the brain to fulfill complicated operations in a short time using these very slowly reacting neurons makes it impossible to explain its function by use of principles of conventional computers.

Furthermore, there are some other properties of the nervous system which underline how they differ from neural networks of the McCulloch type:

- The brain contains neurons and synapses of quite different shapes, properties and functions.
- The neurons in the brain are not connected in a predicted way but connect and disconnect themselves in a process of self-organization controlled by its own function.
- In contrast to computers, the processors, data-stores, data-busses etc. in the brain are not separate elements.
- The loss of neurons which begins at birth in humans is a normal process and does not affect brain function. This process at the beginning of life even runs

parallel with the development of human mental capacity. Even dramatic kinds of brain injury as well as the removal of parts of the brain in some operations mostly do not affect the function of the brain significantly.

In context with these properties, the function of the human brain can be characterized in the terminology of computer science as follows. It is a system

- With parallel information processing,
- With a program hierarchy,
- With self-organizing software as well as hardware (in contrast to computers, hardware and software in the brain cannot be distinguished),
- With non-linear elements, and
- With stochastic properties in some functions.

A number of network theories have already taken these peculiarities into account.

A special problem concerns the large memory capacity of the brain, connected with extremely fast access to stored information. Obviously the memory is not based only on one single mechanism. The discovery of the holographic principle by T. Gabor (1948) and its realization after the development of laser technology raises the idea that the memory in the brain could probably be based on a holographic principle. This has been supported by the fact that a local loss of neurons does not affect the memory, as would be the case in the information storage of technical computers. If information is stored in a holographic way, the loss of a part of the storage elements would only lead to a loss of the resolution of the stored picture, but not to its loss. This is really an intriguing idea, but there is no experimental evidence for this yet, and no proposals for a molecular mechanism.

These considerations show that despite intensive research and some concrete results which are used successfully in computer technology, our models of functions of the central nervous systems are still quite crude.

Further reading: Arbib and Amari 1989; Hertz 1991; Peretto 1992

References

Able, K. P. (1994) Magnetic orientation and magnetoreception in birds. Progress in Neurobiology 42: 449

Alexander, R. McN. (1983) Animal Mechanics. (2nd edn.) Blackwell Scientific Publ., Oxford

Amman, D. (1986) Ion-selective Microelectrodes. Springer, Berlin

Anderson, D. H. (1983) Compartmental modeling and tracer kinetics. Lecture Notes in Biomathematics 50. Springer, Berlin

Arbib, M. A., and S. Amari (1989) Dynamic Interactions in Neural Networks: Models and Data. Research Notes in Neural Computing, Vol. 1, Springer, Berlin

Atema, J., F. R. Fay, A. H. Popper, and W. N. Tavolja (Eds.) (1987) Sensory Biology of Aquatic Animals. Springer, Berlin

Au, W. W. L. (1994) Sonar of Dolphins. Springer, New York

Azuma, A. (1992) The Biokinetics of Flying and Swimming. Springer, Tokyo

Barber, J. (1992) The Photosynthesis: Structure, Function and Molecular Biology. Elsevier, Amsterdam

Barry, P. H. (1998) Derivation of unstirred-layer transport number equations from the Nernst-Planck flux equations. Biophys J 74: 2903

Bashford, C. L., and C. A. Pasternak (1986) Plasma membrane potential of some animal cells is generated by ion pumping, not by ion gradients. Trends in Biochemical Sciences 11: 113

Batenburg, J. J., and H. P. Haagsman (1998) The lipids of pulmonary surfactant: Dynamics and interactions with proteins. Prog Lipid Res 37: 235

Bauer, J. (Ed.) (1994) Cell Electrophoresis. CRC Inc., Boca Raton

Benz, R., and F. Conti (1986) Effects of hydrostatic pressure on lipid bilayer membranes. Biophys J 50: 91, 99

Bereiter-Hahn, J., O. R. Anderson, and W.-E. Reif (Eds.) (1987) Cytomechanics. The Mechanical Basis of Cell Form and Structure. Springer, Berlin

Bernhardt, I. (1994) Alteration of cellular features after exposure of low ionic strength medium. In: Bauer, J. (Ed.) Cell Electrophoresis. CRC Inc., Boca Raton, 163

Bernhardt, J. H. (1988) The establishment of frequency dependent limits for electric and magnetic fields and evaluation on indirect effects. Radiat Environ Biophys 27: 1

Bernhardt, J. H., R. Matthes, and M. H. Repacholi (Eds.) (1997) Non-Thermal Effects of Rf Electromagnetic Fields. International Commission on Non-Ionizing Radiation Protection 3: ISBN 3-9804789-2-0

Bertalanffy, L. von (1968) General System Theory, Braziller, New York

Biewener, A. A. (1992) Biomechanics. Structures and Systems – a Practical Approach. IRL Press, Oxford

Bingeli, R., and R. Weinstein (1986) Membrane potentials and sodium channels: Hypotheses for growth regulation and cancer formation based on changes in sodium channels and gap junctions. J Theoret Biol 123: 377

Blank, M., and L. Soo (1992) Temperature dependence of electric field effects on Na, K-ATPase. Bioelectrochem. Bioenergetics 28: 291

Bloom, M., E. Evans, and O. G. Mouritsen (1991) Physical properties of the fluid lipid-bilayer component of cell membranes: A perspective. Quart Rev Biophys 24: 293

Blumenfeld, L. A., and A. N. Tikhonov (1994) Biophysical Thermodynamics of Intracellular Processes. Springer, Berlin

Britton, N. F. (1986) Reaction-diffusion Equations and Their Applications to Biology. Academic Press, London

Bucala, R. (1996) Lipid and lipoprotein oxidation: Basic mechanisms and unresolved questions in vivo. Redox Report 2: 291

Buettner, G. R. (1993) The pecking order of free radicals and antioxidants: Lipid peroxidation, α-tocopherol, and ascorbate. Archiv Biochem Biophys 300: 535

Cantor, Ch. R., and P. R. Schimmel (1980) Biophysical Chemistry III. The Behavior of Biological Macromolecules. W. H. Freeman and Comp., New York

Cevc, G. (1990) Membrane electrostatics. Biochim Biophys Acta 1031: 311

Charry, J. M., and R. Kavet (1987) Air Ions: Physical and Biological Aspects. CRC Inc., Boca Raton

Christen, H. R. (1968) Grundlagen der allgemeinen und anorganischen Chemie. Diesterweg, Frankfurt

Collins, K. D., and M. W. Washabaugh (1985) The Hofmeister effect and the behaviour of water at interfaces. Quat Rev Biophys 18: 323

Cook, J. S. (1967) Nonsolvent water in human erythrocytes. J Gen Physiol 50: 1311

Cooper, A. (1976) Thermodynamic fluctuations in protein molecules. Proc Natl Acad Sci USA 73: 2740

Creighton, Th. E. (Ed.) (1992) Protein Folding. W. H. Freeman and Company, New York

Deisenhofer, J., and J. R. Norris (Eds.) (1991) The Photosynthetic Reaction Center, Vol. 1, 2, Academic Press, New York

deNoordhout, A. M. (1998) Cortical magnetic stimulation. Clin Neurophys 28: 9

Dertinger, H., and H. Jung (1969) Molekulare Strahlenbiologie. Springer, Berlin

Dessauer, F. (1964) Quantenbiologie. Springer, Berlin

Dieckmann, S. (1984) Einwirkung mechanischer Schwingungen (Vibrationen) auf den Menschen. In: Arbeitsmedizin aktuell, ein Loseblattwerk für die Praxis, Vol. 14, Gustav Fischer Verlag, Stuttgart

Donath, E., and A. Voigt (1986) Electrophoretic mobility of human erythrocytes. Theory and experimental applicability. Biophys J 49: 493

Doucet, P., and P. B. Sloep (1992) Mathematical Modeling in the Life Sciences. Ellis Horwood, New York

Dunin-Borkowski, R. E., M. R. McCartney, R. B. Frankel, D. A. Bazylinski, M. Posfai, and P. R. Buseck (1998) Magnetic microstructure of magnetotactic bacteria by electron holography. Science 282: 1868

Eigen, M. (1971) Selforganization of matter and the evolution of biological macromolecules. Naturwiss 58: 465

Eigen, M. (1992) Steps Toward Life. Oxford University Press, Oxford, England

Evtodienko, V. Y., Y. N. Antonenko, and L. S. Yaguzhinsky (1998) Increase of local hydrogen ion gradient near bilayer lipid membrane under the conditions of catalysis of proton transfer across the interface. FEBS Letters 425: 222

Farhataziz and Rodgers, M. A. J. (Eds.) (1987) Radiation Chemistry – Principles and Applications. Verlag Chemie, Weinheim

Ferreira, H. G., and M. W. Marshall (1985) Biophysical Basis of Excitability. Cambridge University Press, Cambridge

Fleischer, S., and B. Fleischer (Eds.) (1989) Transport Theory: Cells and Model Membranes. Methods in Enzymology, Vol. 171, Academic Press, San Diego

Franks, F. (1985) Biophysics and Biochemistry at Low Temperatures. Cambridge University Press, Cambridge

Friedman, M. H. (1986) Principles and Models of Biological Transport. Springer, Berlin

Fritz, M. (1998) Three-dimensional biomechanical model for simulating the response of the human body to vibration stress. Med Biol Eng Comput 36: 686

Freedman, J. C., and J. F. Hoffman (1979) Ionic and osmotic equilibria of human red blood cells treated with nystatin. J Gen Physiol 74: 157

Fuhr, G., and R. Hagedorn (1996) Cell Electrorotation. In: Lynch, P. T., and M. R. Davey (Eds.): Electrical Manipulation of Cells. Chapman & Hall, New York: 38

Fuhr, G., U. Zimmermann, and S. G. Shirley (1996) Cell motion in time-varying fields: principles and potential. In: Zimmermann, U., and G. A. Neil (Eds.): Electromanipulation of Cells. CRC Inc., Boca Raton: 259

Fukada, E. (1983) Piezoelectric properties of biological polymers. Quart Rev Biophys 16: 59

Fung, Y. C. (1984) Biodynamics. Springer, New York

Fung, Y. C. (1993) Biomechanics (2nd edn.) Springer, New York

Gabelnick, H. L., and M. Litt (1973) Rheology of biological systems. Charles C. Thomas, Springfield

Gabriel, C., S. Gabriel, and E. Corthout (1996) The dielectric-properties of biological tissues. I. Literature survey. Phys Med Biol 41: 2231

Galla, H.-J., N. Bourdos, A. von Nahmen, M. Amrein, and M. Sieber (1998) The role of pulmonary surfactant protein C during the breathing cycle. Thin Solid Films 327: 632

Gause, G. F. (1935) Experimentelle Untersuchungen über die Konkurrenz zwischen Paramecium caudatum und Paramecium aurelia. Archiv Protistenkunde 84: 207

German, B., and J. Wyman (1937) The titration curves of oxygenated and reduced hemoglobin. J Biol Chem 117: 533

George, M. S., Z. Nahas, A. M. Speer, T. A. Kimbrell, E. M. Wassermann, C. C. Teneback, M. Molloy, D. Bohning, S. C. Risch, and R. M. Post (1998) Transcranial magnetic stimulation: A new method for investigating the neuroanatomy of depression. Advances in Biological Psychiatry 19: 94

Georgieva, R., B. Neu, V. M. Shilov, E. Knippel, A. Budde, R. Latza, E. Donath, H. Kiesewetter, and H. Bäumler (1998) Low frequency electrorotation of fixed red blood cells. Biophys J 74: 2114

Giebsch, G., D. C. Tosteson, and H. Using (Eds.) (1979) Membrane Transport in Biology. Springer-Verlag, Berlin

Gimsa, J., and D. Wachner (1999) A polarization model overcoming the geometric restrictions of the Laplace solution for spheroidal cells obtaining new equations for field-induced forces and transmembrane potential. Biophys J 77: 1316

Gimsa, J., and D. Wachner (1998) A unified resistor–capacitor model for impedance, dielectrophoresis, electrorotation, and induced transmembrane potential. Biophys J 75: 1107

Ginsburg, L. R., and E. M. Golenberg (1985) Lectures in Theoretical Population Biology. Pentire-Hall

Glansdorff, P. and I. Prigogine (1971) Thermodynamic Theory of Structure, Stability and Fluctuations. Wiley-Interscience Publ., London

Glaser, R. (1989) Grundriß der Biomechanik. (2nd edn.) Akademie-Verlag, Berlin

Glaser, R. (1996) The electric properties of the membrane and the cell surface. In: Zimmermann, U., and G. A. Neil (Eds.): Electromanipulation of Cells. CRC Inc., Boca Raton

Glaser, R., M. Brumen, and S. Svetina (1980) Stationäre Ionenzustände menschlicher Erythrozyten. Biol Zbl 99: 429

Glaser, R., and J. Donath (1984) Stationary ionic states in human red blood cells. Bioelectrochem Bioenerget 13: 71

Goel, N. S., and R. L. Thompson (1988) Computer Simulations of Self-organization in Biological Systems. Croom Helm, London

Goldman, D. E. (1943) Potential, impedance and rectification in membranes. J Gen Physiol 27: 37

Goldman, St. (1960) Cybernetic Aspects of Homeostasis. In: Comar, C. L., and Bronner (Eds.): Mineral Metabolism. Academic Press, New York

Govindjee (Ed.) (1975) Bioenergetics of Photosynthesis. Academic Press, New York

Graevskij, E. J., and N. I. Šapiro (1957) Sovremennye voprosy radiobiologii. Nauka, Moskva

Griffin, D. R. (1958) Listening in the Dark. Yale Univ. Press, New Haven

Grodzenskij, D. E. (1966) Radiobiologija. Nauka, Moskva

Grosse, C., and H. P. Schwan (1992) Cellular membrane potentials induced by alternating fields. Biophys J 63: 1632

Gruler, H. (1995) New insights into directed cell-migration – characteristics and mechanisms. Nouv Rev Fr Hematol 37: 255

Guzelsu, N., and W. R. Walsh (1993) Piezoelectric and electrokinetic effects in bone tissue. Review Electro- and Magnetobiology 12: 51

Haggis, G. H., D. Michie, A. R. Muir, K. B. Roberts and P. M. B. Walker (1965) Introduction to Molecular Biology. J Wiley & Sons, New York

Haken, H. (1983) Synergetics, Springer Ser. Synergetics, Vol. 1, 3rd. edn., Springer, Berlin

Hall, D. O., and K. K. Rao (1994) Photosynthesis. Studies in Biology (5th edn.) Cambridge University Press, Cambridge

Hallam, T. G., and S. A. Levin (Eds.) (1989) Applied Mathematical Ecology, Springer-Verlag, Berlin

Harris, C. M. (1991) Handbook of Acoustical Measurements and Noise Control. McGraw-Hill Inc., New York

Harrison, L. G. (1993) Kinetic Theory of Living Pattern. Cambridge University Press, Cambridge

Harth, O., and P. Vaupel (1971) Die Verteilung von Na^+ und K^+ unter dem Einfluß von Temperaturgradienten. Pflügers Arch 323: 158

Hassenstein, B. (1966) Kybernetik und biologische Forschung. In: Gessner, F. (Ed.): Handbuch der Biologie. Bd. 1, 629, Akad. Verlagsgesellschaft Athenaion, Frankfurt/Main 1966

Hayashi, K., and N. Sakamoto (1986) Dynamic Analysis of Enzyme Systems. Springer-Verlag, Berlin

Heinrich, R., H. G. Holzhütter, and St. Schuster (1987) A theoretical approach to the evolution and structural design of enzymatic networks: Linear enzymatic chains, branched pathways and glycolysis of erythrocytes. Bull Math Biol 49: 539

Heinrich, R., and St. Schuster (1996) Modeling of Metabolic Systems. Structure, Control and Optimality. Chapman and Hall, New York

Heinz, W. F., and J. H. Hoh (1999) Relative surface charge density mapping with the atomic force microscope. Biophys J 76: 528–538

Hertel, U. (1963) Struktur, Form, Bewegung. Krauskopf, Mainz

Hertz, J. (1991) Introduction to the Theory of Neural Computation. Addison-Wesley, Redwood City

Hianik, T., and V. I. Passechnik (1995) Bilayer Lipid Membranes. Structure and Mechanical Properties. Kluwer Academic Publishers, Dordrecht

Higashi, T., A. Yamagishi, T. Takeuchi, N. Kawaguchi, S. Sagawa, S. Onishi, and M. Date (1993) Orientation of erythrocytes in a strong static magnetic field. Blood 82: 1328

Hille, B. (1992) Ionic Channels of Excitable Membranes (2nd edn.) Sinauer Ass. Inc., Sunderland

Hoffmann, E. K., and L. O. Simonsen (1989) Membrane mechanisms in volume and pH regulation in vertebrate cells. Physiol Rev 69: 315

Hoppe, W., W. Lohmann, H. Markl, and H. Ziegler (Eds.) (1983) Biophysics. Springer-Verlag, New York

Horne, R. A. (Ed.) (1972) Water and Aqueous Solutions. Structure, Thermodynamics, and Transport Processes. Wiley-Interscience, New York

Hotary, K. B., and K. H. Robinson (1990) Endogenous electrical currents and the resultant voltage gradients in the chick embryo. Developmental Biology 140: 149

Hudspeth, A. J. (1985) The cellular basis of hearing: The biophysics of hair cells. Science 230: 745

Hudspeth, A. J. (1989) How the ear's works work. Nature 341: 397

Hug, O., and A. M. Kellerer (1966) Stochastik der Strahlenwirkung. Springer, Heidelberg

International Commission on Non-Ionizing Radiation Protection (ICNIRP) (1988) Guidelines for limiting exposure to time-varying electric magnetic, and electromagnetic fields (up to 300 GHz). Health Physics 74: 494

International Commission of Radiological Protection (ICRP) (1991) Recommendations of the International Commission of Radiological Protection, Publication No. 60. Pergamon Press, Oxford

Israelachvili, J. (1994) Intermolecular and Surface Forces (2nd edn.) Academic Press, London

Israelachvili, J. (1974) Van der Waals forces in biological systems. Quart Rev Biophys 6: 341

Jaffe, L. F. (1979) Control of development by ionic currents. In: Cone, R. A., and J. Dowling (Eds.): Membrane Transduction Mechanisms. Raven Press, New York

Jeffries, C. (1989) Mathematical Modeling in Ecology. A Workbook for Students. Birkhäuser, Boston

Jehle, H. (1969) Remarks on the problems of morphogenetic movement in the development of embryos. Intern J Quantum Chem 3s: 75

Kaandorp, J. A. (1994) Fractal Modelling: Growth And Form In Biology. Springer, Berlin

Kalmijn, A. J. (1997) Electric and near-field acoustic detection, a comparative study. Acta Physiologica Scandinavica 161: 25

Kauffman, S. A. (1993) The Origin of Order. Oxford University Press, New York

Katchalsky, A., and P. F. Curran (1965) Nonequilibrium Thermodynamics in Biophysics. Harvard Univ. Press, Cambridge

Keszthelyi, L., and P. Ormos (1989) Protein electric response signals from dielectrically polarized systems. J Membrane Biol 109: 192

Keynes, R. D. (1994) The kinetics of voltage-gated ion channels. Quart Rev Biophysics 27: 339

Kiefer, J. (1990) Biological Radiation Effects. Springer, Berlin

Kirschvink, J. L., D. S. Jones, and B. J. MacFadden (Eds.) (1985) Magnetite Biomineralization and Magnetoreception in Organisms. A New Biomagnetism. Plenum Press, New York

Knorre, W. A. (1981) Pharmakokinetik. Akademie-Verlag, Berlin

König, H. L., A. P. Krueger, S. Lang, and W. Sönning (1981) Biologic Effects of Environmental Electromagnetism. Springer, New York

Kotyk, A., K. Janacek, and J. Koryta (1988) Biophysical Chemistry of Membrane Function. J. Wiley and Sons, Chichester

Lammert, P. E., J. Prost, and R. Bruinsma (1996) Ion drive for vesicles and cells. J theor Biol 178: 387

Lamprecht, I., and A. I. Zotin (Eds.) (1978) Thermodynamics of Biological Processes. Walter de Gruyter, Berlin

Lamprecht, I., and A. I. Zotin (Eds.) (1985) Thermodynamics and Regulation of Biological Processes. Walter de Gruyter, Berlin

Lauffer, M. A. (1975) Entropy-driven Processes in Biology. Springer-Verlag, Berlin

Läuger, P. (1991) Electrogenic Ion Pumps. Sinauer Associates Inc. Publishers, Sunderland

Leff, H. S., and A. F. Rex (1990) Maxwell's Demon. Entropy, Information, Computing. Adam Hilger, Bristol

Leitgeb, N. (1990) Strahlen, Wellen, Felder. Ursachen und Auswirkungen auf Umwelt und Gesundheit. Georg Thieme Verlag, Stuttgart

340   References

Leopold, A. C. (Ed.) (1986) Membranes, Metabolism and Dry Organisms. Comstock Publ. Assoc. Cornell Univ. Press, Ithaca

Lerche, D., and H. Bäumler (1984) Moderate heat treatment of only red blood cells shows down the rate of RBC-RBC aggregation in plasma. Biorheology 21: 393

Levitt, D. G., and H. J. Mlekoday (1983) Reflection coefficient and permeability of urea and ethylene glycol in the human red cell membrane. J Gen Physiol 81: 239

Leyton, L. (1975) Fluid behaviour in biological systems. Clarendon Press, Oxford

Logofet, D. O. (1993) Matrices and Graphs. Stability Problems in Mathematical Ecology. CRC Inc., Boca Raton

Low, J., and A. Reed (1992) Electrotherapy Explained. Principles and Practice. Butterworth-Heinemann Ltd, London

Lynch, P. T., and M. R. Davey (Eds.) (1996) Electrical Manipulation of Cells. Chapman & Hall, New York

MacGinitie, L. A. (1995) Streaming and piezoelectric potentials in connective tissues. Adv Chem Ser 250: 125

Machemer-Röhnisch, S., H. Machemer, and R. Bräucker (1996) Electric-field effects on gravikinesis in Paramecium. J Comp Physiol A 179: 213

Makarov, V. A., M. Feig, B. K. Andrews, and B. M. Pettitt (1998) Diffusion of solvent around bimolecular solutes: A molecular dynamics simulation study. Biophys J 75: 150

Maniewski, R. (1991) Magnetic studies on mechanical activity of the heart. Critical Reviews in Biomedical Engineering 19: 203

Maret, G., N. Boccara, and J. Kiepenheuer (Eds.) (1986) Biophysical Effects of Steady Magnetic Fields. Springer, Berlin

Markin, V. S., V. F. Pastushenko, and Y. A. Chizmadzhev (1987) Theory of Excitable Media. A. Wiley-Interscience Publication, John Wiley & Sons, New York

Markin, V. S., and A. J. Hudspeth (1995) Modeling the active process of the cochlea phase relations, amplification, and spontaneous oscillation. Biophys J 69: 138

Matthes, R., J. H. Bernhardt, and M. H. Repacholi (Eds.) (1997) Biological Effects of Static and Elf Electric and Magnetic Fields. International Commission on Non-Ionizing Radiation Protection, Vol. 4

Matthew, J. B. (1985) Electrostatic effects in proteins. Ann Rev Biophys Biophys Chem 14: 387

Mazumdar, J. N. (1992) Biofluid Mechanics. World Scientific, Singapore

McLaughlin, St. (1989) The electrostatic properties of membranes. Ann Rev Biophys Biophys Chem 18: 113

Meinhardt, H. (1982) Models of Biological Pattern Formation. Academic Press, London

Meinhardt, H. (1995) The Algorithmic Beauty of Sea Shell. Springer, Berlin

Mohandas, N., and E. Evans (1994) Mechanical properties of the red cell membrane in relation to molecular structure and genetic defects. Ann Rev Biophys Biomolec Structure 23: 787

Møller, H. (1984) Effects of Infrasound on Man. Aalborg University Press

Moore, L. R., M. Zborowski, L. P. Sun, and J. J. Chalmers (1998) Lymphocyte fractionation using immunomagnetic colloid and a dipole magnet flow cell sorter. J Biochem Biophys Methods 37: 11

Morecki, A., J. Ekiel, and K. Fidelus (1984) Cybernetic Systems of Limb Movement in Man, Animals, and Robots. Horwood Ed., Halsted Press

Murray, J. D. (1993) Mathematical Biology. (2. Aufl.). Springer, Berlin

Nèmethy, G., and A. Scheraga (1962) Structure of water and hydrophobic bounding in proteins. J Chem Phys 36: 3382

Netter, H. (1959) Theoretische Biochemie. Springer-Verlag, Berlin

Nias, A. H. W. (1990) An Introduction to Radiobiology. Wiley, Chichester

Niklas, K. J. (1992) Plant Biomechanics. An Engineering Approach to Plant Form and Function. The University of Chicago Press, Chicago

Nuccitelli, R. (1995) Endogenous electric-fields measured in developing embryos. Advances in Chemistry Series 250: 109

Nygborg, W. L. (Ed.) (1985) Biological Effects of Ultrasound. Livingstone New York

Ogston, A. G. (1966) When is pressure osmotic? Federation Proceedings 25: 1112

Overbeek, J. T. G. (1956) The Donnan equilibrium. Progress Biophys Biophys Chem 6: 58

Olsvik, Ø., T. Popovic, E. Skjerve, K. S. Cudjoe, E. Hornes, J. Ugelstad, and M. Uhlén (1994) Magnetic separation techniques in diagnostic microbiology. Clinical Microbiology Reviews 7: 43

Pain, R. H. (1994) Mechanism of Protein Folding. Oxford University Press, Oxford

Palmer, T. (1995) Understandig Enzymes. Prentice Hall, London

Pauwels, F. (1980) Biomechanics of the Locomotor Apparatus. Springer, Berlin

Pearson, K. (1892) The Grammar of Science. Walter Scott, London

Penzlin, H. (1991) Lehrbuch der Tierphysiologie. (5th edn.) Fischer Verlag, Jena

Péqueux, A. J. R., and R. Gilles (1985) High Pressure Effects on Selected Biological Systems. Springer, Berlin

Peretto, P. (1992) An Introduction to the Modeling of Neural Networks. Cambridge University Press, Cambridge

Pethig, R. (1979) Dielectric and Electronic Properties of Biological Materials. John Wiley & Sons, Chichester

Pethig, R., and D. B. Kell (1987) The passive electrical properties of biological systems: Their significance in physiology, biophysics and biotechnology. Phys Med Biol 32: 933

Petracchi, D., and G. Cercignani (1998) A comment on the sensitivity of fish to low electric fields. Biophys J 75: 2117

Pohl, P., S. M. Saparov, and Y. N. Antonenko (1998) The size of the unstirred layer as a function of the solute diffusion coefficient. Biophys J 75: 1403

Polk, Ch., and E. Postow (1996) CRC Handbook of Biological Effects of Electromagnetic Fields. CRC Inc., Boca Raton, 2nd edn

Pollak, G. D., and J. H. Casseda (1989) the Neural Basis of Echolocation in Bats. Springer, New York

Popper, A. N., and R. R. Fay (Eds.) (1995) Hearing by Bats. Springer, New York

Precht, H., J. Christophersen, and H. Hensel (1955) Temperatur und Leben. Springer-Verlag, Berlin

Prigogine, I. (1967) Introduction to Thermodynamics of Irreversible Processes. (3rd edn.) Wiley Interscience, New York

Pullman, B., and A. Pullman (1963) Quantum Biochemistry. J. Wiley & Sons, New York

Qassem, W., M. O. Othman, and Abdul-Majeed (1994) The effects of vertical and horizontal vibrations on the human body. Med Eng Phys 16: 151

Rajnicek, A. M., K. R. Robinson, and C. D. McCaig (1998) The direction of neurite growth in a weak DC electric field depends on the substratum: Contributions of adhesivity and net surface charge. Developmental Biology 203: 412

Rashevsky, N. (1960) Mathematical Biophysics, Physico-Mathematical Foundations of Biology. Dover Publications, New York

Raudino, A., and D. Mauzerall (1986) Dielectric properties of the polar head group region of zwitterionic lipid bilayers. Biophys J 50: 441

Rayner, J. M. V., and R. J. Wootton (1991) Biomechanics of Evolution. Cambridge University Press, Cambridge

Reich, J. G., and E. E. Selkov (1981) Energy metabolism of the cell. Academic Press, London

Reilly, J. P. (1998) Applied Bioelectricity. From Electrical Stimulation to Electropathology. Springer, New York

Reiter, R. (1992) Phenomena in Atmospheric and Environmental Electricity. Elsevier, Amsterdam

Renger, G. (1994) Biologische Wasserspaltung durch Sonnenlicht im Photosyntheseapparat. Chemie in unserer Zeit 28: 118

Renshaw, E. (1990) Modelling Biological Populations in Space and Time. Cambridge University Press, Cambridge

Repacholi, M. H. (1998) Low-level exposure to radiofrequency electromagnetic fields: Health effects and research needs. Bioelectromagnetics 19: 1

Repacholi, M. H., and B. Greenebaum (1999) Interaction of static and extremely low frequency electric and magnetic fields with living systems: Health effects and research needs. Bioelectromagnetics 20: 133

Riu, P. J., J. Rosell, R. Bragós, and O. Casas (Eds.) (1999) Electrical Bioimpedance Methods. Ann NY Acad Sci 873: 1

Roederer, J. G. (1979) Introduction to the Physics and Psychophysics of Music. (2nd edn.) Springer, New York

Rowbottom, M., and Ch. Susskind (1984) Electricity and Medicine. History of their Interaction. San Francisco Press Inc., San Francisco

Rowe, G. W. (1994) Theoretical Models in Biology. The Origin of Life, the Immune System, and the Brain. Clarendon Press, Oxford

Rumbler, L. (1898) Physikalische Analyse der Lebenserscheinungen der Zelle. Arch Entw Mech 7: 103

Schienle, A., R. Stark, B. Walter, and D. Vaitl (1997) Effects of low-frequency magnetic fields on electrocortical activity in humans: A sferics simulation study. Intern J Neuroscience 90: 21

Schmid-Schönbein, G. W., S. L.-Y. Woo, and B. W. Zweifach (1986) Frontiers in Biomechanics. Springer, New York

Schneck, D. J. (Ed.) (1980) Biofluid Mechanics. Plenum Press, New York

Schrödinger, E. (1944) What is Live? Cambridge University Press, London

Schwan, H. (1957) Electrical properties of tissue and cell suspensions. Adv Biol and Med Phys 5: 147

Segel, I. H. (1993) Enzyme Kinetics. Behaviour and Analysis of Rapid Equilibrium and Steady-State Enzyme Systems. John Wiley & Sons Inc., New York

Seidel, H., and R. Heide (1986) Long-term effects of whole-body vibration: A critical survey of the literature. Int Arch Occup Environ Health 58: 1

Shannon, C. E., and W. Weaver (1962) The Mathematical Theory of Communication. University of llinois Press, Urbana

Shipley, R. A., and R. E. Clark (1972) Tracer Methods for in vivo Kinetics. Academic Press, New York

Sigworth, F. J. (1994) Voltage gating ion channels. Quarterly Rev Biophysics 27: 1

Silver, B. L. (1985) The Physical Chemistry of Membranes. Allen and Unwin, Boston

Singer, S. J., and G. L. Nicolson (1972) The fluid mosaik model of the structure of cell membranes. Science 175: 720

Skalak, R., and Shu Chien (Eds.) (1987) Handbook of Bioengineering. McGraw-Hill Book Company, New York

Slivinsky, G. G., W. C. Hymer, J. Bauer, and D. R. Morrison (1997) Cellular electrophoretic mobility data. A first approach to a database. Electrophoresis 18: 1109

Solomon, A. K., M. R. Toon, and J. A. Dix (1986) Osmotic properties of human red cells. J Membrane Biol 91: 259

Spanner, D. C. (1954) The active transport of water under temperature gradients. Symp Soc Exp Biol 8: 76

Stanier, M. W., L. E. Mount, and J. Bligh (1984) Energy balance and temperature regulation. Cambridge Texts in the Physiological Science 4, Cambridge University Press, Cambridge

Stein, W. D. (1986) Transport And Diffusion Across Cell Membrane. Academic Press, San Diego

Stein, W. D. (1990) Channels, Carriers, and Pumps. An Introduction to Membrane Transport. Academic Press, San Diego

Strait, B. J., and T. G. Dewey (1996) The Shannon information entropy of protein sequences. Biophys J 71: 148

Solomon, A. K. (1960) Compartmental methods of kinetic analysis. In: Comar, C. L., and Bronner (Eds.): Mineral Metabolism. Academic Press, New York

Sukhorukov, V. L., H. Mussauer, and U. Zimmermann (1998) The effect of electrical deformation forces on the electropermeabilization of erythrocyte membranes in low- and high-conductivity media. J Membrane Biol 163: 235

Syganov, A., and E. vonKlitzing (1999) (In)validity of the constant field and constant currents assumptions in theories of ion transport. Biophys J 76: 768

Szent-Györgyi, A. (1968) Bioelectronics. A Study in Cellular Regulations, Defense, and Cancer. Academic Press, New York

Takashima, S. (1963) Dielectric dispersion of DNA. J Mol Biol 7: 455

Takashima, S., A. Casaleggio, F. Giulliana, M. Morando, P. Arrigo, and S. Ridella (1986) Study of bound water of poly-adenine using high frequency dielectric measurements. Biophys J 49: 1003

Talbot, L., and S. A. Berger (1974) Fluid-mechanical aspects of the human circulation. American Scientist 62: 671

Tasaka, M., K. Hanaoka, and Y. Kurosawa (1975) Thermal membrane potential through charged membranes in electrolyte solutions. Biochem Chem 3: 331

Thom, R. (1969) Topological Models in Biology. Topology 8: 313

Thompson, and W. D'Arcy (1966) On Growth and Form. University Press, Cambridge

Tributsch, H., and L. Pohlmann (1993) Co-operative electron transfer: Superconduction in biomolecular structures? J Theor Biol 165: 225

Vainshtein, M. B., N. E. Suzina, E. B. Kudryashova, E. V. Ariskina, and V. V. Sorokin (1998) On the diversity of magnetotactic bacteria. Microbiology 67: 670

Varma, R., and J. R. Selman (1991) Characterization of Electrodes and Electrochemical Processes. John Wiley & Sons Inc., New York

Verkman, A. S., and R. J. Alpern (1987) Kinetic transport model for cellular regulation of pH and soluble concentration in the renal proximal tubule. Biophys J 561: 533

Videler, J. (1993) Fish Swimming. Chapman and Hall, London

Vogel, St. (1994) Life in Moving Fluids. The Physical Biology of Flow. (2nd edn.) Princeton Univ. Press, New Jersey

Voigt, A., and E. Donath (1989) Cell surface electrostatics and electrokinetics. In: Glaser, R., and D. Gingell (Eds.): Biophysics of the Cell Surface. Springer Series of Biophysics. Springer, Berlin 5: 75

Volkenstein, M. V. (1994) Physical Approaches to Biological Evolution. Springer, Berlin

Wassermann, E. M. (1998) Risk and safety of repetitive transcranial magnetic stimulation: Report and suggested guidelines from the international workshop on the safety of repetitive transcranial magnetic stimulation, June 5–7, 1996. Electroencephalogr Clin Neurophysiol 108: 1

Waterman, T. H., and H. J. Morowitz (Eds.) (1965) Theoretical and Mathematical Biology. Blaisdell Publ. Comp., New York

Webb, P. W., and D. Weihs (Eds.) (1983) Fish – Biomechanics. Praeger, New York

Westerhoff, H. V., and K. van Dam (1987) Thermodynamics and Control of Biological Free-energy Transduction. Elsevier Science Publ., New York

Wiggins, P. M. (1990) Role of water in biological processes. Microbiological Reviews 54: 432

Wilson, B. W., R. G. Stevens, and L. E. Anderson (Eds.) (1990) Extremely Low Frequency Electromagnetic Fields: The Question of Cancer. Battelle Press, Columbus

Wilson, Ch. E. (1989) Noise Control. Measurement, Analysis, and Control of Sound and Vibration. Harper and Row Publ., New York

Wiltschko, R., and W. Wiltschko (1995) Magnetic Orientation in Animals. Springer, Berlin

Yeagle, P. (Ed.) (1992) The Structure of Biological Membranes. CRC Inc., Boca Raton

Yockey, H. P. (1992) Information Theory and Molecular Biology. Cambridge University Press, Cambridge

Zeeman, E. C. (1976) Catastrophe theory. Scientific American 234: 65

Zhao, M., A. Dick, J. V. Forrester, and C. D. McCaig (1999) Electric field-directed cell motility involves up-regulated expression and asymmetric redistribution of the epidermal growth factor receptors and is enhanced by fibronectin and laminin. Molecular Biology of the Cell 10: 1259

Ziemann, F., J. Radler, and E. Sackmann (1994) Local measurements of viscoelastic moduli of entangled actin networks using an oscillating magnetic bead micro-rheometer. Biophys J 66: 2210

Zimmer, K. G. (1961) Studies of Quantitative Radiation Biology. Oliver and Boyd, Edinburgh

Zimmerberg, J., and V. A. Parsegian (1986) Polymer inaccessible volume changes during opening and closing of a voltage-dependent ionic channel. Nature 323: 36

Zimmermann, U. (1978) Physics of turgor and osmoregulation. Ann Rev Plant Physiol 29: 121

Zimmermann, U. (1989) Water relations of plant cells pressure probe techniques. Methods in Enzymology 179: 338

Zimmermann, U., and G. A. Neil (1996) Electromanipulation of Cells. CRC Inc., Boca Raton

Zimmermann, U., and E. Steudle (1978) Physical aspects of water relation of plant cells. Adv Bot Res 6: 45

Subject Index

Bold-faced numbers indicate pages, where the subjects are defined or explained explicitly.